Sex, Time, and Power

Also by Leonard Shlain

Art & Physics: Parallel Visions in Space, Time, and Light

The Alphabet Versus the Goddess: The Conflict Between Word and Image

Sex, Time, and Power

*How Women's Sexuality
Shaped Human Evolution*

LEONARD SHLAIN

Viking

VIKING
Published by the Penguin Group
Penguin Group (USA) Inc., 375 Hudson Street,
New York, New York 10014, U.S.A.
Penguin Books Ltd, 80 Strand,
London WC2R 0RL, England
Penguin Books Australia Ltd, 250 Camberwell Road, Camberwell,
Victoria 3124, Australia
Penguin Books Canada Ltd, 10 Alcorn Avenue,
Toronto, Ontario, Canada M4V 3B2
Penguin Books India (P) Ltd, 11 Community Centre, Panchsheel Park,
New Delhi–110 017, India
Penguin Books (N.Z.) Ltd, Cnr Rosedale and Airborne Roads, Albany,
Auckland, New Zealand
Penguin Books (South Africa) (Pty) Ltd, 24 Sturdee Avenue,
Rosebank, Johannesburg 2196, South Africa

Penguin Books Ltd, Registered Offices:
80 Strand, London WC2R 0RL, England

First published in 2003 by Viking Penguin,
a member of Penguin Group (USA) Inc.

10 9 8 7 6 5 4 3 2 1

Illustration credits appear on pages 403–404.

LIBRARY OF CONGRESS CATALOGING IN PUBLICATION DATA
Shlain, Leonard.
    Sex, time, and power : how women's sexuality shaped human
evolution / Leonard Shlain.
        p.   cm.
    Includes index.
    ISBN 0-670-03233-6
    1. Sexual attraction.   2. Mate selection.   3. Human evolution.   4. Social evolution.
    5. Sex (Biology)   6. Evolution (Biology)   I. Title.
    HQ23.S45 2003
    306.7–dc21   2002041186

This book is printed on acid-free paper. ∞

Printed in the United States of America
Set in OPTI Administer
Designed by Erin Benach

To my wife Ina, my daughter Kimberly, my son Jordan, and my daughter Tiffany—four very inspiring people who have inspired me.

Preface

There is a female human nature and a male human nature, and these natures are extraordinarily different. . . . Men and women differ in their sexual natures because throughout the immensely long hunting and gathering phase of human evolutionary history the sexual desires and dispositions that were adaptive for either sex were for the other tickets to reproductive oblivion.

—Donald Symons[1]

Error is the inevitable by-product of daring.

—Stephen Jay Gould[2]

Iron/Sex

*J*uxtaposing the words "iron" and "sex" creates an odd couple. The two rarely have occasion to appear together in the same sentence, much less find themselves standing side by side with so little editorial support. In the following pages, I will propose that the first word fundamentally influenced economic matters between men and women and, as a result, profoundly affected the politics of the second word. Along the way, I will present a scenario for how the kaleidoscopic, maddening, exciting, enchanting, and baffling man-woman dance, more commonly referred to as "a relationship," evolved.

This book arose out of a question I posed to a professor when I was a second-year medical student, making rounds on patients in a large ward. Although the incident occurred over forty years ago, I had never really forgotten or accepted his answer.

The sophomore year of medical school represents a major transition for students. They leave the cadavers of the freshman year behind and begin having contact with respiring, perspiring patients. On this particular day, we were being taught how to interpret laboratory results.

Shifting from bedside to bedside, our knot of students listened intently to the professor. Every patient admitted to the ward, he explained, had three basic laboratory tests: a chemistry-26 panel, a urinalysis, and a complete blood count (CBC). The first measured the concentration of twenty-six constituents floating about within the patient's bloodstream.

The amount of a patient's circulating sodium or potassium, for example, provided a snapshot of the health of various internal organs, such as the heart and kidneys. The measurement of every one of the twenty-six constituents was like having a miniature finger figuratively take the pulse of some very important cellular function, which in turn reflected on the state of one or more of the body's organs. The lab reported each value on a slip that was placed in the patient's chart. Neatly aligned, parallel to the patient's results, was a column stating the expected normal ranges for each particular variable. What caught my eye that morning was the list of normals. Of the twenty-six numbers on the chem panel, none distinguished between the values for a man and a woman. And why should there be any? After all, sex has nothing to do with the way a lung or a stomach goes about its business, so I would not have expected any variation between the sexes. The same held true for the normal values reported on the urinalysis.

The CBC, however, was distinctly different in this regard. A complete blood count measures several different parameters of the red and white cells circulating within the body's miles of vascular tunnels. Although the white-cell numbers were the same for the two sexes, I noticed that the red-cell normals for men and women were surprisingly askew. I thought that was very strange.

A red cell's chief function is to pick up oxygen in the lungs, transport it through the blood vessels, and deliver it to every organ in the body. For all complex creatures, oxygen is the staff of life. Deprive them of this most precious element and they will rapidly die of asphyxiation. Yet a man normally has a 15 percent higher concentration of circulating red cells than a healthy woman has. I puzzled over this discrepancy (even after taking into account that, generally, a woman is smaller in stature and has less muscle mass than a man), and it prompted the question that lies at the heart of this book. Why would a woman need less of blood's critical essence than a man?

I raised my hand and asked the professor to explain the reason for this vital disparity. He replied, in a way that suggested he thought the answer was obvious and my question, had I given it more thought, was unnecessary:

"Women bleed and men don't."

Somewhat annoyed that I had interrupted his train of thought, he then returned to the subject on which he had been discoursing. I reddened.

I remember thinking at the time that his curt answer could not be the whole story. Later, as I turned over his reply in my head, I could not think of a plausible explanation for this difference. Why would a species evolve whose females—but not males—operated routinely on less than a full complement of tankers to ferry a fuel as crucial as oxygen? (One could argue that the female has a normal amount and the male has a 15 percent excess capacity. Either way, the inequality begs for an explanation.)

The magic behind a red cell's seamless performance is the protein called "hemoglobin." And, like a set of nested Russian *babushka* dolls, at the core of hemoglobin lies the element iron. Iron and oxygen, because of the salutary arrangement of their outer electron shells, eagerly seek out each other's embrace. Once merged, the two form a molecule called iron oxide. Everyone is familiar with this substance. We call it "rust."

Rust has a signature red color. Iron oxide in rock lends to the Grand Canyon its spectacular chromatic array. It colors the fields in Kenya and the plantations of Hawaii, and rust's distinctive hue made the earth of Scarlett O'Hara's beloved Georgian Tara red. The hemoglobin molecule, with its precious complement of iron atoms, makes blood red and white skin pink, and it tints medium-rare steaks with a distinctive ruby sheen.

Hemoglobin transforms iron's strong affinity for oxygen into a delicate "grasp and release" maneuver, allowing oxygen to be easily acquired in the lungs and readily relinquished farther down the line to the cells. The felicitous combination of iron, oxygen, and hemoglobin is central to the process that facilitates the neurons in your brain to extract meaning from the sequentially aligned squiggles you are reading on this page.

A few weeks after I had posed my red-cell question on rounds, a second incident piqued my interest in the subject. During a rotation on the obstetrics-gynecology service, I was assigned to the prenatal clinic, where wise, experienced nurses instructed me how to care for pregnant women. Detroit's Harper Hospital in 1958 served a large population of poor people who had emigrated from other states, hoping to improve their lives in the city's then booming auto industry. Unfortunately, most women we served were unaware of the prenatal program and did not enter it, if at all, until late in their pregnancy. Playing catch-up, I counseled these soon-to-be moms about the importance of maintaining a healthy diet, for both themselves and their unborn children.

A key element of my care was to monitor each woman's weight, blood pressure, CBC, and blood-sugar levels. My goal was to keep these four cornerstones of prenatal health within a normal range. Specifically, I tried to maintain their hemoglobin level in the 11–12 range, using iron pills to supplement their dietary intake if necessary.* A hemoglobin level below 10 heightened my concern, because it confirmed the diagnosis of anemia.

Consider my consternation and sense of helplessness when I discovered that it was *exceptional* to find one among the late-term women visiting my clinic in possession of a normal hemoglobin count. More often, they ranged

---

*Hemoglobin is measured in grams per deciliter.

from mild (8–10) to severe (5–7) anemia.* My mentors sadly shook their heads, explaining to me that fetal brains would most likely not develop properly if the mother was anemic; at worst the babies would suffer mental retardation, and at best they would fail to attain their full intellectual potential.

Outwardly, these very pregnant women seemed normal, but I soon learned to discern their anemia's telltale signs of apathy and lethargy. Paradoxically, they appeared well fed, and, even more puzzling, many were actually overweight. A peculiar side effect of iron deficiency is the dulling of the taste buds that sense sweets. To compensate, an iron-deficient woman craves carbohydrates.

I was often disconcerted to observe a pregnant woman bringing a carton of cornstarch into my examining cubicle, eating the white powder right out of the box during my interview—so great was her drive to satisfy her carbohydrate hunger. Meanwhile, she was unaware that the real culprit was a lack of a vital mineral in her diet. Increased intake of calories contributed to her weight gain, giving her a deceptively healthy appearance; cornstarch, which in 1958 contained negligible amounts of iron, only contributed to masking the problem.

In the years that followed, I reflected many times on my frustrating experience in that metropolitan prenatal clinic. After I had immersed myself in evolutionary theory, I wondered why Mother Nature would favor the evolution of mothers whose internal circuit breakers did not protect them from endangering their babies and themselves. Humans, like all animals, have sensitive trip wires alerting us whenever salt, water, oxygen, or carbohydrate levels enter the red zone. Why were women in general and pregnant women in particular so blissfully unaware of their susceptibility to iron deficiency?

Human intelligence is measured, for want of a better standard, in IQ points. In the long march from anthropoid hominid toward *Homo sapiens,* the piling up of one IQ point upon another eventually propelled our species to the top of the IQ heap. Low maternal iron depots ultimately subtract IQ points from offspring, thus subverting the entire thrust of human evolution. Our species' females, it would appear, had traded away sufficient maternal hemoglobin levels, but for what? Where, I wondered, was the beneficial offset that could possibly justify such a Faustian bargain?

Curious about this matter, I sought the answer in my authoritative physiology textbook. The explanation proffered was that males led a more stren-

---

*Significant shifts occurring in red-cell and plasma concentrations in the last trimester of pregnancy somewhat confuse the issue of anemia; nevertheless, iron deficiency remains a considerable problem. A fetus's brain may develop normally within a slightly anemic mother's womb, but the crisis that is birth often sharply diminishes the oxygen supply to the fetus's brain. Even slight anemia in the mother can translate into grave consequences for the baby's brain if her delivery is difficult or prolonged.

uous life than females and therefore needed a more robust aerobic circulatory system. Although initially I accepted this answer, as I grew older I increasingly suspected that it could not be the complete story. I doubted that any man engages in a more metabolically taxing activity than trying to form a seven-pound infant over a nine-month period from scratch.* Also, the drain upon the energy stores of a breast-feeding mother are of the same order of magnitude as the demands made upon an Olympic shot-putter during his rigorous training.†

A similar disparity appears in the CBC in some other animal species, such as chimpanzees and gorillas. Hematologists suspect that this differential has something to do with the red-cell-enhancing effect of testosterone and the red-cell-depressive effect of estrogen. (Men treated with estrogen for prostate cancer are at risk to develop anemia.) Compared with these other animals, however, the differences between a man's and a woman's CBC are deeper and wider. This variation gains in significance because the other species that manifest an inequality in the number of red cells between the sexes do not have a brain as large as humans do, and as we shall see, the human brain is very sensitive to any factor that diminishes its ability to gain access to oxygen. Also, anemia is quite rare among pregnant primates, yet it commonly occurs in the human line. Many years later, after learning more, reading more, and experiencing more, I began to harbor a growing suspicion that an important missing piece of the human evolutionary puzzle was hiding somewhere behind the innocuous-looking numbers I had noticed that morning so long ago on a routine CBC slip.

I confess that I had difficulty staying focused on this subject during medical school. An overwhelming avalanche of other facts, lessons, and procedures cascaded down upon me, and distracted me from pursuing the question. Besides, the subject of blood was not my passion in those days. As a teenager, I had stumbled onto Sigmund Freud's *Civilization and Its Discontents*, one of those seminal books that one discovers in youth that rock one's beliefs. My curiosity whetted, I made subsequent explorations deep into Freud's subterranean warren. I was so intrigued by what I was learning there, I became convinced very early on that my destiny was to become a psychoanalyst.

Looking back now, I think it would be difficult for someone who had not lived through the 1950s to appreciate just how profoundly Freud's ideas per-

---

*A woman must consume 80,000 calories above her normal maintenance requirements to accomplish the feat.[3]

†A nursing mother needs 76,000 surplus calories above her normal intake to satisfy the demands of breast-feeding.[4] Behavioral ecologist Hilliard Kaplan estimates that provisioning a child until he or she reaches the age of independence requires 10 million to 13 million calories![5]

meated the intellectual climate then. A rough indicator of his influence was that over 30 percent of my 1957 medical-school freshman class had initially aspired to be psychiatrists. (Many medical students switch from their original choice of specialty, as did a lot of my classmates.) I envisioned that I would become a dauntless spelunker of the mysteries of the mind. Psychoanalysis seemed to be the "open sesame" to the entrance of that dark cave.

As a young boy, I had loved to build model airplanes and play sports. I enjoyed drawing and fancied myself a budding artist. When I rotated on the surgery service, I discovered, to my delight, that this field of medicine provided an outlet for the skills I was incubating. Surgery seemed a perfect fit for me. It was romantic, challenging, and intensely exciting. The drama I witnessed daily, combined with my sensing the enormous gratification surgeons seemed to enjoy, changed my mind about a career choice. I decided to specialize in general and vascular surgery. Though I remained faithful to my aspiration to be an explorer, now I decided I would navigate the interior passageways of the body.

Maintaining adequate circulating levels of red cells in my patients became a central concern of mine. As a young surgical resident at Bellevue Hospital in New York, I experienced an exhilarating frenzy while running with a group of emergency-room nurses and doctors gripping a gurney carrying a gunshot victim. Our goal: Reach the operating room before the patient "bled out."

Throughout my career, I have, on occasions too numerous to count, glanced up from a particularly vexing operative field to monitor anxiously a forest of IV poles, their precious transfusions hanging like ripe fruit. I have had many years during which to contemplate the nature, importance, and vitality of this substance we call blood.

These musings have convinced me that significant differences in iron levels between the sexes were the initial driving force behind many uniquely human cultural innovations. I shall trace the connections of art, calendars, marriage, mayhem, fatherhood, and homosexuality, to name but a few, back to this arcane feature of human physiology. Along the journey, I shall also explore the links between the moon and menses, sex and death, and funerals and paternity.

Throughout my life, I have maintained a lively interest in matters relating to life-forms and evolutionary theory. As a premed undergraduate learning the wondrous stages of embryology or comparing the anatomy of differing species, I recall having moments of pure rapture whenever I grasped the intricacies behind some aspect of the breathtaking beauty of life. The study of its sumptuous diversity is, for me, the quintessential melding of science and aesthetics. I still marvel over the improbability of a caterpillar's metamorphosis, the texture of a calla lily's singular petal, or the amazing grace of a formation of pelicans skimming waves. I will try to infuse this book with my sense of awe and enthusiasm for the processes of life while I elaborate my theories on the bonding glue that holds human relationships together.

Some might wonder why a surgeon would dare to wander so far from his field of specialized expertise to enter the bramble-ridden thicket of human sexuality. The simple answer: It fascinates me, as, I suspect, it does you. And I believe I have some fresh insights to contribute to the subject.

My direct experience with matters relating to physiology, anatomy, biochemistry, and psychology, essential to my training as a physician and surgeon, has served me well in researching this book. My knowledge of anthropology, primatology, evolutionary biology, and archeology is the result of my abiding interest in these fields. However, I must admit at the outset that I am not an expert in all the fields into which I delve.

Sex, Time, and Power is intended for both generalists and specialists. Although it is based on scientific research and theories, I wish to keep to a minimum the standard academic practice of citing the pedigree of a particular idea by listing the numerous authorities who were involved in its lineage. I do not wish to diminish these innovators, but I also do not want the book to read too much like a textbook. I also will resist the temptation, whenever possible, to use arcane zoological names, or to cite daunting statistics.

Space requirements imposed by the publishing world prevent me from presenting every alternative theory to the ones I propose. This does not necessarily mean that I am unaware or dismissive of other possibilities, but, in a bow to the wordsmith's aesthetic, I will try to keep many supporting facts, counterarguments, and authoritative references caged in the footnotes and endnotes.

One of the great pleasures I derive from writing is conjuring metaphors that can translate complex physiological and evolutionary processes into rich images, increasing an idea's accessibility for the nonspecialist reader. A key metaphor that I will employ throughout this work is to treat the process of natural selection *as if* it behaved like an intelligent entity with forethought and purpose. Exercising a writer's prerogative to use poetic license, I will use the terms "Mother Nature," "Natural Selection," and "the Red Queen" interchangeably.*

---

*Evolutionist Leigh Van Valen suggested that evolution resembled the contest the Red Queen describes to Alice in Lewis Carroll's classic *Through the Looking-Glass*. She tells Alice, "It takes all the running you

My use of these terms, however, should not be misconstrued. I am not imputing purposeful design to a supernatural entity. *Natural* selection is a *natural* process. The origin of species does not need a *deus ex machina** to explain how it works.

Darwin's ideas have been put to the rigors of scientific examination for a century and a half, and although there still remain many intriguing questions, overall he has provided scientists with a powerful predictive tool. Whether or not a supernatural entity first set in motion evolution's ingenious processes of natural selection based on random mutations of genes interacting with environmental changes is a religious question better left to each reader to answer.

While on the subject of disclaimers, let me neutralize the contentious nature-versus-nurture debate at the outset. There is no gene-controlled inheritable trait that cannot be altered by the environment. Similarly, the genetic makeup of the organism can overcome the influence of the environment. Each factor can affect and alter the other. Humans enter the world as a work-in-progress. In some cases, the culture or environment into which a person is born more strongly determines his or her responses to the vagaries of life, and sometimes responses are more influenced by the genes he or she has inherited. Nature/nurture is not an *either/or* duality but, rather, represents a *both/and* type of complementarity.

I assume that the reader is reasonably familiar with the essence of Darwin's ideas, but there are two notions I feel compelled to differentiate at the outset. After Darwin electrified the intellectual world in 1859 with his theory of evolution based on natural selection, he refined it further in 1871 by emphasizing the importance of *sexual selection*. For, as Darwin realized, it was simply not enough that an organism survived both the rigors of competition and the hardships imposed by its environment (natural selection), it also had to reproduce successfully (sexual selection). Mate selection became a critical factor in accounting for how sexually reproducing organisms came to be the way they were. Males competed among themselves to see who won the right to mate with females. And females, by picking and choosing among myriad suitors, exerted an enormous influence on which male traits advanced in the

---

can do to stay in the same place." Van Valen proposed that the process of natural selection works similarly. Once either a prey or a predator evolves an advantageous adaptation that foils the intentions of the other, then the other side must adapt to the new situation between them or face the possibility of extinction. After the new adaptations are in place, equilibrium is restored and both parties are generally back to where they started, just as the Red Queen described.[6] Author Matt Ridley titled his very readable 1993 book concerning sexual evolution *The Red Queen*.

*The Latin term refers to having to invoke a miracle to explain a feature of the natural world that confounds reason. The metaphor is based on the common practice among classical playwrights of resorting to the arrival of a god onstage by means of a mechanical contrivance. Writers used this ploy to get past a difficult transition in their plots.

genome. Males, therefore, tended to be what females wanted them to be. Natural Selection is about survival, and Sexual Selection concerns reproduction. I will use the term "Natural Selection" as the generic process moving evolution along, even when there may be elements of sexual selection combined within it.

Another caveat: Because the story I intend to tell unfolds in a linear, sequential narrative, it might appear that I am proposing that first one thing occurred and then that caused another to occur, with cause and effect clanking along in a prescribed sequence. This is not the case. The process of Natural Selection is a to-and-fro, give-and-take, nonlinear whorl. Ongoing, simultaneous feedback loops between local environments and individuals are the driving force compelling species to alter their shape, behavior, and metabolism. This ever-changing, continuous ebb and flow creates a fluid dynamic that linear narrative can never adequately convey.

Because I will be writing at length about matters relating to sex, birth, and death as well as the love between a man and a woman and between parents and children, I feel I should share something of my personal background in addition to my professional qualifications.

I was born in Detroit, Michigan, the youngest of four children, to first-generation Russian immigrants. My father was an extremely hardworking man who brought with him a distinctly Old World view of the place of men and women in society. My mother was gentle and loving, and laughed easily. Their relationship was a typical patriarchal one. They both taught me many valuable lessons worth emulating, and some that I have striven not to repeat. They remained married for sixty-five years and lived into their nineties.

I made the usual teenager and young-adult explorations into parties and dating, then, at twenty-seven, I married, after a tempestuous and passionate four-year courtship in which each of us experienced the dramatic highs and lows of young love. Following a stint in the army in France and a surgical residency in New York, we settled in northern California, where I finished my surgical training. During this time, we had three children one right after another: a daughter, Kimberly, a son, Jordan, and then another daughter, Tiffany.

After seventeen years of marriage, my wife and I divorced in the same manner as we had courted. I remained single for an equal number of years, during which time I had the opportunity to participate in the "dance" a second time around, but this time as an older, marginally wiser, but more observant "dancer."

I have often contemplated the nature of the persistent longing present in the majority of the hearts of both men and women. Persons of each sex, no matter how old, seem to strive to find their respective soul mates. Four years ago, I found mine. Ina and I married, each for the second time.

A judge and a surgeon—some combination. When we cook together, I, in the manner of my professional training, place my hand palm-up without looking away from the slicing and the dicing and bark, "Tomato!" Ina laughingly intones, "Motion overruled." We see ourselves engaged in a grand adventure. We have set out to illustrate that a man and a woman can love each other and mesh both our needs and identities in such a way that the sum of us together is greater than each half alone.

I think of my life as resembling an onion. Each layer symbolizes one of the many roles that I have assumed. Son, brother, lover, husband, teacher, student, father, doctor, writer, surgeon, scholar, and lecturer constitute the main ones. The role I would place at the very core of the onion is the one I have cherished the most: that of a father.

I have known the delicious delight of carrying a freshly bathed, flannel-encased, sweet-smelling, sleepy toddler to his or her bed. I have run alongside three different bicycles each recently divested of training wheels and then . . . let go, to cringe in anxious anticipation of whether I had judged the moment of my release correctly. And that was just the beginning of a whole series of wincing withdrawals.

The military issues campaign ribbons to personnel who have served in various wars and skirmishes to wear on their chests so that comrades-in-arms can instantly identify each other. I recommend that similar insignia should be displayed by all parents who have survived the harrowing teenage years so that they, too, can acknowledge each other's experience. Happy to say, our unit made it through the guerrilla warfare of those years. Those who had been temporarily missing in action are now all present and accounted for. I have lived long enough to watch proudly as my children have grown into high-spirited, accomplished, interesting people.

My children are adults now, forming their own families. Listening to them tell me about their loves and courtships has afforded me the opportunity to observe how this man-woman thing works once again, but in a different generation. The fight for love and glory, it seems, is still the same old story.

And now, let me tell you of my qualifications to write about death. At the age of thirty-seven, I was diagnosed as having a non-Hodgkin's lymphoma. I underwent extensive surgery, followed by many months of health-debilitating radiation treatments. The experience shook me to my very core and changed me as a human being. As Samuel Johnson astutely remarked, the prospect of hanging clears a man's mind wonderfully. I spent many dark, sleepless hours contemplating the meaning of life and the consequences of death—particularly mine.

After over a year of treatments and their inevitable complications, I recovered sufficiently to resume my surgical career. I began to receive many referrals for surgery of patients in the same dire straits that I had recently passed through. I suspect doctors believed that their patients would relate better to a surgeon who had just endured what they must now suffer.

Treatment for cancer, unfortunately, often fails—more so twenty-five years ago than at present. Many times, after I had developed a close connection with a patient, circumstances demanded that I abandon my mission as a healer and assume the role of Charon, the mythical boatman who ferried souls across the River Styx to the other side. My own personal experience and intimate contact with dying patients led me to explore many of the issues I will raise in this book.

All writing, despite authors' best efforts to conceal themselves behind the scrim of objectivity, contains intimations of the autobiographical. Although somewhat unorthodox, I have briefly outlined my personal history so that you, the reader, may know something of the perspective I bring to this work.

I have chosen artwork and other illustrations to accompany my narrative. Please see pages 403–404 for art credits.

Above my writing desk hangs a quote from Franz Kafka urging writers to create books that "can be wielded like a pickax to shatter the frozen sea within the reader's mind." If a book didn't change the way the reader thought about the world, then Kafka deemed it not worth writing. I have taken Kafka's words as my credo. May this book set your mental ice floes grinding against each other.

A well-worn metaphor draws the analogy between an author finishing a book and a woman birthing a newborn. Observing my odd cube-shaped "child" in the form of a neat stack of freshly printed pages sleeping peacefully on my desk this fine morning, I can sense how the Old Testament's Jochebed, the mother of Moses, must have felt.

The moment has arrived to tear my baby from my protective embrace and place it in its basket, preparatory to setting it adrift down the river. Like Jochebed, I, too, fervently hope that the result of my labor will become entangled in the bulrushes and find a hospitable home among accepting strangers.

Enjoy.

Leonard Shlain, 2002
Mill Valley, California

Acknowledgments

Sprinkled among these pages is the distilled, accumulated wisdom of the many people who supported my efforts in writing this book. I sought out some for their exquisitely refined specialized knowledge. I valued others for their critical and curious open-minded opinions. I passed around manuscript pages liberally throughout the multiple stages of this project.

I wish to thank Michael Corballis, Elisabet Sahtouris, Chris Knight, Malcolm Potts, Robert Sapolsky, Linda Clever, Jerry Lowenstein, Michan Afsari, John Locke, Diane Baker, Larry Garlington, Elizabeth Snyder, Don Campbell, Tom Gage, Joan DePaoli, Gary Hunter, Gail Weber, Carl Levinson, Martha Steel, Ralph Wallerstein, Cora Stryker, Ken Bush, Ricki Pollycove, Jeffrey Aron, Sharon Ferret, Mark Cohen, Colleen O'Connor, Wen-I Chang, Evelyn Resh, Garth Petal, Toni Brayer, Barry Gurdin, China Galland, Hal Nash, Barbara Roether, Phil Hockenberger, Vladimir Dinets, Jody Widelitz, Michael and Lynn Braverman, Yoel and Eva Haller, Jack and Pat Futuran, Marty Carr, and Sheldon Levin.

Several readers tackled the task with particular gusto, engaging me in long discussions that helped me refine my ideas immeasurably. To Ernst Simon, David Tresan, Irwin Gootnik, Sam Gray, Michael Trupp, and David Nelson, a tip of my hat to you all.

When the book was in its jig-saw-puzzle stage and the pieces were lying about the floor, I sought editorial advice from Barbara Szerlip, a superb wordsmith who helped me organize and edit the early phase of the manuscript. At a later stage, Woodeene Bricker-Koenig gave the entire manuscript a thorough going-over, adding and subtracting here and there with her very deft touch. The suggestions of my friend Bill Henkin were, as always, right on the mark.

I wish to especially thank both my son, Jordan Shlain, and my son-in-law Ken Goldberg for their excellent edits. Thanks also to my daughters, Tiffany and Kimberly, for their support. I offer my heartfelt thanks to my wife, Ina Gyemant, for her many thoughtful suggestions—especially when she reined in some of my more exuberant expressiveness on the many occasions during this four-year magnificent obsession that I showed undeniable evidence that I was getting carried away.

I appreciate the efforts of Judy Snyder and Robby Gyemant, who typed some of the revisions, and of Aaron Baker, who assisted in organizing the bibliography and footnotes. Matt Hunter performed fancy legwork in securing permissions for the images. I drew upon the artistic expertise of Mark Reynolds, Robert Fox, and Matt Hunter to create the graphics. Lynn and Mara Fritz, thank you for providing me with your wonderful country retreat in which I did some of my best writing.

My masterful agent and old friend Robert Stricker skillfully guided the book through the steep narrows of the New York publishing world. Paul Slovak, my superb editor at Viking with whom I developed a wonderfully rewarding working relationship, did a yeoman's job of shaping the manuscript once it hit his desk. Kate Griggs and her very thorough copyediting associates Terry Zaroff-Evans and Don Homolka attempted to ensure that every "tittle" in the book was correctly "jotted." I wish to wax effusive over the great design team at Viking—Jaye Zimet, Erin Benach, and Michelle Ishay—who turned my ordinary-looking stack of typed paper sheets into a beautifully bound book with a compelling cover. And thanks to everyone whose ear I bent so insistently over the years. You were an invaluable sounding board that helped me shape and hone this complex project from its inception as a foggy idea to its completion as the finished work you now hold in your hand.

Contents

Sex, Time, and Power

During the medieval period, when Christian mysticism uncomfortably yoked ancient beliefs and a newer religion, a strange myth captured the imagination of the people of Europe. At the heart of the Quest for the Holy Grail lay an unanswered question. Several centuries had passed since this magic chalice had disappeared after receiving the blood of Jesus' wound as he hung transfixed upon the cross. The land was sorely desolate, and the people believed that only an unblemished knight could recover the Grail and solve the riddle that would lift the curse. This blessed act for which the devout ardently prayed would make the wasteland bloom with life again. According to the myth, Parsifal, a knight of King Arthur's Round Table, embarked on a hero's journey and after many adventures at last arrived at the gates of the Grail Castle.

Upon entering a great hall, Parsifal saw a procession of young men and maidens holding aloft a spear dripping with blood. Behind them, other youths carried the long-sought-for, sacred Grail, which to Parsifal's astonishment was also brimming with the red liquid. The Fisher King, the Grail's guardian, sat anguished on his throne, trying unsuccessfully to stanch the mysterious wounds in his genitals that bled day and night. When they saw the holy knight, the music ceased and all assembled in the castle stood frozen as if holding their breath. The silence was complete. Each waited for Parsifal to ask the simple question that would lift the curse that both harried their king and devastated the land. But Parsifal was so awestruck that he forgot to ask the obvious question: "This bleeding from the cup, what purpose does it serve?"

Iron, Sex, and Women

Part I

The emergence approximately 150,000 years ago of *Homo sapiens,* an animal that opened a chasm between it and all other species, was as improbable an event as coming upon a nude Caucasian woman reclining on an Empire couch in the middle of the jungle.

Chapter 1

Unknown Mother/African Eve

Sex endows the individual with a dumb and powerful instinct, which carries his body and soul toward another; makes it one of the dearest employments of his life to select and pursue a companion, and joins to possession the keenest pleasure, to rivalry the fiercest rage, and to solitude an eternal melancholy. What more could be needed to suffuse the world with the deepest meaning and beauty?   —George Santayana[1]

The reconstruction of evolutionary history is better regarded as a game than as a science, evolutionary hypotheses should be stated with varying degrees of confidence always keeping in mind that certainty cannot be achieved.
—Sherwood Washburn[2]

She died an agonizingly slow and painful death. She was not accorded funerary rites, nor was her corpse laid to rest in a grave. Her remains constitute but a sliver of debris—a disconnected tooth here, a chip of a fossilized bone there, fragments lost in the strata of bygone ages. At the time of her death, she represented the latest in a line of primates called "hominids" that had begun their evolutionary trial run several million years earlier. If paleontologists ever find her final resting place, we should erect a memorial on the spot in recognition that she did not die in vain. An appropriate name for her marker would be "The Tomb of the Unknown Mother." Her passing heralded the birth throes of a new species.

Imagine that a group of intergalactic anthropologists had been observing these primates from the beginning. When Unknown Mother died, the visitors would have exchanged knowing looks, because they could plainly see that her fate was foredoomed. The hominid line from which she arose had split away from other primates by developing two adaptations destined to collide. Hominids were the only primates to depend on a new means of moving about that required only two limbs instead of four. An upright stance allowed them to clamber down from the trees and seek a living first on the forest floor and later on the open savanna. Because their erect posture greatly increased the possibility that the first creature to stride would end up as "cat food," they needed a crucial second adaptation. Since they could not outrun or outfight predators, they required an enlarged brain capable of outwitting those creatures intent on devouring them.

During the last two and half million years, the hominid brain had tripled in size but the opening in the pelvic girdle through which this rapidly enlarging brain had to pass at birth did not keep pace. These two adaptations—two-leggedness and watermelon-sized heads—were clearly incompatible.

The new engineering imperatives of standing upright had sculpted the hominid's pelvic ring of bone into a new shape, flattening it from front to back. The bipedal pelvis, anatomically dissimilar to its counterpart in four-legged animals, also acquired a novel architectural function. It had to serve as a basin to contain the mass of intestines pressing down from above and prevent them from falling down and out through the rectum. Consequently, the bony hole in the pelvis had to remain relatively small. Only the wide, comparatively horizontal flanges of the human iliac pelvic bones, the narrowness of the pelvic inlet, and the thin sheet of muscles suspending the anus prevented this unusual primate from having the discomfiting experience of being turned inside out while out for a stroll after a particularly heavy lunch—a gravitational hazard that does not pose a problem for *any* other animal.*

These functional constraints prevented the channel in the female's pelvis from enlarging sufficiently to accommodate easily the continually growing size of her fetus's brain during childbirth. Mother Nature devised numerous ingenious sleights of hand to thread the baby through the "eye" of a mother's birth canal.† Despite these clever adaptations, hominid females began to experience increasingly difficult deliveries. The problem became especially acute around 150,000 years ago, at which point the hominid brain

---

*The muscles making up the human pelvic floor formerly served the genial function of wagging animals' tails. Natural Selection urgently pressed what was left of them into a new use in the bipedal hominid. They now served to buttress a potentially lethal defect. Some intermittently upright animals—for example, penguins—have evolved similar adaptations to defend against this problem, but in no other species is the gravitational hazard as serious as it is in humans.

†The soft bones of a human baby's skull resemble tectonic plates. As the infant's head wends its way down the mother's tortuous birth canal, the plates slide and bend to conform to each twist and turn. To assist this molding process, the bony circle of the mother's birth canal relaxes. Under normal conditions, the iliac, pubic, ischial, and sacral bones are welded together by dense bridges of cartilage nearly as rigid as the bones they join. During delivery, however, this tissue undergoes a remarkable transformation, akin to concrete dissolving into Silly Putty. As the fetal head progresses, the pelvic circle, in a complementary maneuver, stretches imperceptibly, its new elasticity conveniently conforming to the mush-skull pushing through it.

had completed a remarkably short burst of rapid inflation that had added one-third to its size. A disaster was in the making.

Eventually, somewhere, sometime, a healthy young hominid had growing within her a new life whose head was simply too large to negotiate the confining walls of her birth canal. During the delivery, her baby became wedged. After a prolonged labor, she died. Her baby died. Those in attendance could do nothing to help. The laws of physics superseded the strength of her uterine contractions. Unfortunately, she was the first of an avalanche of young mothers to die. For the first time in the history of any higher animal, extraordinarily high numbers of healthy females began to die in childbirth; the percentage of stillbirths rose with the number of maternal deaths.

The number of live progeny per mother at the outset of our species was low, because prolonged childhoods forced ancestral women to space their pregnancies far apart. Moreover, one child per pregnancy was the general rule. Young children who lost their mother during a subsequent delivery experienced a catastrophe. Their prospects of surviving without her were bleak. Even a small percentage of mothers dying in childbirth in each generation, especially when combined with factors like disease, drought, or predators, could have placed great stress on a local population.

In a supreme paradox, the leading cause of death for females of the human species became birth. A cursory examination of dates on old gravestones in any cemetery prior to the twentieth century confirms the high mortality routinely associated with childbirth, a condition that does not exist for any other mammal. No female of *any* other species has as much difficulty bearing her young as a human. And no female of any other species routinely solicits and requires help from others to deliver her baby.*

The death of the Unknown Mother signaled the onset of an evolutionary crisis. The loss of a significant number of mothers and their newborns in childbirth was a wasteful reproductive strategy that could have been expected to toll the death knell of the line. Yet it created precisely the kind of crucible in which a species must adapt—or die.

Scientists working in the field of evolutionary biology hypothesize a

---

*There have been sightings of dolphins and whales attempting to assist pregnant females of their kind with deliveries. In one spectacular example observed in captivity, three different species of dolphins were involved. A full-term female was in trouble: Her newborn's dorsal fin was caught in her pelvis. The second dolphin pulled out the baby and assisted the mother to raise it to the surface. While this was happening, the third female delivered the afterbirth, using her teeth. A few land mammals, such as rodents and primates, may also offer limited assistance to a female in labor. Despite these isolated reports, what is clear from observations of many animal births is that no other species' full-term females routinely signal their need for birth assistance. Contrary to popular myths, indigenous women do not simply go into the field and bear their babies alone. In a cross-cultural study of 296 peoples, only 24 reported that a woman on occasion has her baby without assistance. In none of the cultures studied was an unassisted first birth a routine event.[3]

mechanism to explain how a new species often seems to appear all at once in the fossil record. Imagine an isolated local population of an existing species living in harmony with its ecosystem. Suddenly, some new, harmful environmental factor impinges upon the system; large numbers of the local population begin to die. At the eleventh hour, a beneficial random mutation (or mutations) that had previously occurred in the genes of one individual increases its owner's chances of surviving to the next generation.* The offspring of this fortunate individual inherit the gene (or genes) and it quickly spreads. Within the span of several generations, the hard-pressed local population that was on the verge of extinction surges back by evolving an innovative suite of internal metabolic adjustments, physical changes, or modified behavioral responses that allows it to adapt to its new circumstances.

The animal to emerge sometimes differs so significantly from its predecessor that it can be categorized as an entirely new species. Scientists refer to this large dying off of the many so that the few (or even one) can evolve as "passing through a bottleneck." When there is a sudden discontinuity between a precursor species and a new one, some scientists propose that this evolutionary process is due to what they call "punctuated equilibrium."[4]

Many conditions can precipitate bottlenecks. Geologic catastrophes, major volcanic eruptions, abrupt climatic changes (such as the sudden onset of ice ages), pluvials (periods of rains of Biblical proportions), and prolonged droughts can all position a species in the crosshairs of extinction. Epidemics of viruses, bacteria, or parasites can decimate food sources or attack the local population directly.

Approximately 150,000 years ago, in a small region of East Africa, around present-day Uganda, Kenya, and Tanzania, the current countries bordering Lake Victoria, one such bottleneck occurred. A local population of *Homo erectus*, a tool-making hominid, had been living there successfully for over a million years.† Then some yet-to-be-identified event occurred that affected the survival of this particular group of hominids. From this stressed band, a single female known as Mitochondrial African Eve succeeded where Un-

---

*A single gene's sequence of DNA contains the instructions for how to build a protein, which in turn can become an enzyme that further directs the building of an organism. Since there are many variations on the 3-D configuration of proteins and the timing of their entry into the building schedule, a single gene can have an enormous impact on the final form, metabolism, and responses of an organism.

†To avoid a clutter of scientific terms, I will gloss over the subtle distinctions between the fossils referred to as *Homo heidelbergensis, Homo ergaster,* archaic *Homo sapiens,* and many other recent paleontological finds. Each new detail gleaned from studying these ancient bones adds incrementally to our understanding of the evolution of our species, but, unfortunately, to do the subject justice would, I believe, distract from my narrative. Also, a species can evolve in response to a positive development in its environment. A new untapped food source, for example, can prod a species into evolving novel adaptations to take advantage of the bounty. I conjecture that the dominant influence affecting our species was a negative one rather than a positive one, however.

known Mother had failed, giving birth to the new species originally classified as *Homo sapiens sapiens*,* the doubly wise human.

Though the exact birthdate of our species remains uncertain, the scenario that a single woman birthed the modern human species is on firmer scientific grounds thanks to the reliability of the new science of molecular biology.[5] Laboratory tests performed on mitochondrial DNA can accurately measure the genetic variation that exists between members of a species and the differences existing among species. Scientists can then construct "molecular clocks" and calculate how long ago a particular species split away from its precursor.[6] Molecular biology has proved to be the great Rosetta Stone of evolutionary changes. The existence of an African Eve is extremely likely, because the genetic material of all humans alive today is eerily similar.

The genes of chimpanzee communities inhabiting ranges only a few thousand yards apart have more genetic diversity than those of humans separated by oceans. Despite the dramatic differences in the skin pigmentation, eye color, body shapes, and hair types of people from disparate regions of the world, all humans are genetically homogeneous to an extraordinary degree. In fact, there is less than 0.1 percent difference between the gene structure of any one human and another. This suggests that each of us is a not-so-distant descendant of one fairly recent ancestral female. Since we have not had time to diverge very far genetically, our species' birthdate can be calculated backward in tens of thousands of years, instead of millions.

Some dire factor, condition, or event adversely affected the species that lived in the area around present-day Lake Victoria, leading to a population bottleneck. Let us call it Factor X. But what was X? What environmental challenge could have been the catalyst for the radiation of a new species? The geologic and archeological record is relatively silent. Variations in the local climate did occur, but none seems harsh enough to prompt our origin. Scientists have not identified sudden discontinuities in the area's flora or fauna. And yet some extreme condition must have occurred, for African Eve to burst forth like Athena fully formed from the brow of Zeus. Though there are many competing scientific theories, none has managed to gain sufficient support to explain the bottleneck fully.

I propose that the "bottleneck" through which our unfortunate immedi-

---

*This term is now outmoded. With the recent identification of Neanderthal DNA, it is no longer necessary to call us *Homo sapiens sapiens,* except for occasional emphasis. For the rest of this book, I will use the current classification of our species, *Homo sapiens.*

ate ancestors squeezed was actually a *real* bottleneck. Scientists scouring the landscape in search of an external Factor X may have been looking in the wrong place. The precipitating event that pushed a local population of hominids toward the edge of extinction was neither a climate change, a geological force, the arrival of predators or disappearance of prey, nor a shift in the availability of food resources. It was *an internal, anatomical one.**

The unyielding walls of the birth canal, like the alignment of Scylla and Charybdis,† produced the bottleneck that shaped all subsequent hominid evolution. The death of the Unknown Mother and her unlucky baby, and the subsequent dying off of increasingly large numbers of hominid mothers and their newborns, was the stressful Factor X that precipitated the *Homo sapiens* line.

The human female's narrow pelvic canal combined with her fetus's large head led to *Homo sapiens'* "evolutionary bottleneck."

---

*This is the "obstetrical dilemma" first described by Sherwood Washburn in 1960 and elaborated by others, particularly Wenda Trevathan and Karen Rosenberg.[7] The major focus in the literature has been on the effect of difficult labor on child development. The narrowness of the human female pelvis caused infants to be born alitricial—that is, extremely immature. According to estimates based on the size of other primate infants, the length of a human pregnancy should be eighteen months instead of nine. Bringing infants into the world long "before their time" created unique survival problems.

Helpless babies imposed immense child-rearing responsibilities on mothers, forcing a drastic division of labor between the human sexes. And it required women to enjoin men to assist them in raising their offspring, since failure in this endeavor would have fatal consequences for the entire species. It also created novel opportunities for children to have a longer period, called childhood, in which to learn. Much has been written about the consequences of the prolonged human childhood. I wish to redirect the focus away from immature infants to what I consider to be a relatively neglected aspect of the human species' obstetrical dilemma—namely, maternal mortality.

†From a Greek myth, Scylla and Charybdis were two dangerous obstacles between which Odysseus' ship had to navigate during the odyssey.

No other species has as difficult or dangerous a labor as a human does.

Chapter 2

Big Brain/Narrow Pelvis

What men say of women is that we have a peaceful time
Living at home, while they do the fighting in war.
How wrong they are! I would very much rather stand
Three times in the front of battle than bear one child.

—Medea, the wronged wife of
Jason, in Euripides' play *Medea*

The ninety-six hours or so following the onset of parturition (birth contractions) constitute the greatest period of mortality risk that a typical human will ever face.   —Peter T. Ellison[1]

*L*et us call African Eve's adaptation to Factor X "Response W"—a clutch of mutations that remolded us into the most successful and disruptive large animal in the planet's history. Molecular biologists have calculated that we share 95 percent to 98.4 percent of our species' genes with our closest relative, the chimpanzee. (If we count only the active genes, the figure bumps up to 99.6 percent.) The slight reshuffling of the linear code in a small number of her genes must have set in motion changes that would result in innovative behaviors. These few genes transformed a relatively dull, plodding primate into a creature that was "something new under the sun." The subtle reconfiguring of Eve's genome would disperse her species to every continent and large island on the planet. Gracile humans, braving blizzards, would trudge north to inhabit frozen tundras. Others would sail fragile wooden boats across uncharted seas to populate hostile southern deserts.

Our ancestors would then bring about the greatest mass extinction of large animals since the dinosaurs abruptly disappeared sixty-five million years ago. Through their ever-burgeoning technological prowess, humans would plant crops, tend herds, invent writing, build the Parthenon, discover gunpowder, transform the ceiling in the Sistine Chapel, compose the *Eroica* Symphony, and eventually evolve into a biological force capable of influencing the very climate of the earth. Eve's descendants have steadily accumulated the power to destroy each other in an unholy Armageddon and, like sleepwalkers, are shuffling toward a planetary ecological disaster. How could a slight, five-foot-tall, two-legged animal create such sublimity and yet wreak so much havoc in so minuscule an interval of earth's history?

Contrary to what the intergalactic anthropologists witnessing the death of the Unknown Mother and the birth of African Eve would most likely have predicted, high maternal and infant mortality was *not* the beginning of the end. They would have predicted two different outcomes—ineffectively waddling, wider-hipped women, or smaller-brained, less intelligent babies—neither of which occurred. Instead of becoming extinct, we are now six billion strong and continue to multiply at a staggering rate. And yet, until quite recently, women continued to die in childbirth in alarmingly high numbers. How could this be? What possible benefit did having such large-brained babies confer upon the new species that could ultimately offset the significant percentage of fertile female deaths? Moreover, what changes wrought by African Eve's few mutated genes with their minute DNA differences could have opened such an enormous chasm between us and our nearest chimp relatives?

According to the available evidence painstakingly pieced together by paleoanthropologists from archeological sites, Response W was not a highly evolved consciousness, evidenced by the creation of art, musical instruments, burial rites, sewing needles, self-adornment, and sophisticated hunting techniques. These features appear much later. Some posit that it was the sudden maturation of our signature attribute, human language. There is little supporting evidence, however, that freshly minted humans mastered complex speech in the infancy of our species' life cycle.

What we know with certainty is that some subtle skeletal changes appeared: lighter, more delicate bones; a more commodious brain case; a flatter face; and a bigger larynx. But the most striking new feature was our balloon heads. The change in the size of the brain from *erectus* to *sapiens* seems to have occurred in an evolutionary instant. Early specimens of *Homo erectus,* our immediate predecessor, averaged a brain volume of nine hundred cubic centimeters. Our brain occupies, on average, fourteen hundred cubic centimeters. This one-third difference represents a huge increase in brain tissue over the astonishingly brief period of the last several hundred thousand years of *erectus*'s existence.

The human brain is a very expensive organ to operate. Representing a mere 3 percent of total body weight, it can hungrily appropriate up to 25 percent of every heartbeat's oxygen-rich output. The brain demands that its owner expend considerable metabolic vigilance on its behalf. The concentration of ions (calcium, potassium, sodium, chloride, and magnesium) swirling in and around individual brain cells must hover in a very inflexible range. Both the brain's pH and temperature must be narrowly regulated. The body's waste-management system must constantly clear accumulating toxins from the rich soup in which the brain soaks.

So why, we may ask, was such an oversized, extravagant organ installed in this one animal's cranial pan? What was it *doing* at the outset to offset the high maternal death it caused? Many smaller-brained hominids seem to have successfully avoided becoming a big cat's dinner for millions of years. If human language was perfected later, what was the selective pressure that favored a brain capable of such firepower?

The enlarged brain does not appear to have led immediately to greatly superior techniques for acquiring food. That wasn't necessary, for one of the earliest of the *Homo* line, the two-and-a-half-million-year-old *Homo habilis* (Handy Man and Handy Woman), seems to have been quite tenacious using his and her smaller brains. They, too, lasted nearly a million years before going extinct. The archeological record presents scant evidence as to why *Homo habilis* or the later *Homo erectus* evolved their crowning attribute.* Why, then, this big brain? The answer, I propose, has to do with sex, iron, and time.

*   *   *

Besides changing the physical structure of an organism, genetic mutations can also alter an organism's behavioral responses to new environmental challenges. Unfortunately, behaviors leave a fainter imprint on the archeological record than do skeletal remains and are therefore harder to discern. Nevertheless, we can surmise that Response W most likely prodded members of the new species to acquire novel modes of behavior. We can deduce this from the fact that the human brain's major design modification occurred primarily in the frontal lobes—the part of the brain that controls such uniquely human be-

---

Homo habilis first appeared in Africa 2.5 million years ago. This early precursor of the modern human lasted 1 million years before becoming extinct. *Homo erectus* supplanted *Homo habilis* 1.5 million years ago. This hominid had a much larger brain and is a direct precursor to modern humans.

haviors as speech, long-range planning, delayed gratification, and complex puzzle solving.

The behaviors of all organisms can be divided into two basic kinds. The first are those that promote the survival of the individual in response to the dangers and opportunities presented by its surroundings, particularly those concerning feeding, fighting, and fleeing. The second kind—courting, mating, and nurturing offspring—ensures an individual's reproductive success, or, to use the term preferred by evolutionists, its fitness. The first set of behaviors help the organism to attain successfully the age of reproduction; the second set enable the organism's genes to be represented in the next generation. (The term "fitness" can apply to advantageous natural or sexual selection.) If an organism fails in either one of these endeavors, it forfeits its place in the evolutionary game.

There is little evidence that when *Homo sapiens* first made its appearance any major changes in feeding, fighting, or fleeing occurred. The cooperative killing of woolly mammoths would begin much later. But there is much to suggest that the second category, the one having to do with sex and child-rearing, underwent a profound alteration, separating *Homo sapiens* from all the species that had gone before it.

Although *Homo sapiens* displayed features and behaviors quite different from those of its precursor species, it appears that the female of the human species, but not the male, "passed through a bottleneck." She emerged profoundly transformed, displaying some novelties that did not exist in any prior female of any other species. Some of her new features had been present in earlier species, but in the human line they underwent modifications of a significant magnitude. The male, on the other hand, modified his behavior mainly in response to the challenges posed to him by the new female of his species.

This asymmetry in the reprogramming rate between the human sexes drastically altered the way the two related to each other, setting the stage for considerable future conflict and misunderstanding. The relations between men and women would generate great outpourings of emotional anguish that have reverberated down through the seventy-five hundred generations since a few genes of African Eve rearranged the tumblers of her genetic code. The asymmetry also created the conditions for an extraordinary dimension to human love and an ongoing cooperation between members of the opposite sexes unparalleled in its degree and duration in any other species.

Contemporary men and women are living relics of bygone days. In the short span of years that we have existed as a distinct species, insufficient time has elapsed to depart radically from the physiological and behavioral patterns we employed to respond to the conditions we found ourselves in at the dawn of our species. Evolutionist John Bowlby called this the "Environment of Early Adaptiveness."[2] He proposed that the environmental challenges existing during the Pleistocene age molded us into the species we are today.*

Since our genetic makeup has changed very little in the last 150,000 years, I will make the key assumption that the main features of modern men's and women's reproductive life histories do not differ substantially from those present at the outset of our species. There can be no doubt that culture can affect sexual behaviors, but the features that I will be referring to are more basic. For example, I assume that the average length of a contemporary woman's menstrual cycle and that of a current man's obsession with sex are both innate traits that ancestral humans exhibited. (To engage in these speculations, I will make a series of generalizations about men and women. Describing only the crest of the bell-shaped distribution curve of both male and female human behavior, I will not list the many, many exceptions that exist. As a place to begin, let us suspend judgment temporarily and assume these premises are reasonable.)

I will hypothesize that the male's behavior evolved soon afterward in response to the female's lead. In fact, I will argue that *the history of our species could be written from the perspective that males have spent the last 150,000 years trying to regain the power they so emphatically lost to females when we differentiated away from* Homo erectus. By examining the habits of modern human males and females, we can infer the many changes that emerged when the new, improved *Homo sapiens* female debuted in Nature's garden. I will briefly describe these features now, but I will revisit each one in more detail in the coming chapters.

The catalogue begins with the absence in Eve's daughters of some sort of signal that would inform a male that they were ovulating. Unlike the vast majority of other females, the one belonging to the human line does *not* advertise her ovulatory burst. With very few exceptions, other species' females have a distinct period of sexual receptivity, during which they experience a powerful instinctual drive to mate. To the males of her species, a female emanates a distinctive "green light," whether olfactory, visual, auditory, gestural, or some combination thereof. These episodic heights of female sexual desire are exquisitely timed to coincide with her ovulation. Previously uninterested males are alerted by her attention-grabbing signals.

Estrus, as this upsurge is called in female primates, promotes harmony between the sexes. When both male and female are equally excited about mating, it is likely that they will have an amicable and mutually rewarding

---

*The Pleistocene epoch began over a million years ago. During this tumultuous climatic time, the planet was subjected to four ice ages alternating with four interstadials, periods of relative warming in between ice ages. It is in this schizophrenic climacteric that *Homo erectus* and *sapiens* evolved. Recently, the Pleistocene became the Holocene epoch, which covers the last 11,000 years.

encounter. Obviously, a considerable benefit accrues to the species if mating occurs in synchrony with ovulation. Sperm meets ovum, and conception occurs. Eve's daughters, however, lack this most basic sexual semaphore, having replaced it with concealed ovulation. Human ovulation is so cryptic that most women remain unaware when, precisely, their eggs have departed from their ovaries.

Further obscuring the timing of her ovulation, the human female acquired the potential to engage in sex, if she desired, 365 days of the year, during pregnancy, lactation, menstruation, and even after menopause.* An alternative way to state this unusual condition would be to say that the human female does not experience a distinct period of estrus because she is in a state of constant estrus. Precious few other species' females could hold a candle to the human female in this department. No other species has so definitively uncoupled sex and reproduction as the human line. Since sex is so intricately intertwined with reproduction in the other three million sexually active species, what would have been the reason that Natural Selection abandoned this successful strategy in humans?

Another innovation: Some human females experienced a prolonged orgasm capable of multiple sustained repeats. Orgasms, both male and female, are intensely subjective, and we cannot measure with confidence the degree of other female species' orgasms. And yet no nonhuman female, in her observable behavior, comes anywhere near to attaining the heights of sexual pleasure manifested by a woman in the throes of her orgasm.

The male's orgasm, in human and other species, is a necessary component of his ejaculation. It is followed by his rapid withdrawal and prompt disengagement. Only in the human can the female notify the male through vocal or body language, after the completion of his delivery call, that *she* is not finished, and that she expects him to continue until further notice.

Moreover, the variety of sexual positions used in human intercourse exceeds that of virtually all other species. Women became the first land females to habitually copulate face to face with their partners, and they became the first females to increasingly take advantage of an alternative position: mounting a supine male.†

Another feature of human sexuality is the prolonged period of sexual foreplay that occurs prior to penetration. Many other species engage in elab-

---

*Bonobos, a kind of chimpanzee, also engage in sex nearly continually. Nevertheless, females signal through smell and visual displays when they are in estrus, even if that estrus lasts two weeks of their six-week cycle. Ninety percent of bonobo sexual penetrations take place within their estrual period.[3] A few other primates, such as marmosets and tamarins, and an occasional other mammal, such as the porcupine, do not seem to exhibit overt signs of ovulation, and some copulate on any day of their cycle.
†Stump-tailed monkeys and bonobos also use these positions on occasion, but a male mounting from the rear of the female remains their preference.

orate mating and courting rituals. However, when they finally get down to business, sexual foreplay is virtually nonexistent. The human male, in contrast, seems to have grasped the key fact somewhere along the line that it was in his best interests to expend considerable time and effort preparing his partner so that she, too, could experience pleasure. Concern for the pleasure of the female he is preparing to penetrate is not a motive that one would impute to the amatory repertoires of any other species' males.*

Biologists estimate that there are between ten million and thirty million different species of life-forms on earth today. Of these, four thousand are mammals. Only one among the four thousand experiences significant blood loss on a regular basis. If conception does not occur, a fertile human female sheds the lining of her uterus along with approximately forty to eighty milliliters (several tablespoons) of blood every four weeks. A few other mammals—for example, hedgehogs, bats, shrews, and elephants—show signs of menses, but for all of them it is a relative nonevent. Primatologist Alison Jolly estimates that there are approximately 270 different species of primates.[4] Only thirty-one species of primates menstruate. All of these but one, a human, lose an insignificant quantity of blood.

Blood is an essential fluid. What conceivable benefit could female blood loss have conferred on the survival of our species? Something as dramatic as monthly menstrual bleeding must somehow be accounted for in the evolutionary scheme of things.

Although it is possible that menses is an incidental spandrel related to some other adaptation, it looms so large as a constant in the life of every woman that it is unlikely to be a mere accident of nature.† (The point raised by some anthropologists that ancestral women rarely menstruated because they were either pregnant, lactating, or menopausal will be addressed in detail in a later chapter.)

Adding to the enigma of human menses is the human females' propensity to coordinate their menses with other women, a feature rarely observed

---

*Again, male bonobos also exhibit this behavior to a limited degree, and chimpanzee males will groom a female who is pregnant or lactating in the expectation that when she comes into estrus again she will be more receptive to him.

†A spandrel, a term Stephen Jay Gould and Henry Lewontin borrowed from architecture, refers to a feature of evolution that did not evolve for the purpose of enhancing an organism's survival or reproductive abilities. Spandrels, they point out, are the extraneous manifestations necessitated by an earlier adaptation that may have evolved to perform an entirely different function. The authors advocate that biological evolution is the result of a series of highly improbable events that do not have any manifest direction, and not every physical feature or behavior pattern organized by the genome has a beneficial purpose. They caution researchers not to fall into the teleological fallacy of attributing a "purpose" to every feature of an organism. Many sociobiologists—for example, Leda Cosmides and John Tooby—believe that Gould and Lewontin have carried their argument too far and there are, in fact, few spandrels. Most adaptations, they are convinced, are present in the genome because they somehow advanced the survival or reproductive potential of an ancestral individual of a species.

in nonhuman primate females. Like the tines of disparate tuning forks all resonating to the same note, women involuntarily tend to synchronize their periods when they live or work together in offices, convents, dormitories, and large families. Ancestral women lived in tight-knit clans numbering around twenty to thirty fecund women. It is likely that these women, too, synchronized their periods. One suspects it is a very old adaptation that once served an important function in our species' development.

An unexpected component of human females' synchronized menses is that the conductor orchestrating this harmony is an inert 81,000,000,000,000,000,000-(eighty-one-quintillion) ton object located 250,000 miles out in space. The moon is the metronome that sets the tempo for cycles that begin to vibrate in unison in the dark interiors of billions of women's pelvises each month. It is likely that ancestral women, too, entrained their menses with the lunar orbit.

During the past century, technology has flooded our lives with artificial light. Fluorescent and electric lightbulbs have illuminated what had previously been darkness. Moonlight is no longer the critical illumination it once was. Most people are unaware of the current status of the moon's phases. Yet most women continue to coordinate their menstrual cycles with lunar ones.

Another anomaly of the human female's sexual life cycle is her menopause. A woman stops ovulating at an earlier point in her life than any other female mammal, while coincidentally acquiring the distinction of becoming the longest-lived terrestrial mammal.* If she avoids maternal mortality and other causes of an early demise, a woman can on rare occasions achieve a life span exceeding a hundred years. The human female was clearly built to last.† At present in the United States, she outlives her male counterpart by an average of six years, while attaining an average life span of eighty-three years.

A postmenopausal woman possesses a longer period of life during which she is incapable of conceiving a new life than any other female mammals, even though she remains quite vigorous for most of these years. With very few exceptions, other mammalian females ovulate right up to the day they die. And a woman stands in stark contrast to a man, who, despite advanced age and many infirmities, usually can generate viable sperm far into his dotage. Another baffling feature of human menopause: Despite the early cutoff in their reproductive faculty, some menopausal women report an increased libido.[5] If the purpose of sex is the continuation of the species through reproduction, why, only in the human line, did early cessation of ovarian function combine with longevity and increased libidinous desire?

---

*Some bowhead whales have been estimated to live to 150 years.
†This is not an artifact of medical advances. Strong evidence suggests that people in traditional societies can also live long lives if they avoid the common killers of humans.

The innovations distinguishing the human female from other mammalian females mentioned thus far pale when compared with her most spectacular new feature. She became the first female of any species who possessed the willpower to refuse consistently to engage in sex around the time she was ovulating. For that matter, she was the first animal of either sex, of any species, capable of deciding to remain celibate if she so desired.

This resolve is at the *heart* of Response W. This is the gift Natural Selection bestowed upon her for having to endure Factor X, high maternal mortality and painful childbirth. It is something that had heretofore never existed in the animal kingdom. Philosophers call it Free Will. And herein lies the crux of relations between the sexes. African Eve and her daughters developed the determination to choose consciously a course of action that overrode the instinctual circuits that drive every other species' females to copulate when they ovulate. Females of some other species may be able to choose which male among multiple suitors upon which they wish to confer their favors; an occasional female of any species may decide not to mate with anyone or at any time. But the human species was the first in which *all* the females evolved the capacity to decide *consciously* to refuse to mate during any one ovulation or *all the time*.

A major overhaul of the human brain was necessary before a female could acquire the requisite mental equipment to exercise Free Will. The radical new configuration encompassed three key additions. First, when the brains of both men and women rapidly increased in size, they acquired considerable heft. The majority of the new nerve cells were primarily located in the outer covering of the brain, called the neocortex. Second, the frontal lobes, the ones located directly behind the forehead, greatly enlarged. Third, the hemispheres of the human brain became highly specialized. Each half dramatically seized control over opposite but complementary functions. (These three trends can be discerned in earlier hominids' brains, but they greatly accelerated in humans.)

These unique developments led to the creation of special areas within the brain solely dedicated to human language, facial recognition, musical appreciation, rationality, and self-conscious thought. Once activated, the presence of novel mental skills allowed the possessor of one of these new turbo-charged brains to override the circuitry that demanded obeisance to the sexual urges. Unexpectedly, the female of the human species gained considerably greater control over these basic urges than did the male.

She developed an ego-consciousness capable of disengaging from the *be-here-now* mentality used by all other animals. There came into existence within her brain "a room of her own" high above the hurly-burly of her insistent instincts and hormones. For the first time ever, a female *had the time* to reflect in depth on the consequences of mating. She became the first female of any life-form to *understand* the connection between sex and pregnancy. Among the grunting, heaving, thrusting biomass of creatures periodically engaging in sex, it is extremely doubtful that any one of them has the faintest idea what this strange activity's ultimate purpose is.*

To make the connection between the pleasure of sex and the intense, painful contractions of childbirth nine months later, an ancestral woman needed to become aware of time. "So that's what caused my belly to swell," mused the first female who figured out this essential link. Her crucial insight marked a sharp line between all the organisms that had evolved before this event and the solitary one that evolved after it. The new *Homo sapiens* female acquired the ability to arch over the present in order to connect the past with the future, a skill that does not exist to such an extraordinary degree in the mind of any other animal.

Haltingly at first but with increasing assertiveness, this new female was finally able to refuse the hard-wired commandments that demanded she mate when she ovulated. The first woman who achieved such veto power must have mumbled, in the proto-language in use at that time, the equivalent of "Free at last, free at last."

To many women, the most liberating sexual event in history was the invention of the Pill. However, 150,000 years earlier, women acquired a more important advantage: the resolve to say *No!* The story of her release from the slavery of brute instinct is the preface to the tale of how modern humans, both men and women, came to be the way we are. *Homo sapiens* means "Wise Man." So much greater were the changes in the female of the new species than those of the male that it would have been more accurate for scientists to have named our genus and species *Gyna sapiens* rather than *Homo sapiens*.† Throughout the rest of this book, I will acknowledge what I believe to be these most critical adaptations by using *Gyna sapiens* when referring to the ancestral females of the species *Homo sapiens,* genus *Homo.*

The female—not the male—underwent a major transformation, because it was the female—not the male—who was dying in childbirth. It was the female—not the male—who confronted an evolutionary crisis. The dictates of Natural Selection would predict that she, rather than he, would evolve novel

---

*Wenda Trevathan has collected many human observations of various primate deliveries. Across the spectrum of primates, first-time mothers, experiencing the initial pangs of birth contractions, are described by their observers as bewildered. They do not appear to have any understanding of what is happening to them. A human-mother will enter delivery with anxiety, but only on rare occasions would a human be ignorant that her birth pangs were the direct result of her earlier sexual activity.[6]

†*Gyn-* is the Greek prefix root for "female," as in the English "gynecology."

adaptations to the challenge. By wresting control away from her sexual urges, Eve and her daughters exerted discipline over the process of conception. It was small compensation for the increased risks she exposed herself to whenever she became pregnant. If she was to be the one who died in childbirth, then it was she who had best be able to choose when, where, how, and with whom she would become intimate.

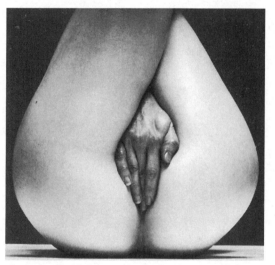

The human female became the first female to override her instinctual sex drive and gain the power to refuse sex when she ovulated.

In Georges Lacombe's wooden bas relief *Isis,* 1895, the artist depicted the fluid issuing from the woman's breasts the color of blood. He uncannily portrayed one of the six major pathways a human female can lose the vital mineral iron.

Chapter 3

Red Blood/White Milk

Child raising is the first industry of every species, and if that industry fails,
then the species becomes extinct.   —Sir Arthur Keith [1]

The social relations of all mammals are determined primarily by the physiology of reproduction.   —Sir Zolly Zuckerman [2]

*I*n Genesis, the Serpent, a reptile, hisses into Eve's ear how she might acquire self-awareness and escape from her reptilian brain. She, not Adam,
takes the first bite of the forbidden fruit, and it is she who then teaches him
what he must do to acquire the Great Gift. The heavy price Eve subsequently
paid to possess this knowledge was the grave sentence God leveled against
her and her daughters: "I will greatly multiple thy sorrow and thy conception; in sorrow thou shalt bring forth children" (Genesis 3:16).

The Genesis story contains all the elements that I suggest actually converged at the dawn of our species, though in a somewhat different sequence.
Maternal mortality was the primary cause, not the punishment, for the emergence of the first woman's ego-consciousness. Eve did not commit the Original Sin; rather, the initiating event of our species was her exercise of Original
Choice. The first man was not prodded by a threat to his existence comparable to what a female had to experience nine months after she stopped bleeding. He evolved cognitively in the area of sexual relations because she
presented him with a threat that he could not ignore.

Adam confronted a knotty problem no other male of any other species ever had to contend with—a female with a mind of her own. She could refuse to mate with anyone, anywhere, anytime. Just as legend has it, that before Eve there was Lilith.*

A quirk of genetic linkages called pleiotropism spurred the first man's rapid evolution. A human has twenty-three pairs of chromosomes. Only one of the twenty-three is considered the sex pair. As long as the gene controlling a trait is not on the female's X or the male's Y chromosome, the mixing of maternal and paternal chromosomes at conception guarantees that an attribute evolving in one sex, such as a nipple or a larger frontal lobe, evolves simultaneously in the opposite sex. As the female was gaining the mental grit necessary to assert her independence, she forced the male to respond to her act of will, because his individual fitness and the survival of the entire hominid line was at stake.

Eve's independence compelled Adam to hastily assemble enough mental wattage to formulate the Brobdingnagian† question "What does a woman want?" Stripped down to its essence, what a man really wants to know is: "What must I do to convince her to let me have sex with her?"

Like Oedipus' encounter with the Sphinx, death awaited a man's genes if he failed to solve the riddle an aloof woman posed. Furthermore, the riddle was a multiple-choice question with protean answers, some of which might be correct in one situation but inappropriate for another. The enigma was complicated enough to force a man to evolve a big brain with large frontal lobes capable of dealing with so complex a mystery.

Of course, if he could not persuade her to join him willingly, he could, as a last resort, overpower her with his superior size and strength. However, aggressive males forcing themselves on unwilling females was not a satisfactory solution to the psychosexual emergency *Gyna sapiens* precipitated with her first adamant *No!* Sheer aggression between males and females has never, *ever* been used as the standard sexual strategy for any other species. From an evolutionary point of view, intersex mayhem would be too costly and dangerous to maintain. Rape in the wild remains an oddity, rarely observed. To install it as a mainstream adaptation in the hominid line would run

---

*The story of Lilith first appears in first-century-A.D. rabbinical commentaries and tells the story of a woman who preceded Eve. Lilith would not bend to Adam's will, spurned his advances, and left him. God supposedly punished her impudence by exiling her. For revenge, she became the killer of the babies born by Eve's more compliant daughters. Lilith's story is a thinly disguised inversion of Eve's story—obviously contrived by men to frighten women—and represents another dreary attempt by men to gain control over women's reproductive resources and blame women whenever something goes wrong.

†Brobdingnag was a country of giants in Jonathan Swift's novel *Gulliver's Travels*.

counter to the tendency toward increasing cooperation that had been building steadily among the highly gregarious and social primates.*

With rape relegated to a rare and hazardous option, the human male found himself at a significant disadvantage as long as the female retained veto power over sexual congress. Nevertheless, the use and threat of rape in historical cultures has introduced a bitterness between the sexes that has poisoned male-female relations.

Since it was literally a matter of life or death to her, the woman retained the upper hand. Fortuitously for the frustrated male, Natural Selection stepped in and gave him an unexpected assist. Sophocles warned, "Nothing vast enters the life of mortals without a curse." Her power to refuse sex was undeniably vast. The curse: Human females began to leak the crucial element iron at persistently alarming rates from a variety of avenues throughout their entire reproductive life.

The depletion of woman's iron stores balanced her veto. Together, these two new adaptations—the ability to say *No!* and chronic iron loss—shaped the course of the many diverse human cultures that flowed from these two remarkable evolutionary developments. In combination, they also provide the answer to the timeless question "What do women want?" At its most fundamental level, the level present at the dawn of our species, what every woman wanted then was the substance that bestowed health and vigor on her and ensured that she birth smart babies. Ancestral women wanted iron. To understand better why iron is so critical and how it undergirds the structural I-beams of a human's pre-eminent attribute—intelligence—we must make a brief digression.

We owe our superior mental agility to our big brain. Brains run on a mixture of two fuels: oxygen and glucose. The latter can be extracted efficiently from either fats, proteins, or carbohydrates courtesy of an enzyme system employed by the liver called the Krebs cycle. Every morsel you eat can be converted to a molecule of glucose.

During a crisis, such as illness or starvation, after your liver has exhausted its readily available reserves, it grimly sets to work cannibalizing the structural components of your body. Fat, skin, sinews, organs, and muscles all become grist for the mill as the liver grinds them down into simple grains of glucose. Implacable in its single-mindedness, the liver has a mandate to sacrifice everything to keep the brain operational, regardless of the debilitating consequences for any particular organ or structure. The liver "knows"

---

*Though rape has been observed in orangutans, right whales, dolphins, and some birds, as well as a few other species, it is not a primary mating strategy among any of them.

what the community of other organs "know"—"If our brain dies, we are as dead as a doornail."

In contrast to glucose, which is a foodstuff, oxygen is an element wafting in the air. Inspired through the lungs, it is transported through arteries, capillaries, and veins to a range of distant tissues by red blood cells. These dinner-plate-shaped discs are unique packets crammed with the spherical protein hemoglobin. A single hemoglobin molecule is huge compared with the average molecule. Yet the mighty engine of hemoglobin cannot function without its tiny spark plugs of iron atoms.

Blood loss from any source leads to diminution of the body's iron stores. Unless replenished, this drop in the level of available iron leads inexorably to an iron-deficiency anemia. An anemic person suffers from lethargy and has a lowered resistance to many diseases. Iron-deficient pregnant women commonly deliver low-birth-weight infants, who more often fail to thrive, are at greater risk for mental retardation, and die earlier and more frequently than those birthed by mothers with adequate iron stores.

Human menses is an evolutionary mystery. Similar to the riddle that confronted the tongue-tied Parsifal standing in the Grail Castle hall, the question suspended in midair awaiting an answer is: This blood the woman sheds, what purpose does it serve? Why would a species evolve that profligately discarded so much of the indispensable liquid, especially when it does not seem to be a particularly important design feature of the vast majority of other females occupying the same mammalian phylum? Millions of females in the other phyla do not lose so much as a drop of menstrual blood during the process of reproduction. A contemporary human female will lose, on average, *forty quarts* of blood during her lifetime of menses.[3]

As puzzling an adaptation as human menses is, it is only the first of six major sources of iron loss that shadow a healthy woman.

The second is the transfer of a mother's iron stores to her fetus during gestation. The brain grows the fastest in utero. Demand for oxygen is exceedingly high in this crucial formative stage. Iron, therefore, plays a critical role during the normal development of the mammalian fetal brain. Among mammals, the human fetus attains the largest brain-to-body size. The only source for fetal iron is the mother's iron stores. The implications for a pregnant woman are portentous. Not only must she keep up with the oxygen requirements of her own brain, but she must also supply the iron necessary to satisfy the demands of her unborn child's mushrooming brain. In the last month of pregnancy, the fetal brain appropriates *three-quarters* of all the energy streaming in from the umbilical cord in the form of oxygen and glucose.[4]

The average daily dietary intake of iron is approximately one milligram per day in an adult man, slightly higher for a woman of reproductive age. A pregnancy causes the transfer of approximately 350 milligrams of iron from mother to fetus, or the amount equivalent to a year's worth of the iron she would need to absorb if she was not pregnant.*

The third cause of iron loss occurs during delivery. Among mammals, there is no more difficult or dangerous labor than that experienced by a human female. The passage of a human life through a life, to enter life, is attendant with more blood, sweat, and tears than is the birth of any other mammal. Not for nothing is this travail called "labor" for the mother and "birth trauma" for her infant.

At birth, the exceedingly large mass of the human fetus's head takes on the function of a battering ram. It stretches the mother's woefully outmatched vaginal outlet beyond its limits. Not uncommonly, significant tears in the channel's lining occur, especially during a woman's first delivery. To prevent delivery lacerations of any degree, obstetricians can perform an episiotomy—a surgically controlled tear in an area away from the all-important anal-sphincter muscles that control fecal continence. This incision's purpose is twofold. First, it enlarges the vaginal outlet, allowing the baby's head to exit more quickly and with less resistance, thereby reducing the stress time on its vulnerable brain. Second, an episiotomy reduces the chances that the baby's head will breach the integrity of the vaginal sheath.† This latter event, though rare, can lead to health-debilitating long-term complications.

The increased blood supply present in the pelvic tissues of a full-term female will result in brisk bleeding from a break in the lining due to any cause. No other mammal except a hyena has such a difficult time extruding its fetus through its vaginal channel.‡ And no other mammal experiences the severity and high probability of tears and lacerations of any kind in the process of birthing its young as a human.

The fourth major cause of iron loss also occurs during childbirth. Mammals bring forth living young that have been nurtured within the mother's womb, unlike the eggs that birds and reptiles lay. Unique among the phyla, a mammalian mother succors her fetus by transferring nutrients in her bloodstream to the bloodstream of her developing fetus. The organ at the interface of this vital transmission is the placenta.

Several different types of mammalian placentas exist. Most have fairly definitive barriers of layered cells protecting the maternal bloodstream from the

---

*Enzymatic pathways in the wall of the duodenal intestine increase the absorption of dietary iron in pregnant females to compensate for the considerable transfer of her iron to her fetus. During pregnancy, a woman absorbs nine times more iron than in her nonpregnant state! For this mechanism to be maximally effective, however, there must first be sufficient bioavailable iron in the mother's diet.
†Recent research suggests that episiotomies have been used too liberally in the past. Judiciously performed, however, they remain a valuable procedure under certain circumstances.[5]
‡More about hyenas later.

fetal one. The most efficient placenta, from a fetus's point of view, exists in higher primates. Here, only the most gossamer of tissue divides the placental inter-face from the uterine one. Tiny blood vessels from each side create mini-lakes of blood separated only by a single porous membrane. As at an open border between two countries, goods and wastes move back and forth with ease.*

Among placentas, a human mother's is the gold-medal winner. It can transfer to her fetus more nutrients more quickly than that of any other pri-mate. If laid end to end, the absorptive surface of the blood vessels of the hu-man placenta would cover thirty miles![6]

The piper, however, must be paid. The critical fourth major source of iron loss occurs when the placenta separates from the uterine wall. Closely en-twined blood vessels on both sides of the maternal-placental divide dis-engage, and a significant number of them are torn in the process. After separation and before the placental delivery, there are critical moments when the divot in the uterus resembles a large raw wound. Between half a pint and over a full pint of blood typically escapes from the mother during this interval. The uterus's charge is to clamp down immediately and com-press the open blood vessels to stanch the bleeding.†

Although this is nature's effective method to minimize blood loss, it is not instantaneous. And with each successive pregnancy, the force of these post-delivery contractions weakens. In spite of the womb's best efforts, the gush of blood from the placental separation can be alarming. Anyone witnessing a live human birth comes away deeply impressed with how bloody mother, baby, sheets, floor, and the hands and gown of the deliverer become.

"Lochia" is the name of the slight bloody discharge that continues for sev-eral days after a human delivery. Although minimal, it is not inconsequential. Its amount and duration in a human mother exceed that of any other mam-malian mother. The combination of tears, lacerations, episiotomies, placental separation, and lochia constitutes a loss of blood, and therefore iron, far in excess of any other mammalian species—five hundred to a thousand mil-ligrams of iron, or the equivalent of one or two pints of blood. One-eighth to one-tenth of a mother's entire iron supply exits at delivery, having been ab-sorbed into the fetus and the placenta or squandered profligately during de-livery.

The fifth major cause of iron depletion in *Gyna sapiens* is not so obvious as the previous four but nonetheless significantly increases her risk of devel-oping an iron-deficiency anemia. The transfer and loss of iron associated

---

*The placental barrier is not completely porous. Red blood cells cannot cross it from either side.
†Administration of modern drugs by obstetricians has significantly reduced the quantity of post-delivery bleeding. Ancestral mothers could not avail themselves of this immense advantage, so their blood loss during delivery would have been considerably greater.

with gestation and birth exist to a lesser degree in other mammalian mothers but still pose a problem. To counter it, Mother Nature equipped females of the other mammalian species with a vital instinct—an urgent hunger driving them to consume their offspring's placenta. A plump soufflé of meaty iron, amino acids, and essential fats, the placenta is the consummate first meal a mother should partake of immediately after the ordeal of delivery. It is the perfect replacement for the very nutrients she lost just minutes earlier, because a freshly expelled placenta contains the iron equivalent of one or two blood transfusions.

Gyna sapiens has lost her craving for this delicacy. Our closest relatives, chimpanzees, dine with gusto on their afterbirth immediately after delivering their infants. In contrast, nurses whisk away the placenta before a mother can even catch a glimpse of it, so that it can be discarded in the trash. Present-day hospitals label the placenta "toxic waste" and issue strict regulations governing its disposal. Among many other human cultures, placentas are thrown to dogs, who, recognizing a valuable resource, gratefully devour it.* Failure to ingest the placenta constitutes the fifth significant source of iron loss in the human female (albeit a passive one).

The sixth and last major source is perhaps the most pernicious. Newborns spend the majority of their time suckling and sleeping. They should. Infants are a complex, incomplete building project feverishly under construction; no wonder they are exhausted most of the time. During the period of breast-feeding, the baby's brain grows rapidly. In the first year of life alone, it more than doubles in size as it steadily lays down layer upon layer of new brain cells.

To get the job done, newborns require a constant energy source. Now that they are breathing on their own, they need to put in place quickly their own elaborate transport system to carry oxygen from lung to organ. Their most demanding organ is their rapidly enlarging brain. And where, we might ask, does an infant derive *all* the iron needed for its shiny new red cells? Again, the mother is the *only* source of this crucial element.

Human breast milk constitutes another significant source of iron loss for a woman. Even though it is not especially rich in iron, the duration of its loss makes up for its concentration. Unlike the sudden exodus of red cells at de-

---

*Though the large majority of mammalian mothers consume the placenta, it is not a universal behavior. Some researchers in the field conjecture that the mother quickly ingests it so that its odor will not give her offspring's location away to a predator. Mares do not eat their afterbirth immediately after giving birth but quickly abandon the spot. The majority of researchers posit that the instinct to consume the afterbirth is primarily driven by its nutritional value, because herbivores eat it as well as carnivores. Humans diverge from most other mammals in their distaste. Among a cross-cultural study of over 300 human cultures, there were *none* in which the placenta was regularly consumed. In many, it was, however, ceremoniously buried, indicating that many people honored its centrality in the birth process.[7]

livery, lactational iron loss occurs steadily over a span of several years. Observations among hunter-gatherer lactating women, such as the !Kung San of the Kalahari, reveal that on average they breast-feed their young for an average of two years and eight months.[8]

Anthropologists use the customs of these and other hunter-gatherer peoples to extrapolate back in time and make approximate predictions concerning ancestral women. Breast-feeding for almost three years represents a steady but incremental drain on a lactating mother's iron stores. Again, changes in her digestive system rapidly increase her gut's ability to absorb iron from foodstuffs, but iron-rich food must be available to her before she weans her baby. During the time a mother is breast-feeding, complex hormonal adjustments in her system forestall menses in all but a few cases.

Within a month or two after separating her toddler from her breast, just as night follows day, the mother begins to menstruate again. But a new factor will cause her to lose more blood with each menses than she did prior to her pregnancy. Once a woman has had her first child, the increased size and more robust vascularity of her uterus will cause a small but significant increase in menstrual loss of iron. Each subsequent pregnancy will be followed by a slightly heavier menstrual flow, and this trend will continue for the rest of the woman's reproductive life.[9]

With the cessation of lactation, a woman's menstrual cycle begins anew. *One constant remains: By one avenue or another, a woman is always losing iron.* Over a lifetime, the average woman loses the equivalent of approximately *fifteen gallons* of iron-rich blood due to menses, pregnancy, delivery, birth trauma, placental loss, failure to consume the expelled placenta, and lactation. If she has access to sufficient dietary replacements, she can easily make up for this loss. However, a crisis occurs if she encounters a problem in either her health or her diet which interferes with her acquisition of iron.

Recent research carried out by the increasing number of female anthropologists entering the field suggests that modern women lose more blood from more frequent menses than did the women of ancestral times. Studying the indigenous Dogon tribes in Mali, anthropologist Beverly Strassman found that the fertile Dogon women are in a near-continuous state of pregnancy or breast-feeding. Also, the onset of menarche occurs later than it does in modern cultures.[10] Strassman calculates that the average Dogon woman experiences 110 periods during her reproductive life, compared with over four hundred for women living in the industrialized West.[11]

Reading the literature on hunter-gatherer reproductive statistics might leave one with the impression that as soon as a mother weans her toddler, she immediately begins another pregnancy, leaving no space for menses to squeeze in between lactation and the next conception. But this is not the case. The Dogon practice agriculture and they are not representative of pure hunter-gatherers. Anthropologist Sarah Blaffer Hrdy documented in her book *Mother Nature* the nearly universal practice among hunter-gatherers to practice abortion and, if necessary, infanticide as methods of family planning. Women in hunter-gatherer societies generally try to space their children at least four years apart. Attempting to care for, feed, and nurture too many children too close together places great stress on both the mother and her tribe. A number of menstrual months occurs between the end of one lactation and the beginning of the next pregnancy. Nevertheless, some women in extant hunter-gatherer tribes recall having few menses during their reproductive life.

There are additional factors. Several obstacles unique to humans exist that make serial pregnancies chancier for them than for other mammals. Reproductive physiologists estimate that it takes, on average, three menstrual cycles of concentrated lovemaking among young, healthy human couples to conceive. The reasons for the delay are several.

The incidence of miscarriages is 40 percent in the first trimester of pregnancy. Many of these occur before a woman even has an inkling that she is pregnant. A woman experiencing such a loss would have to rejoin her partner in another ninety-day cycle of lovemaking to conceive again.[12] There is little reason to doubt that ancestral women experienced the same percentages of lost fertilized ova. (The possibility cannot be dismissed, however, that modern culture has introduced a number of artificial environmental toxins that may be affecting the odds of conception and the percentage of miscarriages—factors with which ancestral women did not have to contend. But this would have been offset by earlier women's much higher parasitic loads.)

Any deficiencies in the health of either member of the pair, when added to the factors elaborated above, make conception in humans a hit-or-miss affair. During the not inconsiderable number of months when a sperm and an ovum were trying to link up, a woman would experience several menses, each requiring iron replacement. Sarah Hrdy in her study of !Kung women estimated that one-half of them died childless.*[13]

Strassman's research on the Dogon women, however, does *not* mean that the fewer menses among her subjects translate into less iron loss than among contemporary women. The iron drain from the other five major

---

*Many cultural factors differing from one culture to another could be used to explain this statistic. Also, there is considerable hazard in trying to extrapolate from an extant culture the behavior of an ancient one.

sources for blood loss exceeds the not insignificant amount of blood ejected with each of nine periods she missed by being pregnant. The iron loss from lactation also exceeds the amount that would have been lost in menses over the two years and eight months of breast-feeding.

Whether a !Kung woman is menstruating, gestating, or lactating, she is constantly losing iron. Although they had less frequent menses than modern women, fertile ancestral women still needed to establish a safe and effective way to replenish their iron stores. (And this does not take into account the woman's body's urgent requirement to restock its warehouses of proteins, fats, vitamins, and minerals.)

*   *   *

Along with the six major iron-leaching causes, there exist four minor, variable bloodletting pathways that can harass women of reproductive age. The first two—polyps and fibroids—are created by the unique cellular composition of a woman's uterus. Imagine a TV weather personality delivering nightly forecasts for each of the body's regions. Weather reports for kidneys, eyes, hair, bladder, and toenails would be monotonously boring, because each area experiences fairly constant climes. The "weather" in the uterus, in contrast, is constantly in a state of flux. Furthermore, a destructive storm regularly lays waste to the landscape. No other organ manifests such a striking change in its appearance during each month, and none has so dramatic and repetitive a cycle of growth, maturity, and death.

Bones and teeth, for instance, once they have stopped growing, remain essentially unchanged structurally for the rest of life.* The heart is certainly dynamic in function, but though it constricts and relaxes seventy times a minute, little change occurs in the life cycles of its component cells. White blood cells and the mucosal† cells that protect the lining of the intestinal tract turn over every few days. Their individual cell cycles, however, are conveniently out of phase with each other. When some cells are young, others are mature, while others are dying. Because of their asynchronicity, the appearance of these organs does not vary from day to day.

A very different situation takes place in the uterus, whose tissues take their orders from hormones secreted by distant glands that convey widely conflicting directives. Most cells lining the uterus respond in lockstep to

---

*At the atomic level, there is considerable constant activity, as old calcium atoms move on and new calcium atoms replace them.
†"Mucosa" describes the inner lining of an organ.

these monthly comings and goings, compelling the uterus to fluctuate in both its internal structure and its appearance.

Cells find it easier to escape from the normally strict biological controls regulating the manner in which they grow if they live in a whipsawed environment. The uterus is the Wild Western frontier of the body. Cellular lawlessness is common. In the course of a woman's reproductive life, she is likely to develop several abnormalities that exacerbate cervical and uterine bleeding.

Polyps and fibroids are overgrowths of cells that have rebelled against the body's strict cellular-growth regulators. These sporadic rogues' independence movements can occasionally transmogrify into cancer. Fortunately, the vast majority of both polyps and fibroids remain benign. Nevertheless, they can cause local problems.

Polyps are heaped-up mounds of exuberant cells arising from the cervical and uterine lining. Older medical textbooks called polyps "proud flesh" because they stand up and out from the surrounding normal tissue. These slightly abnormal cells are very fragile, tend to bleed easily, and are one of the most common causes for "spotting" between periods.

Fibroids are benign tumors consisting of uterine muscle cells that have experienced runaway growth. Stimulated by estrogen and progesterone, they are usually multiple in numbers. Though slow-growing and usually painless, some can reach prodigious proportions. Fibroids arising close to the interior of the uterus can cause a thinning of the delicate inner lining. Eventually, the tumor stretches the lining to the breaking point, causing persistent "spotting" from the exposed raw surface of the tumor. In some cases, bleeding fibroids can produce a life-threatening blood loss. The surgical removal of the uterus will definitively stop the bleeding from any abnormal condition, but this recent medical advance was not available to ancestral women.

We have a record of just such blood loss in the New Testament. Luke, a physician, relates the story of a woman who sought Jesus' help because she was plagued by "an issue of blood for twelve years" (Luke 8:43–48). "Menorrhagia" is the medical term for her abnormal uterine bleeding. Her anguish is evident in the passage. It is reasonable to assume that uterine bleeding from pathological conditions also vexed ancestral women. Physicians treating a patient with menorrhagia routinely prescribe iron supplements. Only after menopause does a human female's daily iron needs stabilize to mimic those of a male. Even then, her total body reserves remain below male levels.

In addition to polyps and fibroids, the third minor source of iron loss plaguing women (which can also beset men) is the dreaded hemorrhoid. A uniquely human condition, it was the direct result of our hominid ancestor's fateful decision to stand up and walk on two legs instead of four. In virtually every other mammal, the anus and heart occupy the same horizontal plane. Venous blood returning from a hind end can flow lazily back to the heart, usually downhill.

Upon assuming an upright stance, a hominid positioned its heart approximately two feet above its anus in a vertical plane. This created a bedeviling hydraulics problem. Arterial blood flow hurries along on its appointed tasks, urged onward by the propulsive force of each systolic heartbeat. Once on the venous side of each capillary, however, the movement of blood slows to a sluggish crawl, since there is no pump propelling it back to the heart. In humans, venous flow returning to the heart from the pelvis must defy the force of gravity, heading straight uphill. As a result, the veins underlying the delicate lining of the distal rectal tract remain permanently engorged, because there is a several-foot-high column of blood pressing down from above.

Exacerbating the problem of hemorrhoids is a most curious behavior. Humans are the only animal in all the phyla that must learn at a very early age to maintain tight control over their anal sphincter.* To the ever-constant delight of small children visiting the zoo, animals generally poop whenever the urge moves them.† Having recently suffered through the intense period dreaded by both parent and child known as "toilet training," it is no wonder that children marvel at all other animals' lack of inhibition.

Controlled defecation must have developed relatively early in our evolution. Hominids were the first mammals to congregate at a semipermanent home base, to which they brought food to prepare. Our ancestors discovered through trial and error that it was imperative to train their young to control their sphincters to prevent the spread of disease.

One can accurately gauge the level of anxiety concerning this subject in human culture by the hearty laughter and continual fascination it evokes in children and many adults. It could be said with assurance that, despite the plethora of varying cultural traditions throughout the world, losing control of one's bowels in public ranks in all of them as the most humiliating of social gaffes. A physiological side effect of this idiosyncratic human cultural convention is that the human anal sphincter is in a state of near-perpetual constriction. The small circular muscle squeezes the veins around it, engorging them further, making them more susceptible to bleeding.

Pregnancy significantly exacerbates the problem. Hemorrhoids tend to recur even after the woman has delivered. Women, more than men, suffer from

---

*Some pets and domesticated animals can be "trained," but it is not a natural condition in the wild.
†Many animals do so only in certain locations within their individual territory, using excrement as means of territorial marking.

bleeding hemorrhoids. Blood loss from hemorrhoids alone can cause a significant chronic anemia. Physicians treating a patient with bleeding hemorrhoids, short of surgical intervention, routinely prescribe iron supplements.

A fourth precipitating cause of female anemia from iron loss is the tiny hookworm parasite, which is in fact the leading cause of iron-deficiency anemia throughout the world for both sexes. Until the era of modern medicine, parasitic infestations were unwelcome hitchhikers common to all human societies. Even today, the World Health Organization estimates that a quarter of the world's population is infested. Hookworm can tip a marginally anemic woman over to one who suffers the full manifestations of the disease.

One other important factor adds to a woman's iron woes. A slight anemia translates into heavier menses and more blood deficits from delivery loss. This sets up a vicious cycle difficult to break. The more anemic a woman is, the more blood she loses, making her more anemic and causing increasing iron loss.

When taken as a whole, the six major sources of blood loss in a woman and the four minor ones are quite puzzling. No other female animal possesses what can only be characterized as a suite of traits that predispose its possessor to the threat of ill health. The question arises—why? Of what conceivable benefit could it be to the fertile female to lose iron constantly and skirt the edge of anemia? How did this piling of one seemingly negative adaptation after another serve to advance the overall fitness of the species? And why would this deleterious iron drain occur in only one sex of that one species?

*   *   *

Health authorities in the First World have masked a woman's need for iron. Recognizing the role the red metal plays in her health and the health of children, manufacturers have fortified most packaged foods with iron supplements in a form that is easy to absorb. Scrutiny of the fine print on any cereal box will confirm the presence of the additive. This simple act has greatly reduced the incidence of iron-deficiency anemia among women fortunate to live in such societies. To avoid making comparisons to women living in a technologically advanced society, the point is worth emphasizing. At the dawn of our species, women did not possess this luxury. Once humans left equatorial Africa for climates more extreme, their access to dietary sources of plant iron, especially in winter, became restricted. In order for both them and their babies to be vigorous and mentally sharp, women became highly motivated to find a way to ensure a reliable source of bioavailable iron for their diets.

The foodstuff highest in absorbable iron is blood. Ironically, there exists a powerful taboo in almost every culture in the world against drinking blood. The vampire myth, for example, is among the most frightening in many cultures, and probably arose to forestall the practice of cannibalism.* The food resources second highest in iron are liver and bone marrow, and the third and most plentiful is meat. Consuming any of these animal products is the easiest and fastest natural way to replenish depleted iron stores. So how would an ancestral woman have guaranteed that she would have ready access to the iron she and her offspring so desperately needed?

---

*Not all people are so squeamish. During their long rides across Asia and Europe, Mongol horsemen sustained themselves by regularly nicking their horses' neck veins to drink their mounts' blood. Contemporary Masai tribesmen of Africa routinely bloodlet their cattle and consume the fresh blood.

Of the 270 species of primates, only one consistently hunts other animals.

Chapter 4

Plant Iron/Meat Iron

There is no question that a reconstruction of ancestral mating systems will be an essential part of any satisfactory account of human origins.

              —Martin Daly and Margo Wilson[1]

Some [indigenous] women in North America avoided meat during pregnancy out of their very practical desire to keep their developing fetus small, to avoid complications or death in childbirth.   —Judy Grahn[2]

*M*other Nature is spectacularly elegant. And She loves the color green—all shades of it. The world wears green because plants possess the tiny but mighty molecular engine, chlorophyll. A photon hurled outward from the superheated surface of the sun speeds toward the earth in the form of a minuscule fireball covering 186,000 miles in a second. Try to intercept one and—splash!—it explodes in an energy wave that dissipates as heat in an instant. Chlorophyll somehow manages to catch these speedy messengers from the sun in slow motion, thereby harnessing the sun's light and converting its energy into the mass of the plant. All stems, trunks, flowers, and leaves are the living proof of this wondrous transmutation.

    The core of chlorophyll, the marvel behind the process of photosynthesis, consists primarily of hydrogen, carbon, oxygen, and nitrogen atoms arranged in a ring pattern. At the exact center of this complex molecule rests a single atom of magnesium—a dull-gray metal that gives much of the world its leafy-green color. Substitute that single magnesium atom for one of iron—and presto!—a chlorophyll molecule magically transforms to resemble closely the core of hemoglobin. Iron ore, another dull-gray metal, plays an important role in coloring the innards of all complex animals red. Green for plants; red for animals. So beautiful, so simple.* In the preface, I spoke of experiencing moments of pure rapture when I learned certain details concerning the workings of life. Encountering the complementarity between chlorophyll and hemoglobin was one such moment of awe. Evolutionary nutritionist Michael Crawford, struck by this similarity, comments, "The hemoglobin of the red cell is so close in structure to chlorophyll . . . it makes one wonder if the genetic information of the red cell came from a mutation of the chloroplast data."[3] Because the early part of this book's thesis hangs on the hook of iron, how an iron atom makes its way into the center of hemoglobin is pertinent to the story I plan to unfold.

Iron is a fundamental element, one of the 118 found in the periodic table. It must be eaten, because it cannot be manufactured from constituents in the body. The iron atom has two stable forms: ferrous (Fe^{++} or Fe_2) and ferric (Fe^{+++} or Fe_3). Since iron is a major constituent of soil, and plants grow in soil, iron is a structural component of many fruits, grasses, grains, and vegetables. Unfortunately for humans, most plant iron is unavailable to us, because both forms of the iron atom are tightly bound to the plant's organic molecules (such as phytates, tannins, and many others) in a process called "chelation." Plant-eating animals possess digestive processes that can readily strip the metal away from the chelating molecule's steely grip to incorporate the vegetable's iron into their hemoglobin. Mysteriously, the human digestive tract lacks the common but critical enzymes that facilitate the rapid absorption of the iron from many vegetal foodstuffs.†

The Fe^{+++} in plants is more difficult for a human digestive tract to absorb than Fe^{++}. Spinach, famously rich in iron, contains primarily its chelated ferric form. (Popeye, contrary to expectations, would not have received much of a jolt from his famous spinach gulp.)

These two nutritional impediments, one major (chelation) and one minor (iron in its chelated ferric form), are the principal reasons why the majority of iron contained in vegetables passes through the human intestinal tract unabsorbed.

Animals, in contrast, are composed of tissues that abound with iron in an easily absorbed ferrous and ferric form, known as heme iron. Bone marrow,

---

*Some sea creatures—crabs, for example—use the organic molecule hemocyanin to oxygenate their blood. Remarkably similar to human hemoglobin, hemocyanin uses copper at its core instead of iron, and copper colors the blood green. Perhaps this is the reason that monsters in film and computer games seem always to have green blood.
†Some vegetables are good sources for easily absorbable iron: for example, carrots, beets, pumpkins, potato skins, broccoli, and cabbage.

liver, and muscle tissue (meat) brim with heme iron. Aiding absorption is the quirk of our digestive tract's metabolism that allows the cells lining the stomach to transfer heme iron more easily into the interior of blood vessels if it is in the presence of other animal proteins. Ancestral humans must have intuited that eating meat was the fastest, easiest way to absorb iron and maintain vigor. Though they did not understand biochemistry, they surely would have developed a native intelligence on matters concerning health.

Another consequence of the change in the human diet from vegetarian to carnivory: The more meat we ate, the shorter our gut became. The shorter our gut became, the more oxygen was available to nourish a brain that grew in size.* Most animals allocate the greatest portion of their available oxygen to their alimentary tract. Digesting plant food is hard work and requires a high-energy source. Upon switching over to a diet high in animal products, the gut no longer needs so much oxygen, and the excess that is freed can be diverted to the brain. A few nonhuman primates—for example, chimps, capuchin monkeys, and baboons—eat meat but not as a major food. Those few that do are the most intelligent nonhuman primates. In general, carnivores are smarter than herbivores; foxes are slyer than donkeys.

If humans were becoming increasingly dependent on oxygen-consuming brains, one might predict that obstacles to the digestion and the absorption of iron would diminish. Instead, they increased. Why did a human lose the genetic instructions that create the enzymes to take in plant iron easily? If prodigious brains depend on the timely delivery of great gasbags of oxygen, why would impediments to the absorption of the key to its transport, iron, coevolve with our increasing dependency on iron? And why would there be a marked gender disparity in the frequency of iron deficiencies that seriously affect health?

Along with pigs and hippopotami, humans are hearty omnivores. They can derive sustenance from a wide variety of both plant and animal foods. But they must partake of each food group. A *vegetarian* easily avoids dietary deficiencies by eating dairy products. A *vegan* is distinguished from a vegetarian by abstaining from eating *any* animal-related food, including eggs, milk, and cheese. A man adopting a vegan diet is at low risk for developing a serious anemia. A woman in her reproductive years without access to modern nutritional knowledge who makes the same choice *cannot avoid* anemia. She must ingest some supplementary form of iron or risk injuring her health and the health of her unborn baby. Natural Selection seems to have rigged the human's digestive tract so that humans, especially females, would have to acquire iron from the flesh of another animal. Before discussing how women

---

*This idea, the Expensive Tissue Hypothesis, is advocated by Leslie Aiello and Peter Wheeler.[4]

solved this dilemma, let us examine what happens to the iron that does make it across the intestinal barrier into the bloodstream.

An atom of iron successfully transiting the human stomach and intestinal lining is rapidly chelated by apoferritin, a specialized deliveryman protein bobbing along in the bloodstream waiting for a shipment.* Once it has taken iron atoms on board, ferritin, as it is now called, transports its charges and deposits them primarily into the bone marrow. When not engaged in making its daily rounds in a red cell's hemoglobin, excess iron is also stored in the liver.

When red-cell levels fall—for example, as a result of bleeding from any cause—sensors within the bone marrow and the interior of the kidneys note the deficit and issue orders to ramp up production of both the hemoglobin protein and new red cells. Once the bone marrow ascertains that red cells have reached maturity, it releases these compact discs into the bloodstream. There each cell begins its nonstop work shift, shuttling back and forth between lung and distant cell, carrying its indispensable hod of oxygen. A red cell has a life span of approximately 120 days. As it nears decrepitude, specially designed sentinels in both the spleen and the liver identify and then yank these seniors off the line. They crack open the red cell's membrane, break down its hemoglobin into simpler components that can be used for other purposes, and extract the senescent red cell's semiprecious iron ore.

Recycled back to the bone marrow, a single atom of iron will later be incorporated into a new red cell, and another 120-day cycle will begin. Iron atoms never tire or wear out; they are indestructible—the perfect reusable part. Nearly all vertebrates, up the evolutionary ladder from fish, depend on hemoglobin and its reliable atom of iron to transport oxygen.

When an animal dies, its body decomposes and its iron dissipates and eventually settles into the soil. There the worms churn it. A seed, blown by the wind, lands gently nearby. Soon roots of what will become a new plant are tentatively exploring the subterranean neighborhood. A root's tentacle sucks the iron in the soil into its substance, and the iron atom finds itself an integral part of the new plant's infrastructure. And then, one fine day in spring, a shy green tendril nudges aside the loam to find its place in the sun. As the plant grows, the iron that began in the soil moves into the upper stories of the plant's leaves. Eventually, an herbivore may come along and eat the plant. Perhaps a carnivore will then eat the animal that ate the plant, and the iron will find itself in yet another silo. This cycle will go on and on as long as the sun continues to shine and the earth abides.

---

*The human digestive system distinguishes between iron chelated in plants and iron chelated in animals—the latter being readily absorbed, in contrast to the former.

*   *   *

When biochemistry became advanced enough to measure the concentration of the various vital elements circulating in a human bloodstream, researchers discovered that women had significantly lower levels of circulating red blood cells, hemoglobin, and iron stores than did men. Unable to explain this discrepancy completely, the clinicians who set the standards for what is normal decided, somewhat tentatively, that it was normal for a woman's red-cell and hemoglobin parameters to be 15 percent below those of her male counterpart.

Implicit in their determination is that women should be able to get by normally on less hemoglobin than men do, or that women's metabolism is more efficient.* No evidence exists that cellular oxygen transfer in a woman differs from that of a man. Basic metabolic enzymatic processes, as a rule, make no distinctions between sexes. Current conditions among contemporary women living in technologically advanced cultures that routinely fortify food with iron should not obscure the fact that ancestral females living in the Pleistocene would have constantly teetered on the edge of developing a chronic iron-deficiency anemia. A woman's best defense against this pathological condition's subtle and deleterious effects on both her health and the intelligence of her unborn children would have been to consume foods rich in iron in its readily absorbable heme form. (Or, alternatively, to live in an environment that contained rich sources of plant iron in easily gathered, abundant quantities.) Men do not need supplementary iron, because losing it is not part of their physiological makeup. Women, on the other hand, have to have it.

---

*Human metabolism contains wide margins of safety for most of its parameters. No diminution of function is detectable until hemoglobin levels fall below 30–35 percent of normal. The 10–15 percent difference between male and female hemoglobin levels does not appear to be clinically significant. Its major impact is on a pregnant female's fetus if she begins to challenge her margin of safety when her hemoglobin levels drift below her normal levels.

Gyna sapiens' reproductive life cycle was a dramatic departure from all that had gone before.

Chapter 5

Gyna Sapiens/
Gyna All-the-Others

The female sexual response was evolved and perfected many millions of years ago ... certainly in creatures far simpler and more primitive than ourselves. The reason they make so little fuss about it is precisely because it *had* been perfected; it was as simple and effortless as eating, and the reward in terms of pleasure was as automatic. So the question to which we ought to be addressing ourselves is not, How and why did the human species evolve this frightful complicated and mysterious female mechanism? It is rather, How on earth did the human species come to lose, mislay, and/or generally louse up such a simple straightforward process?   —Elaine Morgan[1]

The psychosexual pattern in man looks very much like the end product of a biological emergency.   —Alex Comfort[2]

*T*he two factors that advance a sexually reproducing organism to the next evolutionary round are its ability to survive to reproductive age and its success in mating with the opposite sex. Jackpot bells ring throughout the evolutionary casino when half the organism's DNA drops down into the next generation in the form of a healthy offspring. The new organism, having received genes from both a fit mother and a fit father, then begins the arduous competition of surviving long enough to reproduce. And so it has gone—for millions of years. *Gyna sapiens* experienced the most dramatic changes in the genes responsible for her reproductive cycle. *Homo sapiens* underwent the greatest genetic revolution in the genes controlling his survival strategies.

When attempting to understand why a particular human trait evolved, an evolutionary biologist seeks to discover the benefit the trait confers upon individuals within the species in increasing the odds of either survival or reproductive success. When an inherited trait seems at first glance to be deleterious, then the researcher must look beyond short-term disadvantages and search for a less obvious long-term benefit.* Confusing the issue is the possibility that the trait under scrutiny might have nothing to do with enhancing survival or reproduction, and instead might simply represent a spandrel of no evolutionary consequence. Spandrels, remember, are evolved traits having no particular impact, either positive or negative, on survival or reproduction.

Evolution's most useful metaphor is a branching bush. Our two closest relatives, *Pan troglodytes* (common chimpanzees) and *Pan paniscus* (pygmy chimpanzees or bonobos), are linked to us by a yet-to-be-identified Last Common Ancestor, from which we three apes evolved around five or six million years ago. The hominid line branched into us, and the other fork split again into two end twigs, on the tips of which at present sit our two slightly different chimpanzee cousins. Jared Diamond titled one of his books *The Third Chimpanzee,* reminding us that we humans were the third member of this small family. Yet, despite our close chimp kinship, a human female's sexuality differs significantly from that of her close cousins and varies even further from other nonhuman females.†

The disparity between *Gyna sapiens'* reproductive life history and the females of her nearest relatives is so great that a veritable army of Ph.D.s seek to explain the reasons for her considerable distance from what constitutes the crest of the reproductive bell-shaped curve for other, nonhuman females. Yet a consensus among interested scientists has thus far failed to gel. Part of the reason *Gyna sapiens'* evolutionary picture remains so opaque is that each of the revolutions in her reproductive life story, examined individually, seems to have either mightily or slightly disadvantaged her in the competition for resources, survival, and reproductive fitness. Let us examine each of her un-

---

*The peacock's fabulous tail is an example of a trait that would seem to hinder the survival of any male saddled with having to display this apparently useless appendage. A predator could more easily capture and eat a peacock with very large tail feathers than one with those less imposing. Amotz and Avishag Zahavi, Israeli evolutionists, proposed a theory they called the "honest handicap" to explain why. A peacock adorned with a magnificent tail ensemble preening before a peahen announces that he is healthy, strong, and smart. Otherwise, how could he have survived this long lugging around his enormous train? His splendid fan is an honest handicap because a male cannot fake it. He cannot deceive a peahen that he is more fit than he is. Once the first peacock developed the mutation for a bigger tail feather and survived to reproduce, peahens began to use this feature of masculinity to judge the quality of their suitors. The process of sexual selection then took over. Discriminating peahens increasingly favored males with large, gorgeous tails. Male genes for large tail feathers advanced over male genes for smaller tails. The trait the females chose became grotesquely exaggerated in a process evolutionist Ronald Fisher called "runaway selection." Each generation of peahens keeps selecting the males with the biggest, most colorful tails, and with each successive generation the tails grow larger and more useless.[3]
†Gorillas and orangutans, our more distant cousins, are also apes.

usual adaptations with an eye to evaluating its potential benefit against its possible harm to her reproductive fitness.

One unusual trait in human females that has baffled scientists for years is cryptic ovulation. A male of the human species does not receive any reliable clues as to the moment when a female ovulates. So secret is her ovulation that she, too, is generally unaware of the moment of the Great Launch. Because cryptic ovulation is intimately associated with both *Gyna sapiens'* loss of estrus and her potential for continual sexual receptivity, I will discuss all three of these interrelated features together.

In the majority of primate species, a female in estrus displays spectacular physical manifestations as ovulation approaches. Her vulvar sexual skin (best seen from behind by the male) blushes a flaming red and swells with edema fluid, sometimes impressively. The sight of a female in such a state stimulates the male tumescently. Excitatory pheromones waft from her vulva, further agitating the troop's males.

In case the male does not express the appropriate interest in this sight-and-smell show, the female primate will actively solicit copulation by presenting him with an enticing, up-close rear view. In a few species—for example, the gray langur—the female will position herself directly in front of a male and perform a shuddering movement of her head and/or stroke the male with her hand in an imploring manner. Still others emit a distinctive call. Singularly or in some combination, the females of 269 primate species use call, smell, sight, touch, and/or gesture to announce their sexual receptivity. One does not. Her secretive ovulation separates *Gyna sapiens* from other mammals as well. Reproductive evolutionist Margie Profet declared, "The only mammal whose ovulation is known to be truly concealed is *Homo sapiens.*"[4]

Gyna sapiens does not emit a consciously detectable odor or distinctive sound when she ovulates. Her facial expression does not change, nor does her body stance. She does not gesticulate in a come-hither fashion. The patch of dense pubic hair unique to humans obscures any hints of a color change in her vulvar skin.* Outer clothing worn by most women in both indigenous and advanced cultures serves to increase a male's uncertainty.†

A few women, particularly young women, do receive an unwelcome alert that announces the precise moment of their ovulation. *Mittelschmerz,* a Ger-

---

*The hair follicles associated with pubic hair are a rich source of sexual pheromones. Their odor attracts the opposite sex. But the human sense of smell is so weak that pubic aromas do not play a significant role in the initial stages of sexual attraction.

†In a few exotic tribes, such as the Yanomami of the Brazilian rain forest, the females cover their pubic area with a thin cord that does not leave much to the imagination. These exceptions aside, public exposure of a female's vulvar area is a rare feature in human societies.

man term meaning "middle-of-the-month pain," is a well-recognized medical
syndrome. Occasionally, a small amount of internal bleeding accompanies
the bursting of the ovarian bubble follicle containing her ovum. Blood irri-
tates the peritoneal lining, causing localized abdominal pain. Many a young
woman has been misdiagnosed with appendicitis because of her physician's
confusion in trying to tell these two nearly identical clinical presentations
apart. As a woman ages and her ovary's husk becomes more pliable, *Mittelschmerz* markedly diminishes in both severity and recurrence.

Some women are exceptionally attuned to their bodies and can intuitively monitor changes in their circulating hormonal levels and use these
fluctuations to inform them of the key moment when their eggs are ripe. For
the majority of women, however, ovulation is truly cryptic. Family planning
requires more reliable information on this issue, and lack of certainty
prompts many women to purchase thermometers and cervical-mucus density kits to pinpoint accurately their internal cycle's apogee.

In every other species, it is extremely advantageous to survival and reproduction to have females signal males that the time is right. What possible
benefit would there have been to losing so crucial a primal instinct? This
stark chasm dividing a woman from the majority of other primate females—
as well as virtually every other species of insect, arthropod, fish, reptile, bird,
and mammal—has been the source of endless speculation. The most commonly accepted theory is that a woman's cryptic ovulation, year-round sexual receptivity, and loss of estrus solidifies the bond between her and her
mate, convincing him to stay and help her raise their children. Cryptic ovulation, according to this line of reasoning, increases the need for abundant sexual intercourse, and this in turn heightens intimacy.

Jared Diamond calls this the "daddy-at-home" theory. Citing the work of
Richard Alexander and Katherine Noonan, Diamond argues that a woman
had to conceal her ovulation; otherwise her husband would only stay with
her when she was exhibiting signs that she was fertile.[5] The rest of the time,
he would be out trying to find other women, who were exhibiting signs that
they were sexually ready. His absence would be detrimental to his children,
and by concealing her ovulation, a woman convinced a man to stay by her
side and make love to her throughout the month, so that he could be sure he
was fathering the children she bore.

Several flaws and inconsistencies, however, weaken the argument that
promotes sex as the glue holding human relationships together. If sex served
the purpose of ensuring the durability of the human parenting commitment,
then parents should become more ardent in their lovemaking following the
birth of a baby. Instead, the opposite occurs. Both parents routinely report a
sharp fall in their respective libidos following the blessed event.[6]

While a mother focuses on tending to her babies, other females, who are also enjoying year-round sexual receptivity, can surreptitiously compete for the affections of the father. Given his hyperkinetic sex drive, it could be reasonably argued that cryptic ovulation, loss of estrus, and year-round sexual receptivity work to a new mother's disadvantage and are just as likely to precipitate marital dissolution as promote marital bliss. The peccadilloes reported in the daily newspapers provide ample corroborating evidence for this observation.

Year-round female receptivity is not necessary to strengthen the sexual relationship among any other monogamous species. Wolves, geese, coyotes, gibbons, albatrosses, and prairie voles mate for life without the female's having to lose her "season" to entice the male not to stray.

Many other inventive theories attempt to explain why ovulatory signals disappeared in the human line, but none of them, in my opinion, propose an advantage so sterling that it offsets the abandonment of the premier gene-replicating method used by millions of species over millions of years—females signaling males their sexual readiness at the same moment they are ovulating. The few species of primates that do not ostensibly signal ovulation are distant from the human line on the evolutionary primate-branching bush. A few other species' females also do not appear to advertise their ovulation—for example, some birds, such as mallards—but the problem of knowing for sure is exacerbated by the fact that human observers can't ask the males of these species whether or not *they* are aware of the females' ovulation.

Among the more ingenious explanations for why sexual signaling disappeared in humans is the one proposed by anthropologist Nancy Burley. Women lost the ability to monitor their ovulatory moment, according to Burley, because those who grasped the connection between sex and pregnancy realized that pain, possible death, and taxing demands were also part of the deal. A woman so enlightened, Burley theorizes, might prudently decide to abstain from sex.[7] Celibates do not leave offspring. Selection pressure would, therefore, favor those women who were unaware of their ovulation.

Primatologist Sarah Blaffer Hrdy proposes that infantile murder coaxed human cryptic ovulation into being in a theory known as the "many fathers." Having carefully documented the horrifying fact that male primates sometimes killed infants, Hrdy posits that the loss of external signs of ovulation ultimately protected newborns by keeping all parties guessing concerning the issue of paternity, especially males. Unsure whether an infant was due to his copulatory efforts, a male would be less inclined to kill it. Hrdy worked primarily with hanuman langurs, but other studies, including Jane Goodall's chimpanzees in the wild and Alison Jolly's ring-tailed lemurs, have confirmed that males commit infanticide in these species. To date, thirty-five species of primates have been identified in which strange males kill infants.[8]

Social scientists Margo Wilson and Martin Daly found suggestive evidence that this abhorrent practice exists among humans. Surveying crime statistics, they noted that when an adult male murders a child he is sixty-five times more likely to be a stepfather or live-in boyfriend than the child's biological father.[9]

Though I have great respect and admiration for Hrdy's work, I wonder whether her theory is the whole story. If infanticide was so great a threat to the continuation of affected primate species, why did only the human line adopt the evolutionary strategy of cryptic ovulation to solve the problem? Chimpanzees, for whom infanticide is a serious problem, have not evolved anything resembling cryptic ovulation. Bonobo males have never been observed to engage in killing infants, yet bonobo female primates come closest to mimicking the human female's reproductive model of loss of estrus and increased sexual receptivity.

The observation that strange males do the killing is in keeping with evolutionary theory. Alpha maledom often does not last very long. A strange male who achieves dominance must make hay while the sun shines. By killing the sucklings of his new group, he can precipitate estrus among the distressed mothers and thus increase his chances of spreading his genes. Among human populations, it has been a common strategy of conquering armies, after beheading the losing side's warriors, to turn their attention to killing the infants who the conquerors know were fathered by those warriors. This slaughter of the innocents has been amply recorded at different times in disparate locales throughout history. Cryptic ovulation has rarely protected infants of the conquered women from being killed by strangers.

If keeping the male uncertain concerning paternity increases the life span of children, what would be the advantage to the female of remaining in the dark on such a vital issue as her own ovulation? Another question: The male primates that engage in the practice of infanticide do not seem to care one fig about the offspring they do sire. Since the majority of infanticide is carried out by recently arrived males that are strangers to the group, it would be safe to assume that these animals are equipped with an instinct to kill the infant of any strange female. Hrdy's theory rests on the assumption that a male primate is capable of making the causal connection between sex and birth, either instinctually or consciously. There is minimal scientific evidence to indicate that this quantum leap in logic has occurred in the mind of another species besides a human.* Another problem: Since knowing when a female ovulates is critical to a male's fitness, why did not the human male

---

*A male lion that has just defeated an alpha lion to take over a pride has an instinct enabling him to know whether newly pregnant lionesses are carrying his offspring or those of the old alpha. Somehow he knows

develop a compensatory counteradaptation to detect the female's ovaries' subterfuge?

Many other theories abound. Donald Symons suggests women use their year-round sexual receptivity to seduce philanderers in exchange for gifts.[10] L. Benshoof and Randy Thornhill propose that cryptic ovulation allows a woman to mate by stealth with a superior man without alerting her husband.[11]

Though the theories outlined above may have been contributing factors in reprogramming *Gyna sapiens'* reproductive cycle, they do not seem to offer sturdy enough reasons to explain the origin of such drastic changes in her life strategy. The rarity of concealed ovulation among the other three million sexually reproducing species suggests that cryptic ovulation is not a mainline solution to any of the problems posed by the various theories.

The primary consequence of cryptic ovulation—the need for increased sexual contact to coincide with ovulation's propitious moment—would appear to be highly disadvantageous, evolutionarily speaking. In the cold calculus of energy conservation, copulation is both dangerous and a very expensive metabolic activity. An ancestral couple in flagrante delicto would have been very vulnerable to a predator. Sex consumes time, calories, resources, and mental effort that might better be used for survival. With a few notable exceptions, other creatures expend minimal time and energy copulating. The mating act of most birds and mammals can be measured in seconds.* The human investment, in terms of time spent thinking about sex, planning, wooing, and actually engaging in the act, exceeds that of any other creature. After their strenuous coitus, humans generally require a longer recovery interval than any other animal.

Additionally, without a visible or olfactory lodestar, men and women have found it necessary to engage in frequent, capricious copulations throughout the year to increase the likelihood of pregnancy. The uncertainty of conception, both for those who yearn for it and those who don't, has been among men's and women's most consistent causes of stress, anguish, and anxiety down through the generations. Evolutionary processes do not care whether an organism is happy or not. Nevertheless, stress tends to diminish an organism's fitness.

Nonestral females of other species, with rare exception, do not appear to begrudge the attentions estral females receive from excited males. When *Gyna sapiens* lost estrus and gained the ability to engage in sex anytime

---

to kill the cubs of the latter at birth. Males of a few other species act as if they know whether offspring are the result of their copulatory efforts, but in all cases save the human, this information is processed at an instinctual level and not consciously. This issue will be explored in depth in a later chapter.

*In a few tiny primates, such as galagos (called bush babies), copulation lasts over an hour. Primatologists suspect that this has more to do with mate guarding than it does with pleasure.

throughout the year (if she so desired), the nettlesome problem of sexual jealousy among women reared its ugly head. The green-eyed monster consumes a staggering waste of spirit and is virtually unknown among other species.

Cryptic ovulation and year-round sexual receptivity also greatly increased the amount and degree of jealousy among many men. Societies have had to construct draconian legal, social, religious, and cultural barriers to regulate members' sexual competition and minimize the outbreak of violence. Duels, dogmas, eunuchs, taboos, so-called honor killings, chastity belts, and female genital mutilation are just a few of the rituals and devices that attest to the difficulty men have had in dealing with women's robust sexual capability.

Desmond Morris in his book *The Naked Ape* makes the argument that humans are the first species to elevate sex to the status of a recreational activity.* Morris speculates, "The vast bulk of copulation in our species is obviously concerned, not with producing offspring, but with cementing the pair bond by providing mutual rewards for the sexual partners."[12]

According to this argument, endorsed by many others besides Morris, we are *Homo ludens* (the Playful Ape) and we have liberated sex from the depths of the Minotaurean labyrinth† in the brain's primitive limbic system of instinctual drives. By elevating sex to the brain's higher, neocortical planes, according to these authors, we have created a new kind of sex. Proponents argue that the pleasure we so derive and the love that enhances the deep human commitments more than offsets any of the disadvantages of drastic changes in sexual programming.

I, for one, am not so sure. The many downsides of cryptic ovulation, loss of estrus, and a woman's continual sexual receptivity do not seem to balance the purported advantages. The reason that this unusual constellation of female reproductive changes constituted an incontestable plus for humans lies in an area far removed from sex itself. Before I make that case, we must continue to enumerate several other features of the *Gyna sapiens'* life history that separate her from her nonhuman sisters. Only after all the pieces of the puzzle have been identified can we begin to see how they all fit together.

*   *   *

Another feature indicative of *Gyna sapiens'* radical reproductive makeover is the mysterious way a close-knit group of women synchronize their

---

*Bonobos appear to be the only other species to have made the crossover.
†The Minotaur was the half-man, half-bull that guarded the subterranean labyrinth below the ancient citadel of Knossos. Theseus, with the help of the princess Ariadne, successfully killed it in a Greek myth.

menses. Nobel Prize laureate Barbara McClintock demonstrated that unconsciously detectable pheromones emitted by the alpha female in the group
mediated this mysterious process.[13] Women who menstruate together must
by necessity also ovulate together. Contemporary societies typically comprise large numbers of unrelated strangers constantly in flux who rarely
congregate in one place for any extended period of time. In contrast, ancestral women often lived most of their lives in close association with female
members of their tribal group or clan whom they knew quite well. Our modern disruption in the living arrangements under which women lived for over
99 percent of our species' history has profoundly attenuated and obscured
what was formerly a transformative evolutionary reproductive trait.

Among most other creatures, circadian rhythms synchronize ovulation
to create mass breeding seasons, timed to coincide with optimal environmental conditions. Primates are a notable exception to this general rule. Both
bonobos and common chimps cycle in and cycle out at various times of the
month, and their pattern is the norm among other apes and monkeys. The
ring-tailed lemurs of Madagascar, a primate species many limbs and
branches away on the primate order's family tree from humans, are one of a
handful of *Gyna sapiens'* primate sisters to coordinate their reproductive
phases instinctually. Typically once a year on the night of a full moon, ring-
tailed lemur females ascend in lockstep to the pinnacle of estrus. All the
males, as one would expect, go wild. What follows is a lunar mass sexual
orgy.*

Biologists do not know at what point in our evolution *Gyna sapiens*
evolved the unique feature of menstrual harmony.† As with any major novel
adaptation, there would most likely have been multiple reasons why *Gyna
sapiens* synchronized menses, but one subtle effect was to accelerate her lib-
eration from the grip of male tyranny.

The majority of species manifest a disparity between the size of the male
and the female. Biologists call this trait "sexual dimorphism." Among inverte-
brates (insects, worms, etc.), females tend to be larger than males; among vertebrates, especially birds and mammals, males tend to be bigger. Ethologists‡
use a species' dimorphic ratio to predict its mating pattern. Whenever the
male is disproportionately larger than the female, he uses his superior strength
to fight other males for possession of the females, whom he overbearingly

---

*Under certain conditions, the females of some other nonhuman primates can also synchronize their
menses, but it is not their natural rhythm in the wild. For example, when a strange male takes over a langur troop, he often kills the young. All of the females then promptly come into estrus.[14]
†Anthropologist Richard Wrangham argues that all the uniquely human female reproductive features
were in place 1.9 million years ago.[15]
‡Ethology is the study of animals in their natural habitat.

protects but keeps in sexual thrall. When males are nearly equal in size and weight to females, then cooperation and monogamy tend to be the norm.

Silverback gorillas and hamadryas baboons are two extreme examples of primate sexual dimorphism. The alpha males are more than twice as large as the females, and they thoroughly dominate the members of their respective troops and herds. The alpha male, along with his close allies, brusquely fends off lower-ranking males who attempt to mate with "his" females. In some species of monkeys, powerless males never gain the opportunity to have sex at all, and the despotic alpha male's hard-won monopoly deprives females of choice. Females of sexually dimorphic primate societies do not synchronize their estrus or menses but, rather, phase in and out of sexual receptivity in a random fashion, a perk not lost on the alpha male, who jumps on every chance to mount one estral female after another.

A group of females cycling in unison, however, dramatically alters the equation. Menstruating in synchrony means that all the females also coordinate their estrus and ovulate nearly simultaneously. Imagine an alpha gorilla's or an alpha baboon's astonished confusion if all the females in his group simultaneously approached him, each presenting her hind, demanding relief. His initial delight over this novel situation would soon turn to consternation. A solitary male, despite his well-advertised sexual prowess, would discover that it was exceedingly problematic to satiate insatiable females if they all demanded sex from him at the same time. Should he try, he would most likely find himself exhausted and depleted, lying flat on his back, nursing a sore penis. In these circumstances, he would presumably welcome the assistance of lower-ranking males.[16]

Through a clever adaptation, females would now have a much greater selection of males, other males would finally get a chance to discover what all the hoopla was about, and, in general, the entire troop would be happier and more content. By menstruating in sororal harmony, ancestral *Gyna sapiens* broke the back of the system that rewarded the strongest, fiercest, males with a sexual monopoly. In short, synchronized menses enhances female equality.*

Underscoring this last statement: Female ring-tailed lemurs, one of the few nonhuman primates to habitually synchronize their menses, belong to the *only* primate species, other than humans, in which a female can completely dominate a male. Allison Jolly observed on many occasions that a diminutive female approached a large male busily eating, snatched the food away from him, cuffed him on the ear for good measure, and then sauntered

---

*This is a very condensed version of the theory Paul Turke proposed to explain human ovarian synchrony.[17]

off to enjoy her ill-gotten gain in privacy. No other female of any other primate species (except *Gyna sapiens*) would routinely dare so brazen an act.[18]

Anthropologist Chris Knight has proposed that women synchronized their menses in order to organize a sex strike and force men to go hunting to bring them meat. He wrote in his comprehensive 1991 book, *Blood Relations:*

> For babies to be conceived, the sexes had to come together. For efficient hunting to take place, they had to separate. If both hunting *and* conception were to occur, the sexes had to alternate between conjunction and disjunction. Periods of sex strike and marital togetherness had to alternate. I assumed that this alternation must have been socially synchronized, rather than a matter for individuals to decide autonomously within couples.[19]

Backing up his arguments with a wealth of ethnographic data, Knight makes the case that menstrual rituals were among the first and most important rites performed by early humans. The liberal use of red ochre for body adornment in these rituals became the basis for many human cultural innovations that followed.

Menstrual coordination played a key role in the evolution of our species. When this was joined with cryptic ovulation, loss of estrus, and potential year-round sexual receptivity, conditions began to coalesce that fueled the explosive rocketing of the human species to a position far ahead of all the others in the competition for resources. Several additional, extremely unusual components that occurred in *Gyna sapiens'* life cycle require investigation before the overall pattern emerges. Like an emulsified photo from a Polaroid camera forming a picture before our eyes, additional enhanced details are necessary before the frame comes into focus. When all the components of her reproductive life history are recognizable against the background of the snapshot, I will propose in chapter 13 a theory to explain how the picture's many diverse parts interrelate.

Because of the superb olfactory sense of most predators, menses would have imperiled a ground-dwelling
human female.

Chapter 6

Periods/Perils

Greater than his fear of death, dishonor, or dismemberment has been primitive man's respect of menstrual blood. The measures he has taken to avoid this mysterious substance have affected his mealtimes, his bedtimes, and his hunting season; and primitive woman, unable to separate herself from her blood, knew that upon her tabooed state depended the safety of the entire society.

—Janice Delaney, Mary Jane
Lupton, and Emily Toth[1]

None of the known features of female mammalian reproductive physiology require menstruation as an ineluctable by-product. If menstruation were both costly and functionless, natural selection would have eliminated it long ago.

—Margie Profet[2]

*A*round the time that *Gyna sapiens* was abandoning the outward signs of ovulation, her menses, initially of minimal consequence, evolved into a copious and potentially dangerous blood loss. On examination, there appear to be many drawbacks to this feature without a single obvious compensatory advantage. Physiologists would protest, pointing out that human menses is the inevitable end result of the extraordinary preparations the uterine lining must undergo to implant a fertilized ovum successfully. But 3,999 out of 4,000 other mammalian species' females pump out litters with as many as ten kits, cubs, pups, and piglets without having to discard significant amounts of blood and protein-rich tissue monthly. Some mammals, such as dogs, bleed during their period of heat. But nature has equipped them with the instinct and the anatomical flexibility to lick themselves, thus allowing them to recycle their iron and prevent deficits.

Menstrual discharge emanates a distinctive odor. Most predators rely heavily on their olfactory sense. The smell of menstrual blood does not pose a threat to the other thirty species of menstruating primates. Discharged by a high-flying monkey or an acrobatic langur, the minimal dribs and drabs fall through the tree branches below, spattering in dense foliage. Most droplets do not even reach the ground. Should a leopard, snuffling along the forest floor, happen to pick up the scent of a menstruating monkey, the weakness and inconsistency of the signal would be of little assistance to it in locating dinner. Even if the leopard could follow the scent, this would not necessarily result in a kill, because primates have few natural enemies, and even fewer that can catch them in treetops.

Given the conditions under which our ancestral hominids lived in the Pleistocene, however, menstrual blood's distinctive odor would pose a much more serious threat for a bipedal hominid female. Judy Grahn in her 1993 book, *Blood, Bread, and Roses,* recounts the following incident.

> Once when I was living in an all-female collective some visiting dogs "went wild" for a few hours one afternoon while we were out. They broke into laundry baskets and ate the crotches out of a half a dozen pairs of Levis and dress pants! . . . That was the day I began to imagine the hazards of being a menstruating woman in the vicinity of blood smelling predators, in those long eras before we learned to use fire, weapons, or houses. How cautious one would need to be. How silent. What to do? Hide, bury yourself in the sand, climb a tall tree?[3]

Nonhuman primates can easily escape from any ground-dwelling predator by handily scampering up a tree trunk. Evolution redesigned the bipedal foot for walking on flat surfaces. During its extensive overhaul, the human foot relinquished the flexibility and grasping power typical of every other primate. In a climbing contest among all the species of primates, *Homo sapiens* would come in dead last.

Men can shinny up trees better than women. Their greater upper-body strength somewhat compensates for their feet's inferior tenacity. A man, much less a woman with small children, cannot outrun any major mammalian or reptilian land predator. Menses and bipedalism would have been a dangerous combination in the forest primeval and an often fatal one out on the nearly featureless savanna. A menstruating woman's leakage of even the most minute amount of blood would lay down a convenient trail of spoor that a predator could easily track, leading it to both her and her young.*

Some scientists have argued that the previous discussion is not pertinent, because predators instinctually do not like to dine on human flesh: It is

---

*Kathryn March reported that bears become aggressive and disposed to attack when they smell menstrual blood. White-tailed deer, a prey animal, display an aversion to menstrual odor and shy away from it.[4]

not particularly flavorful. The significant number of humans attacked and eaten by bears, tigers, lions, and crocodiles in historical times invalidates this argument. The more likely reason that predators shy away from humans is that they have acquired an instinct ethologists call "flight distance." Humans have become so dangerous to other animals, including predators, that only those that have learned to avoid us have survived. But at the outset of our species, when the changes I speculate took place, predators would not yet have been so cautious.

The radical hormonal flip-flop that initiates human menses creates conditions within the breasts, ovaries, and uterus that over time predispose the cells making up their linings to cancer. No other species' females in the wild experience high cancer rates in their reproductive organs, yet these malignancies rank among the leading causes of death for young women today.* There is little reason to doubt that ancestral women did not experience relatively high cancer rates in their reproductive organs. Admittedly, most did not live long enough to be affected, but the loss of even one old wise woman in a tribe to a reproductive organ's cancer could have seriously diminished a group's chances for survival.

Jared Diamond recounts his experience working with isolated South Pacific islanders.

> When I began pestering my middle-aged Rennellese informants with my questions about fruit edibility, I was brought into a hut. There, in the back of the hut, once my eyes had become accustomed to the dim light, was the inevitable, frail, very old woman, unable to walk without support. She was the last living person with direct experience of the plants found safe and nutritious to eat after the hungi kengi [cyclone], until people's gardens began producing again. The old woman explained to me that she had been a child not quite of marriageable age at the time of the hungi kengi. Since my visit to Rennell was in 1976, and since the cyclone had struck sixty-six years before, around 1910, the woman was probably in her early eighties. Her survival after the 1910 cyclone had depended on information remembered by aged survivors of the last big cyclone before the hungi kengi. Now

---

*Domestic dogs suffer a fairly high rate of breast cancer, but they do not approach the high incidence prevalent in humans. Artificial neutering contributes to this problem and is not a factor in the wild.

A major factor in human breast cancer is whether or not a woman breast-feeds. Uterine and ovarian cancers are relatively unaffected by lactation.

the ability of her people to survive another cyclone would depend on her own memories, which fortunately were very detailed.[5]

Had this one particular woman died of a reproductive cancer earlier in life, the consequences for the entire tribe would have been calamitous.

Another deleterious effect of menses concerns Free Will. Humans are extremely proud of their unique ability to override their instincts, drives, and hormones and rise above the urgent demands of thirst, hunger, sleep, fatigue, and sex. We can voluntarily forgo eating even to the point of starvation. We enter convents and monasteries pledging to live a celibate life. We can work hours without sleep. We can control (up to a point) our bladder and bowels. Yogi adepts can regulate their heart rate and breathing to a remarkable degree. We admire heroes and heroines who control their destinies through their forceful characters, delaying gratification to achieve majestic goals. We marvel at the discipline and willpower necessary to become a virtuoso pianist, an Olympic gold medalist, or a Nobel Prize winner.

We are exceptionally proud that we can rein in the part of us that we consider our "creatureliness." However, one bodily function resisting volitional control, except under the most extreme circumstances, is the timing of a woman's menstrual cycle.* It is for naught that a woman brings to bear mental fortitude and concentration determined to change the appointed day.

Her inevitable resignation to the power her monthly cycle exerts upon her will contributes, I believe, to a woman's receptivity to accept "fate," a passive mental attitude more commonly associated with the feminine than the masculine. Whether this is a disadvantage or an advantage is debatable, but a man's greater reluctance to accept matters as they stand is, in no small part, due to the fact that he does not experience a monthly bodily function, comparable to menses, over which he cannot impose his will.†

One of the most detrimental side effects of human menses is the prevalence of the gynecological condition of endometriosis. This distressing disease

---

*Under certain extreme conditions (e.g., fasting, or strenuous physical activity such as training for a marathon), a woman can influence the timing of her menses.

†There has been much speculation among sociologists as to the reason why young girls undergo a drastic reduction of self-esteem during their middle-school years when compared with boys of the same age. Girls have nearly as much self-esteem as boys do until puberty. Then there is a dramatic falloff. It would be worth investigating what the effect of monthly menses is on the psyche of young girls. Could their lack of control over this dramatic bodily function be a major contributing factor in the unexplained puzzling drop in self-image?[6] Some women might protest, recalling how their menarche followed by regular menses boosted their confidence, because they felt more grown-up. This benefit, however, does not seem to translate into an increasing sense of self-esteem compared with the drop reported in the majority of studies.

plays virtually no role in the well-being of *any* other species' females. One in ten women of reproductive age in the industrialized nations suffers from it, and it at present constitutes the most common cause of female admissions to hospitals.[7] The volume of blood and tissue lost during each human menstrual cycle surpasses that of any other animal. The uterine contractions required to expel it are, therefore, proportionately more energetic. In some cases, their propulsive intensity nearly equals the force of uterine contractions delivering a full-term fetus.[8] These monthly spasms can cause uterine tissue to relocate to other areas within the woman's abdominal cavity.

When a woman's period begins, her menstrual contents must exit via the narrow tunnel in the cervix, the tubular outlet of the uterus that leads into the upper vault of the vagina. During menses, the opening in the cervix enlarges to accommodate this exodus. In some women, however, not all of the compressed blood and tissue can escape through this preferred channel; some of it is forced to exit through two alternative openings into the uterine cavity, the entrances of the horn-shaped fallopian tubes, high on either side of the uterine apex.

Normally, the fallopian tubes function as follows. At midcycle, the ovary releases an egg into the interior of the peritoneal cavity. Nearby, beckoning the ovum on the fallopian tube's trumpet-shaped extremity is a delicate circular fringe resembling the fingers of a sea anemone. These waving, frondlike fimbriae gently scoop up the egg and funnel it into the tapered portion of the fallopian tube's narrow passageway. Through gentle, undulating waves, the ovum is then milked slowly toward the womb's great chamber. The fallopian tube's design allows peristaltic movement only in a wombward direction.

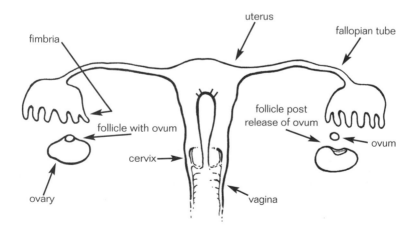

A schematic of the human female reproductive pelvic organs.

The force of the human uterine contraction during menses, however, can cram a small amount of menstrual blood and cellular debris along the length of the pencil-sized fallopian tubes. Overpowering the fallopian muscular contractions, uterine contractions can occasionally propel menstrual contents to the tube's fimbrial opening, spilling some of the waste products out into the peritoneal cavity.

The specialized sentinel cells making up the pelvic peritoneal lining react vigorously to the foreign material invading their pristine landscape. An aroused peritoneum responds, activating pain reflexes, and the bedeviled woman experiences abdominal discomfort different from that associated with normal menstrual cramps. Locally, the peritoneal membrane mounts an inflammatory counterattack. White blood cells, the immune system's military arm, rush to the scene and, along with other reinforcements, slay, disrupt, and absorb what is left of the foreign invaders. When the conflict is over, the peritoneal lining resembles a scarred and roughened battlefield. Adhesions composed primarily of collagen, the body's first phase in trying to heal the area, appear. The blood vessels supporting their enterprise permanently enlarge.

On occasion, some of the cells belonging to the uterine lining that were near death manage to escape the detection of the vigilant white-cell patrols, and they possess sufficient vitality to latch on to a contiguous organ or transplant onto the newly perfused peritoneal lining. Like a weed blown into a fertile garden, these errant uterine cells take hold, insinuating themselves in their new hospitable surroundings. Patches of uterine lining begin to sprout where no uterine lining was ever intended. Bowel, bladder, ovary, and the outer surface of the uterus all become potential hosting sites. The displaced cells then transform into true parasites, drawing all their nutrition and oxygen from the reluctant hosts. The cruelty of their incursion soon becomes apparent. The doleful symptoms of endometriosis now unfold.

Uterine-lining cells are very responsive to a woman's reproductive cycle's hormones circulating throughout her entire body. Molecules of estrogen and progesterone cannot distinguish whether their target cells are in the uterus or growing on the peritoneum. Displaced endometrial tissue increasingly swells and grows turgid with blood in the days following midcycle, exactly mimicking the response of cells inside the uterus. At the end of a cycle, the diaspora cells suffer the same fate as intrauterine cells if conception does not occur—they die. But endometrial carpetbaggers have nowhere to go and must die in place. There is no mechanism by which the body can rid itself of this small but significant toxic load.

Endometrial cells' necrotic implosion initiates another cycle of peritonitis, inflammation, scarring, and adhesions. A woman afflicted with endometriosis will experience periodic pelvic pain and, on occasion, severe discomfort during the sexual act, called dyspareunia. Infertility is another baleful consequence.

There is no reason to believe that ancestral women had weaker uterine contractions or a lesser volume of menstrual effluvia requiring expulsion. I conjecture that they, too, suffered from endometriosis.* Ancestral women, however (because of their more frequent pregnancies and lactation), experienced relatively fewer cases of endometriosis than contemporary women. But the few debilitated women suffering from this condition constitute another dramatic reason why heavy menses would have been an undesirable addition to *Gyna sapiens'* life story.

*   *   *

If a wounded animal bleeds excessively, a hunter anticipates that it will soon collapse. Animals that bleed intuit instinctively that they have been injured, and will retreat to the back of the cave or burrow to lick their wounds. A predatory hominid male would be acutely aware that copious bleeding in a wounded animal is an event preceding its death.

Imagine, then, the awe, fright, and confusion that men experienced when they furtively caught sight of a woman's menses. Women bled, but they did not grow weak. They bled, but they were not injured. They bled, but they did not die. Sexual relations with a menstruating female would conclude with the male's withdrawing a blood-smeared member. Feeding many males' innate castration fears, this disturbing sight would tend to cool a man's ardor and make him believe that a menstruating woman possessed a power beyond his ken. Menses would seem to him to be some sort of magic. Perceived supernatural powers induce fear, and men began to fear women. This in turn led men to resent women, because, even though they were bigger and stronger, men were afraid of the otherworldly supremacy they imputed to women.

The sheer inconvenience of monthly bleeding also creates hygienic problems, especially to ground-dwelling nomadic females. Keeping clean while on the move, at present problematic for women living in developing nations, would have been extremely difficult in the Pleistocene. Fabrics were nonexistent, large hides were not yet plentiful, and appropriate plant substi-

---

*Endocrine disrupters, chemicals in the environment that mimic female hormones, have been implicated as the cause of the current upsurge in endometriosis.

tutes were not always readily available. Remaining at a home base instead of traveling daily presented a menstruating woman with a different set of problems. Special handling of menstrual effluvia became imperative. Dependent on the purity of local water supplies for drinking, washing, and bathing, men and women would have had to address the issue of menses communally.

Extrapolating from what anthropologists have gleaned from male attitudes toward menses in extant hunter-gatherer cultures, early *Homo sapiens* would most likely have insisted on measures to isolate the menstruating women in the group as he became increasingly aware of *Gyna sapiens'* menses. Fastidious women would have already disengaged from the others during their monthly times and sought out other women similarly indisposed. Perhaps men and women voluntarily supported these restrictions for the good of the tribe. Later, these strictures, passed down through generations by ritual and tradition, became menstruation taboos, the vast majority of which, over time, have served to strip political power and autonomy from women.

Among the Bribri Indians of Costa Rica, a menstruating woman was considered unclean. She was forbidden to come near a male's hunting weapons or touch any food or utensils. The only plates she could use for her food were banana leaves, which she had to abandon in a sequestered spot because it was believed that if a cow ate them the animal would die.[9] In the Jewish Orthodox tradition, the period of *niddah* circumscribes five days of menstruation followed by seven days free of discharge. During this period, a woman is considered "impure." According to one rabbi's interpretation of the *niddah* laws, "One should therefore be careful that the beds are far enough apart to ensure husband and wife will not touch each other when they sleep. He may not sit on her bed even when she is not present. He is forbidden to take care of her when she is ill unless there is no one else, and she is in great need of it."[10] The first-century-A.D. Roman pseudo-scientist Pliny warns men that a menstruating woman, by her touch, can "blast the fruits of the field, sour wine, cloud mirrors, rust iron, and blunt the edges of knives."[11] During the witch craze that convulsed Renaissance Europe, accused women purportedly had the magical powers to do the very things that Pliny's menstrual superstitions attributed to a woman having her period.

Initially, elders may have instituted menstrual restrictions for rational hygienic reasons. Menstrual blood is a rich medium, encouraging the growth of pathogenic organisms. Its proper disposal had to be regulated. Predators had to be kept at bay. Prey must not be alerted to the presence of hunters by whiffing the scent of a menstruating female who might be among them. When dealing with something as mysterious as menses, however, the irrational quickly trumped the rational. Men put women in an untenable position by insisting that it was their responsibility to assuage men's menstrual fears. The powerful taboos associated with menses among indigenous peoples, as well as many that persist in modern cultures, are testament to the tenacity of these beliefs—which, in many cases, are nothing more than misguided superstitions.

Anthropologist Camilla Power put forth the idea that menstruation was beneficial to the human species because it provided men with the only reliable clue that they could use to calculate whether a woman was *imminently* fertile. In a population of women who were either pregnant or lactating, the occasional menses would be an important signal. "Pleistocene males who were attentive to recently menstruating females in an effort to increase their mating prospects," Power wrote, "should therefore enhance their fitness. No male could afford to ignore this signal."[12] According to Power, menstrual rituals evolved to become complex coalitionary attempts by all the tribe's women to confuse the men as to the timing of a woman's menses, which in turn would strengthen the power of women's veto over sex. In every other species, the female signals the male she is potentially fertile by advertising her ovulation as estrus, heat, rutting season, or any of the many other terms used to describe a female animal's heightened sexual receptivity. Power's explanation of why Mother Nature went to the trouble of saddling human females with concealed ovulation and then undid Her efforts by further burdening women with a dangerous blood loss for the purpose of alerting men that pregnancy was a possibility is ingenious but, in my opinion, too complex.

Not all women experience menses as a "curse." Some premenstrual women report an upsurge in positive feelings. Paula Nicholson, after reviewing interviews with many women, reported, "The premenstrual phase of the cycle is frequently accompanied by heightened activity, intellectual clarity, feelings of well being, happiness, and sexual desire."[13]

And not all the taboos and rituals associated with menses detracted from a woman's well-being. Some women would have welcomed these restrictions, which afforded them an opportunity to create zones that men dared not trespass. Camaraderie and freedom from the daily grind allowed congregating women to create, tell stories, and rest. In the apocryphal Red Tent, women could pass along valuable lore and woman wisdom from one generation to another. But these are, remember, cultural advantages that occurred long after the human line developed sophisticated cultures. The original question I posed was: What evolutionary advantage did menses provide to advance the fortunes of individual females of the human species? Why did it evolve in the first place?

Hippocrates, the classical-Greek father of medicine, believed that menses flushed "poisonous" humors monthly from a woman's body, a necessary cleansing, he was sure, that prevented the accumulation of toxins. Menses supposedly promoted a woman's health. Galen, the influential second-century Roman physician, endorsed Hippocrates' theory, and this unsubstantiated belief has persisted in one variation or another right up to the present day.*

Margie Profet, an evolutionary biologist, proposed in 1993 that menses purges the uterus of possible microscopic pathogens that may have hitch-hiked on the sperm of the previous month's suitors.[14] Profet arrived at her controversial conclusion because she was convinced that something as detrimental as menses must have a benefit to the species—otherwise it would have been culled from the genome long ago.

Skin is an impermeable membrane that covers and protects the body. When skin becomes denuded, as happens in the case of a third-degree burn or a scraping injury from trauma, the raw tissue beneath becomes exposed. Raw surfaces stripped of their protective lining and then coated with blood provide the ideal conditions for the overgrowth of pathogenic bacteria. If the surface is an internal one kept at an optimal incubating temperature of 98.6°F, the bacteria will multiply even faster.

The monthly denuding of the uterus's protective lining at the height of menses would leave it more vulnerable, not less vulnerable, to disease. Other mammalian females mate with multiple males, and venereal diseases do not seem to present a problem for them. Why would the lining of their wombs seem so much more resistant to disease than the human one?

Beverly Strassman has challenged Profet's theory, arguing that menses is a metabolically sound way to rid the body of the endometrial lining, and disputed Profet's main contention: that the sloughing of the uterine lining provides women with protection.[15] The gynecologic literature overwhelmingly supports Strassman on this point. Menses is a time when uterine infections are the most common. Toxic-shock syndrome occurs because a tampon has prevented the egress of menstrual blood. Harmful bacteria, then, take advantage of the uterus's hospitable environment, and overpower the woman's immune system. And not only bacteria: In one study done in the historical span

---

*Only recently did the medical profession finally abandon the unscientific belief that menstrual blood contained unidentified "menotoxins" (a fancy way of saying "bad humors"). And only in the last few years have reproductive physiologists openly questioned whether it is necessary for a woman to menstruate at all!

when paralytic polio was a scourge, over three-quarters of adult women diagnosed with the virus were infected during their periods.[16]

In general, menses is a vulnerable time for women, and they intuitively know that it is usually an inauspicious moment to initiate ambitious projects. I suspect this is nature's way of protecting an organism during its passage through a period of relative defenselessness. Even if one were to grant the validity of Dr. Profet's claim that menses was an evolutionary gain for women, the protection from an occasional unsavory rider accompanying a man's sperm would not be a weighty enough reason to offset menses' other serious disadvantages.

One valuable contribution of Dr. Profet's theory is her raising the question of menses' cost-benefit ratio. What *was* the reason it became so florid and nettlesome in only one species out of thirty million?

The question becomes more insistent when one considers that, incrementally, the negatives of menses pile up, and it can be said with conviction that, to date, gynecologists have been unable to identify a single uncontested benefit of human menstruation.* On the contrary, it seems to have had a harmful impact on the survival odds of the individual and ultimately the species. Mother Nature rarely burdens a species with a vexatious attribute unless She balances the scales with a positive offset of greater value. Negative adaptations tend to be bred out of a population quickly, within several generations. In chapter 13, I will propose a theory to explain the benefits of human menses that more than counterbalance its many downsides.

---

*Delaney, Lupton, and Toth, in their book *The Curse,* point out one advantage of menses not typically cited in the medical or physiological literature: "If a woman does not wish to engage in sexual intercourse, her period is one legitimate way out. . . . Using 'the curse' as an excuse, many a woman has enjoyed a dinner date free from the bothersome knowledge that she herself might be the dessert."[17]

The human female orgasm was an extraordinary addition to the human species' reproductive life history.

Chapter 7

Her Climax/His Climax

And what about us? Free, we say, yet the truth is they get erections when they're with a woman they don't give a damn about, but we don't have an orgasm unless we love him. What's free about that?

—Doris Lessing[1]

Now there you have a sample of man's reasoning powers, as he calls them. He observes certain facts. For instance, that in all his life he never sees the day that he can satisfy one woman; also, that no woman ever sees the day she can't overwork, and defeat, and put out of commission any ten masculine plants that can be put to bed to her. He puts those strikingly suggestive and luminous facts together and from them draws this astonishing conclusion: The Creator intended the woman to be restricted to one man. . . . Now if you or any other really intelligent person were arranging the fairnesses and justices between a man and a woman, you would give the man a one fiftieth interest in one woman, and the woman a harem. Now wouldn't you? Necessarily, I give you my word, this creature with the decrepit candle has arranged it exactly the other way.

—Mark Twain[2]

*B*esides the dramatic new reproductive adaptations reconfiguring *Gyna sapiens'* life cycle covered in the last two chapters, Natural Selection bestowed several others that proved to be equally momentous. Of these, her orgasm is so extraordinary that it deserves a chapter of its own.

Humans are at present unable to communicate across the interspecies barrier and therefore can never know for sure what other animals are feeling. The human female orgasm is intensely personal, and the question whether or not other female animals experience a similar degree of pleasure during sex must remain uncertain. Nevertheless, astute animal-handlers can gauge with a high degree of accuracy when one of their charges is in distress, hungry, anxious, in pain, or content. Extensive research has revealed that other animals share with humans remarkably similar, and in most cases identical, neurochemical transmitters and hormone physiology, and the same neuroanatomical structures that mediate emotions.

Extrapolating from the physiological changes evident in human females when they are at the height of orgasm, reproductive physiologists have been able to identify very few female animals that mimic the extraordinary respiratory and heart-rate increases, profuse sweating, wild gestures, contorted facial expressions, and audible noises associated with the human female's orgasm. Hormones drive female nonhuman mammals to want to mate, but once the act begins there is little outward evidence that the female experiences the degree of pleasure that a woman enjoys.

I believe it would be safe to assume, when observing a doe leisurely munching on some tasty clover while a buck busily thrusts at her backside, that this animal does not experience an orgasm of the same order as a human one. For the majority of species' females, the consummation of the sexual act is followed immediately by a behavior that can best be characterized as nonchalance. Little observable evidence indicates that for the nonhuman female "the earth moved." From all the available data, it would appear that the intensity of the human female's orgasm is a unique adaptation. A few female nonhuman primates—most notably, the chimpanzees and stump-tailed monkeys—exhibit physiological signs indicating that they are experiencing an orgasm, but, again, their outward manifestations pale in comparison with those of a woman attaining her apex.[3]

Female orgasm was a taboo subject for centuries. Male physicians sporting impeccable credentials pontificated that the very concept was a female fantasy born of fevered minds and could not exist. With the rise of modernity, the floodgates have opened and a deluge of descriptions of female orgasm, advice to women on how best to achieve it, and debates as to its relative merit have poured forth from many different quarters. A casual perusal of the shelves in the women's-studies section of any bookstore confirms this observation.

Evolutionary biologists waded into the fray in the late 1970s, contributing a number of possible explanations to the question that interested them most: What was the selective pressure that would have stimulated this unusual trait to hypertrophy so in the current female version of the hominid line? Many features of the human female's orgasm suggest it is a recent hominid addition and still a work-in-progress. For instance, why is female orgasm so unevenly distributed? Some women experience it all the time, some experience it some of the time, and some never experience it. Some women achieve orgasm quickly, whereas others may take a prolonged time to reach a climax. And some women easily attain multiple sustained repeats.* Evolutionary theorists begin with the premise that orgasm is an adaptation that evolved in response to a specific environmental challenge. They reason that it seems to be too central to the life of *Gyna sapiens* to have been an accident or a spandrel.

However, the possibility also exists that the whole subject is so much threshing of straw, and any attempt to discern an evolutionary reason for the human female orgasm may well be an exercise in futility simply because there is no "reason." At present, three theories vie with each other for top honors. Sexologists and interested anthropologists have whimsically given them the names Pole Ax, Upsuck, and Cuddles.

The Pole Ax Theory proposes that orgasm had to be installed in human females to prevent the species from simply leaking away to extinction. When *Gyna sapiens* stood up, stretched, and started walking, she created a plumbing problem no other species' females had ever encountered. Bipedalism radically re-engineered the lower half of the body to accommodate this new and unusual form of locomotion. Along with the innovative shape of her pelvic bony ensemble, *Gyna sapiens'* vaginal opening moved from high in the rear to low in the front. Positioning the vaginal orifice (or its anatomical equivalent) high in the behind provides easy access for a mounting male and is the favored location in virtually all arthropods, amphibians, reptiles, birds, and other mammals.† The rear-entry position has been used with unerring success for millions of years by millions of other coupling species.

Because of the relative youth of our species, the vagina's present position may represent a halfway station on its journey to its intended final resting place, somewhere slightly below the navel (said with tongue in cheek). If indeed this is its ultimate destination, sexual intercourse will be considerably more comfortable for future cuddling couples but probably less interesting, and certainly less challenging. Such an anterior position would most likely eliminate a great deal of embarrassment and markedly reduce the humiliating fumbling of the inexperienced. Moving the vagina nearly halfway between front and back has not only introduced the need for a certain degree of contortionism between lovers, but also created a confounding evolutionary problem.

The vagina's anatomic migration, necessitated by bipedalism, markedly altered this vital sheath's spatial orientation. When a woman is standing erect, the axis of her vagina aligns close to vertical. In every other mammalian

---

*Numerous studies, both within contemporary culture and cross-cultural, confirm these observations.[4]
†Some sea mammals, such as whales and dolphins, are the notable exception, and they are the only other animals that habitually perform ventro-ventral intercourse besides humans.

female on the move, the vagina lies in a horizontal axis. A flat vagina makes perfect evolutionary sense. Upon completion of a male's delivery call, sperm can begin the sprint of their short lives on a level playing field.

If a *Gyna sapiens* resumed normal activity immediately after intercourse, as is the practice of other species' females, she would stand up and begin ambling away. *Homo sapiens'* sperm, having just left the starting line, would discover to their collective dismay that, despite their earnest efforts to advance, they would keep slipping backward. If they looked over their metaphorical shoulders, they would witness with horror that they and millions of their competitors were descending toward a precipice from which there was no return.

Mother Nature had two options to prevent this inevitable occurrence. She could either equip each sperm with some sort of miniature piton and hand-ax device to hang on and climb upward, hence eliminating the basic streamlined design She had perfected, or She could flatten the female by creating an experience that would temporarily disincline her to arise from a horizontal position. Mother Nature decided in favor of streamlined sperm. She therefore invented the human female orgasm.

A woman who has just completed satisfying sex luxuriates in repose. While she is recumbent, the most vigorous sperm inside her take advantage of this respite to charge past the cervical opening. Once inside the womb, they are free from the vagina's deadly gravitational hazard. Those laggards still remaining in, for them, the Tunnel of Terror would also receive the gift of precious postcoital minutes during which they could advance higher, thereby reducing their danger of spilling out.

From a single sperm's point of view, a female's orgasm has the same effect on its quixotic quest as if a postcoital woman had been hit over the head with an ax handle (hence the name Pole Ax Theory). Her orgasms serve a similar but far more pleasurable function by stunning her into a temporary paralysis followed by a period of sweet lassitude.

The second of the trio of main theories concerning female orgasm is inelegantly named the Upsuck Theory. Robin Baker and Mark Bellis, using both sophisticated monitoring equipment and uninhibited volunteers, measured the movements of the uterus during and just after female orgasm. They also roughly quantified the amount of sperm entering the uterus in women who had experienced an orgasm as compared with those who had not. The uterus, contrary to what most people assume, does not lie in a straight line with the axis of the vagina but is oriented nearly at right angles to it. During intercourse, the upper vaginal vault distends significantly when a woman experiences an orgasm. These actions serve to form a large reservoir in which sperm collect if the coupling woman lies supine.[5]

At the moment of orgasm and for a short period afterward, the cervix, now positioned directly above the concentrated collection of sperm and driven by the rhythmic contractions associated with orgasm, repeatedly dips its cone deeply into the posterior pool of sperm, exerting a gentle suction and drawing many of them up into the womb. In cases of rape, or if a woman is not particularly receptive to a male, reverse mechanisms serve to protect the woman from pregnancy by increasing the difficulty for those sperm trying to penetrate the cervical opening.[6] Unfortunately, these maneuvers to aid in conception or to prevent its occurrence are not always reliable.

The third theory, Cuddles, postulates that Mother Nature bestowed on *Gyna sapiens* a multisynaptic orgasm to help a couple more thoroughly bond with each other through mutual great sex. Female orgasm encouraged women to engage in sex more often, because they found it to be so pleasurable. This in turn made them feel sexier and increased their efforts to attract men. Men were drawn to women who displayed outward signs of availability and interest in sex. The ultimate result was that more babies were born. Because of their mutual interest in sex, couples remained together to help raise their children to the point where their children had children. Despite the warm, fuzzy appeal of this hypothesis, there are several features of human orgasm that do not square with the facts of life.

Other organisms that mate monogamously do not require the female to experience an exciting orgasm to cement their relationship. Moving sex away from the nuts-and-bolts function of merging an overeager, ovulating female at the height of her estrus with a sperm-laden, lusting male places sex in the arena of a sporting activity. This transfer would seem to be a strange and risky adventure out of keeping with Natural Selection's normally conservative agenda. Remember, evolution is driven by an organism's twin needs to survive long enough to reproduce. One or both eyebrows must elevate when sex begins to disconnect from reproduction.

Then there is the problem of the extraordinary mistiming of the male and female orgasms. Men's capacity to ejaculate, and quickly repeat, peaks in early adolescence and then suffers an inexorable downhill slide. The slope of his decline accelerates after a man reaches his mid-forties and, like Lewis Carroll's Cheshire Cat, slowly fades until nothing remains but his feral leer. A woman typically does not experience orgasm until she is in her late teens or early twenties. Her peak arrives in her mid- to late thirties, many years *after* her optimal years to gestate a baby have passed.

Was Natural Selection playing a cruel joke? Perhaps Nietzsche was on to something when he claimed, "God is a comedian performing before an audience that is too frightened to laugh." What conceivable evolutionary benefit would there be in having women and men so jaggedly out of phase with each other? Ideal sexual compatibility would demand that thirty-something women should pair up with postpubertal boys. Such an arrangement, however, would completely upend the norm of family, child-rearing, and societal organization. In general, women, the world over, prefer older men, and men, the world over, prefer younger women. Yet both sexes' orgasmic barometers are set awry against these universal predispositions.

If Natural Selection provided orgasm as a mutually pleasurable event to encourage sexual union, why not enable the two sexes to synchronize easily? Adolescent boys can set off their supersensitive detonator in their sleep, in class, or while daydreaming. All this hyperactivity occurs at an age when it is unlikely that any of their sperm will ever be called upon to swim anyplace useful. Yet most women of the same age don't experience such hypersexuality. In fact, many women never know what all the fuss is about throughout their entire lives.*[7]

There are further incongruities. Why do many women who can achieve orgasm, sometimes (or in some cases always) experience multiple sustained repeats? The man lying on top of (or under) her is spent. His reload time increases with his years, just as her ability to do it over and over again increases. How can this mismatch be a feature that guarantees a man and woman will mutually desire to re-engage in an activity that can often be so unsatisfying for either participant? The disharmony of male and female orgasm is the source of much contention and frustration in the bedroom. Again, it would seem to make little sense, from an evolutionary point of view, to have these two physiological events occurring in tandem instead of simultaneously. For many new couples, on that uncommon occasion when their two orgasms occur in synchrony, the surprised pair effusively congratulate each other on their extraordinary feat, as if they had both climbed to the top of Mount Everest, arriving from different directions but reaching the peak together.

Many women, frustrated by their encounters with premature ejaculators, turn to masturbation, discovering that they can achieve satisfaction much more easily by themselves. Many men, exasperated by the directives of a hyperorgasmic (or hypo-orgasmic) woman, turn to masturbation, discovering they can save themselves the embarrassing feelings of inadequacy by playing solitaire. Mother Nature would have to have Her head examined if She purposely installed a feature in the human genome that encouraged individuals to desire sexual release without a partner. What is the evolutionary point She was making when She initiated so much distressing mischief?

---

*Considerable debate surrounds this issue. Many psychologists maintain that failure to achieve orgasm is largely due to cultural mores and lack of education. Many reproductive physiologists claim it is more a function of genetic programming.

Another quirk further jangles relations between the sexes: Men, as a general rule, achieve speedier ejaculations more often with unfamiliar women than with women with whom they have been intimately familiar over a long period of time. Sexologists call this phenomenon the Coolidge Effect, named after an incident that occurred between the taciturn thirtieth president of the United States and his wife.

The president and his entourage were visiting an agricultural state and were on the obligatory farm tour when husband and wife were taken to different areas. While they were being shown the henhouse, the guide informed Mrs. Coolidge's group that a single rooster could service all the hens. Mrs. Coolidge, incredulous, asked the farmer, "You mean to say that one rooster is able to keep all these gals satisfied?" When the farmer confirmed with a "yup," Mrs. Coolidge smiled knowingly at the other women in her party and then instructed the farmer, "Go tell *that* to President Coolidge."

When he caught up with the president, the farmer relayed Mrs. Coolidge's amazement about the sexual prowess of the rooster. Coolidge, without a moment's hesitation, asked the farmer if the rooster repeated his studly performance always with the same hen, and the farmer was quick to add, "Oh no, Mr. President, it is always with a different hen."

To which Silent Cal deadpanned, "Go tell *that* to Mrs. Coolidge."[8]

Montaigne observed, "I have put out to stud an old horse that could not be controlled by the scent of mares. Familiarity presently sated him toward his own mares, but let a strange one pass his pasture, and he returns to his importunate neighings and furious heats as before."[9] The French also express the same sentiment in another way: "Fresh mount, fresh courage," men would say to each other with a wink and a nod. How does this feature of male orgasm fit into the scheme of survival of the species through sustained joint parenting by means of great mutual sex?

And why is it that women who have birthed a child are more likely to experience an orgasm than women who have not? The capacity to have multiple sustained orgasms is awarded most often to women who have experienced multiple births. It is uncommon in childless women. Shouldn't orgasm, especially among the childless, serve as the carrot encouraging them to keep at it? Why bestow the benison on those who have already performed their duty to their species?

Many arguments and counterarguments could be made to the points raised above. Taken as a whole, however, there appear to be too many inconsistencies to embrace wholeheartedly the notion that female human orgasm evolved to entertain couples while they unwittingly increased the species' numbers. Perhaps there are other, more subtle reasons for why this feature evolved.

All three of the commonly proposed theories pass the test of plausibility. Selective pressures that encourage an adaptation to evolve are frequently multifactorial, and it may very well be that each of the three current theories I discussed above played a part in wiring *Gyna sapiens'* nervous system for her extraordinary orgasm. I would, however, propose adding three alternative reasons to explain why Mother Nature propelled *Gyna sapiens,* alone among the feminine myriad, to the heights of ecstasy.

The first reason: Female orgasm directly addressed the biological emergency that struck our species when African Eve realized that mortality was intimately related to sex. The veto power she gained over sex as a result of her insight endangered the viability of the species. A sober ancestral woman facing the odds that she might *die* because she had intercourse may have reasonably decided to forgo engaging in the activity. Mother Nature found it imperative to reward *Gyna sapiens,* thus countering her newfound caution. The reward was orgasm. If a woman knew that she could revisit the most amazing rush she had ever experienced, then she would be substantially more willing to risk confronting the Grim Reaper nine months hence, believing the game was worth the candle.

The second and subtler reason for the evolutionary value of woman's orgasm addresses the nettlesome problem of a human female's mate-selection process. A woman had to be far more discerning than a man before deciding to say *Yes!* to sex. The gravity of her assent weighed heavily on her, because she understood, as no other species' females ever did, that she had to determine *in advance* whether a male prospect would be an excellent protector, provider, and companion for many years into the future. She also needed a means to gauge his suitability as a father to the children she had discovered were the real purpose behind sex.

Human language, as we will see in a later chapter, provided her with a major assist in this department, but it was not foolproof. A slick-talking male could deceive her as to his true intentions. Some men desire only to satisfy their lust, some to achieve a "conquest," and some to prove their virility. She needed to predict who would abandon her and who would stay. By creating marked differences between the quality and timing of a female's and a male's orgasm, Mother Nature provided *Gyna sapiens* with the quintessential measuring rod to take the measure of her lover's future behavior—one that superseded in value the information she garnered from mere words.

In general, a man has straightforward, no-nonsense sexual goals. These are (1) easy access; (2) control over the pace, pressure, position, and duration of sex; (3) rapid withdrawal at the conclusion of his ejaculation; followed by (4) his unencumbered departure to attend to other business. Among the many reasons a man frequents prostitutes is his desire to satisfy his sexual wish list without having to feel guilty for not considering his partner's feelings, needs, or desires.

Unlike a man's experience, a woman's orgasm does not always confer its beneficence every time she copulates.* She learns through experience that, for her to achieve it regularly, several preconditions help.

For optimal results, most women prefer to be immersed in the trappings of romance prior to engaging in sex. The man with whom a woman contemplates an assignation should have considered the setting, lighting, and music leading up to their union, because romantic seduction increases her lubricity, which in turn predisposes her to a more pleasurable sexual experience.

She desires that he exhibit patience by engaging in a lengthy period of foreplay, consisting of his lambently stimulating her multiple erogenous zones. She hopes that he will show consideration and postpone his entry. Once he is inside, she wants him to prolong his movements and delay his ejaculation so that they can arrive at their respective pinnacles together. She prefers to guide him gently in matters concerning pace, pressure, position, and duration. Propelling her along the steep ascent to orgasm would be abundant amorous feelings toward the man with whom she is intimate.

Clearly, a man's and a woman's lovemaking agendas are in sharp opposition to each other. Compromise is necessary after a couple comes together if both parties are to feel "togetherness." But a woman learns, to her surprise and dismay, that very little is required of her to satisfy the majority of her male partners.

In matters relating to *her* orgasm, however, a man cannot fake his concern or allow his attention to wander. Having reached orgasm sooner than his partner in most cases, a man must steadfastly remain at his station and continue to provide her with the most efficacious stimuli that he has learned over time will assist her in achieving a similar degree of pleasure. A man willing to do this reveals much about his character. By noting his willingness to sacrifice many of his original fast-sex, short-term goals, a woman can gauge how likely a man is to give her pleasure in the future.

A man's response to the possibility of a woman's orgasm decisively separates the wheat from the chaff. The major traits a woman hopes to find beating behind a man's rough exterior include: health, humor, patience, empathy, kindness, stamina, honesty, creativity, reliability, and generosity. All can be

---

*Some men have difficulty achieving an orgasm all the time, and some men intermittently have difficulty achieving orgasm, at different times in their lives. In general, however, male orgasm is much easier to attain than is a female's.

ascertained by evaluating the nuances of how a man makes love. These traits also predict what kind of provider, protector, companion, leader, and, most important, father this particular man will be. (This statement does not negate the experience of both men and women that sex with someone who may be entirely unsuitable as a prospective mate can often be the most exciting and satisfying for the very reason that it is flirting with danger.)

An old saw states, "As the courtship goes, so goes the marriage." To that I would add, "As the lovemaking goes, so goes the union." Women determined after much experimentation that a man "with a slow hand and an easy touch who did not come and go in a heated rush" had the right stuff to see her through life's inevitable travails.*

A man, too, learned, through many embarrassing trials and errors, that if he desired to obtain her nod of assent in the first place, and later to keep her interested in him, he would have to adjust drastically the short-term goals of his fantasized sexual style. The long-term benefits accruing to a man living with a sexually satisfied woman outweighed the extra effort he expended on his partner's behalf.

The third alternative reason for the evolution of female human orgasm concerns the mysterious G spot. In 1944, German gynecologist Ernst Grafenberg identified a location in the superior wall of the vagina, just behind the pubic bone, that, when stimulated correctly, produced a vaginal orgasm distinct from a clitoral one. He became convinced of the existence of this accessory pleasure zone because many of his patients confirmed its presence by recounting their experiences. He published his report in an obscure medical journal and offered no scientific evidence to back up his claim.

His revelation, however, was not new. Similiar observations had appeared in sexual manuals of antiquity, and seventeenth-century anatomists identified an excitatory zone in the same location. Alice Ladas, Beverly Whipple, and John Perry seized upon Grafenberg's report and published a book in 1983 naming the area "The G Spot."

Scientists and lay sexologists have since attempted to verify Grafenberg's observation, but, despite a vigorous debate about the issue, there remains little scientific consensus as to the exact whereabouts or even the existence of the elusive G spot. Couples, too, in the privacy of their bedroom, merrily joined the hunt. Freud, without ever actually doing any experiments, surmised that a vaginal orgasm existed distinct from a clitoral one. Masters and

---

*This sentence is a paraphrase from a Pointer Sisters song.

Johnson meticulously measured vaginal responses to stimulation and debunked Freud's claim, and in the process sank Grafenberg's, too.[10] To many male reproductive physiologists, the G spot resembles the mythical unicorn, the horned white stallion that some women can easily see and know exists. For some women and the majority of men, despite a diligent search, the magnificent creature continues to remain elusive.

Gynecologist Terence Hines recently proclaimed in the prestigious *American Journal of Obstetrics and Gynecology* that the G spot did not exist. Period. He dismissed the concept, calling it "a gynecological UFO" with many sightings but no hard proof. Hines based his conclusion on his exhaustive review of the medical literature, which failed to turn up any substantive or credible scientific evidence. He dismissed the numerous anecdotal reports from women who were quite certain, based on their own personal experience, that the G spot was indeed real.[11]

However, Hines and other skeptics should be reminded that absence of evidence is not evidence of absence. D. T. Suzuki, the great Buddhist teacher of the last century, noted that the primary characteristic of an ecstatic experience is its *authenticity*. Once someone has experienced ecstasy, no amount of logical argument can convince that person that what he or she experienced did not happen.

Missing from the debate on whether the G spot is or is not present is an explanation for why such a spot would have evolved in the first place. Form follows function. The tip of the clitoris has crowded together in one small area over eight thousand nerve endings, making it the most concentrated collection of sensitive dendrites anywhere in the male or female body.[12] It does not appear to have any function other than to provide its owner with pleasure. If the clitoris evolved in humans only to confer upon a woman exquisite delight, what would compel Mother Nature to create a secondary pleasure center in an inaccessible region that is difficult to find and unlikely to be stimulated in the course of prosaic lovemaking?

Although scientific evidence for a G spot remains murky at this time, there could be a sound evolutionary reason for this secondary pleasure zone. Could its existence be more related to the pain of childbirth than to the pleasure of sex? The final stage of human delivery pits the largest skull among animal newborns (relative to body size) against the vaginal orifice's diameter. Despite the vagina's rapid circumferential dilation during delivery, every woman who has ever birthed a baby can vouch that the contest between the two—infant's head and distended vaginal opening—is a terrible mismatch.

If Natural Selection did not step in and conceive of a way to ameliorate the pain a mother endures in the last phase of delivery, the searing memory of it might traumatize her forever. Her ordeal during her baby's difficult pas-

sage could exert a powerful influence over her psyche, inclining her to promise herself that if she survived this experience she would never—*ever*—repeat
it. Based on innumerable hours of observations of other female animals, it
would appear that none experiences as painful a delivery, nor do any others
possess the capacity to hold the memory of it indefinitely and, moreover, act
on this information far into the future. If *Gyna sapiens* exercised control over
whether or not she would engage in sex, then an excessively painful childbirth could tip the scales toward her decision to abstain permanently.
African Eve's acquisition of Free Will could then become a double-edged
sword, placing the survival of the human species at risk.

Enter the semi-mythical G spot. Let us begin by assuming that the G spot
does exist. Grafenberg postulated that this on-off button lay just behind the
pubic bone, far outside the outer limits of the vaginal sheath, hidden among a
dense plexus of pelvic nerves. Serendipitously, some women discovered that
deep pressure on the G spot during intercourse can activate its reflex, adding
considerably to their pleasure. The most advantageous position for its stimulation occurs when the woman straddles a man facing backward. In this
position, his member deeply strokes the area just behind the pubic bone.
The position is awkward, and infrequently used for a variety of reasons, not
the least of which is that it misdirects sperm from their intended target, the
cervix, which lies in an entirely different direction. It hardly seems reasonable to conjecture that Natural Selection intended the G spot to promote female sexual pleasure if at the same time it inhibited conception. Its obscure
location would seem to diminish the argument that it is present to provide
its owner with an alternative jolt of pleasure during intercourse. If this were
the G spot's raison d'être, why would not Mother Nature position it on the
kneecap, or the elbow—someplace easy to get at?

The only time in the course of a woman's life span when the G spot's remote dense knot of nerves would ever have rhythmic, extreme pressure applied to it would be in the final push of birth. As the baby's head traps the
upper wall of the vagina, compressing it against the unyielding junction of the
pubic bones, the G spot's fortuitous reflex would be set in motion. The function of the G spot, in my scenario, would be to flood the delivering woman's
brain with endorphinlike substances. Many women report an almost out-of-
body experience accompanying the birth of their child, and in a few instances women have claimed they experienced the *best* orgasm of their life.

In her book, *The Crack in the Teacup*, Canadian storyteller Joan Bodger
recounts the events surrounding the birth of her third child:

> Lucy was born on February 1, 1956, in San Pedro Hospital, the very one
> where my mother had almost died in 1929. Like my doctor in Nyack, the

doctor present had never participated in a "natural" childbirth, a birth where the mother *chooses* to be conscious. This time, John was allowed to stand just outside the delivery room, separated from the scene by a flimsy curtain. In the first stages of birth, I talked to him, but further along I was too involved in the birth—and with something else that was happening to me. Along with the pain, I was experiencing the greatest orgasm ever to roll over me, through me, from the roots of my hair to the tips of my toenails. I must have indicated something of the sensations that were flooding my body. My yells of pain were punctuated by cries of sexual ecstasy. The staff seemed embarrassed, not wanting to believe what they had seen and heard. (Years later, I discussed my experience with a gynecologist. I was blushing as I spoke, but I had to know whether I was the only woman in the world who . . . The doctor said he had read that some primitive women experienced birth orgasm, but he himself had never before talked to a woman who was willing to admit such a thing had happened to her. Of course, most of the women he dealt with had been under anesthesia.)[13]

I have had the pleasure of meeting Ms. Bodger, who in my opinion is a very reliable reporter. Her experience is not an isolated instance. Other women have also described experiencing birth orgasm.

Let us consider why it would have been very beneficial to the survival of the human species for a woman in the throes of childbirth to experience an orgasm. Or, if unable to attain this extreme height, she would have, at the minimum, neurotransmitters tickling her brain's synapses, diverting her attention away from the painful crush occurring between her legs. How convenient for Natural Selection to arrange to have an internal morphine pump alleviating the pain of childbirth. The switch for it, abutting the back side of the pubic bone, would be indented "on" only when heavy pressure was applied—the kind of pressure that could happen routinely during birth, or in odd positions during sex. In a few women, the pump releases other psychoactive substances that can stimulate a woman to orgasm.

Nepenthe was a Greek nymph whose name means "surcease from sorrow." The narcolepsy named after her is associated with the endorphin/morphine fog that often leaves women with only a hazy memory of the birth experience. Many soporific women are vaguely aware of Nepenthe's ephemeral presence in the delivery room. Mothers tell worried first-time-pregnant women the comforting saw, "Yes, the pain is awful, but you tend to forget it." Perhaps the G spot is the physiological reason for the relative amnesia. William Wordsworth might just as well have been referring to the mother as well as the newborn when he wrote, in his ode on "Intimations of Immortality," "Our birth is but a sleep and a forgetting."

If Nepenthe was a midwife, another character from Greek mythology also stands by as the woman's internal anesthetist. Morpheus was the Greek god of sleep, and the opiate morphine was appropriately named after him. Morpheus had a twin brother whose name was Thanatos, the god of death. Natural Selection enlisted the powers of Morpheus to help a woman edge past the most dangerous encounter with his brother that she would routinely experience in the course of her entire life.

The role hormones play in inclining a mother to repeat her pregnancy emerges from the recent discovery that, the moment a newborn begins to nurse, this sucking action activates a reflex that ensures that the mother will fall in love with her baby. Breast-feeding causes nerve impulses to leap away from her nipple and travel to her brain's limbic system. This stimulus causes the amygdala, a major component of the emotional brain, to release a flood of the hormone oxytocin, which then inundates all the synapses of her nervous system. Oxytocin is the "love hormone," and high levels are associated with the bliss of profound attachment. It cannot be a coincidence that among oxytocin's major physiological effects is to cause the uterine muscles to contract sharply following the expulsion of the placenta. Women bond with their babies just as they markedly limit postpartum blood loss. What a cleverly designed system!

I propose that the human female orgasm was an evolutionary prize awarded to women because *Gyna sapiens* became the first female to learn of copulation's onerous and even deadly price. The intense pleasure of her orgasm gave her a powerful incentive to keep re-engaging in sex. Also, by enabling women to gain access to an important alternative means of evaluating a man's character, orgasm provided them with a valuable tool in making better choices among potential mates. A secondary pleasure center evolved deep in the pelvis for the purpose of reducing the agony of childbirth by ingeniously crowding out the pain.

The human female orgasm is, by all accounts, a many-splendored thing. The multifactorial mix of possible reasons for its evolution should not obscure the reality that its presence is a distinguishing trait separating *Gyna sapiens* from *Gyna all-the-others*.

Three generations.

Chapter 8

Grandmothers/Circumcision

No spring, nor summer beauty hath such grace
As I have seen in one autumnal face   —John Donne

There is no greater power in the world than the zest of a post-menopausal
woman.   —Margaret Mead[1]

My own preference, if I had the good fortune to have another son, would be to
leave his little penis alone.   —Benjamin Spock, M.D.[2]

*T*he organs and systems of the human body resemble the interlocking cogs
and gears of a well-crafted grandfather (or, for the purposes of this chapter, let
us call it a grandmother) clock. Every part is designed at the outset to work in
harmony with the others for the duration of the life of the person. Some parts
are more susceptible to the impairments of aging than others, but brain, kid-
ney, heart, lung, and colon come with a warranty claiming that, with proper
usage, each organ system should approximate the allotted life span of the individual. Not covered are misuse, abuse, and Acts of God. When an organ or
a system malfunctions prematurely, specialists in the culture assigned to attend to such matters diagnose the condition as a disease.

One system, however, affecting only one species, and targeting just one
sex of that species, *always* winds down long before the others. That vital
function is a woman's fertility. The paired organs primarily responsible are
her ovaries. The directive to shut down operations considerably before closing time comes from her brain's tiny pituitary gland, and its commands are
programmed into a woman's genes when she is born.

The out-of-sync aging of *Gyna sapiens'* ovaries sets in motion the cascade of events known as menopause. Typically beginning in a woman's late forties or early fifties, the constellation of signs and symptoms are the result of the ovaries' ceasing to mix, shake, and stir the complex cocktail of hormones that drive the menstrual cycle.

As previously emphasized, evolution creates a variety of living forms each of which must survive long enough to reproduce. Survival and reproduction are so central that some species, like salmon, survive *only* long enough to reproduce. Adult ocean salmon embark on a long and grueling journey by swimming up rivers, jumping rapids and waterfalls if necessary, all for the purpose of finding the creek where they were born. Upon arriving at this improbable destination, the exhausted adults then spawn the next generation before expiring from their Ulyssean effort. Since reproduction is so basic to all species, the benefit to the human species of *Gyna sapiens'* early and abrupt end to her fertile years, when set against the potential of her exceedingly long life, must be explained in Darwinian terms.

The onset of menopause is marked by a series of notable events. Ovaries lose their effervescence. Eggs no longer bubble in slow motion to the surface before bursting forth. Those that remain trapped within slowly disintegrate. The curtain rings down on the womb's long-running monthly drama, which had featured a compelling story of rebirth, wild youth, ripe maturity, and sudden death punctuated periodically by nature's grandest showstopping denouement—birth. Menstrual bleeding ceases.

The complex infrastructure enabling a breast to produce milk involutes. The mammary glands, the distinguishing feature that led the eighteenth-century taxonomist Carolus Linnaeus to name our phylum "mammals," lose their previous resiliency.

Secondary sexual characteristics defining the essence of the feminine rapidly begin to recede. Lips lose their fullness, skin loses its turgor, hair loses its luster, eyelashes lose their length, and the *labia majora* lose their plumpness. The robust corrugated lining of the vagina thins, and the former ease with which it lubricated under sexual stimulation attenuates.

Highly dependent on estrogen and progesterone, the branch of the autonomic nervous system responsible for regulating the size of the skin's tiny capillaries goes wobbly. Hot flashes, night sweats, and facial blotches are the involuntary, embarrassing, and discomfiting results.

The rhythm of the menstrual tides washing in and out of a woman's internal seas affects nearly every organ system in the body. The cessation of the monthly music often produces psychic dissonance, manifested by mood swings and, on occasion, severe depression. Less discernible physiological changes also accelerate. Arteries harden, muscles weaken, bones become more brittle and less calcified. Libido can suddenly drop. Conversely, because the individual response to estrogen and testosterone levels varies considerably, some women experience a rise in their libido following menopause. The evolutionary paradox of a mammal having an increased appetite for sex at a time when conception is impossible and the bloom of youthful beauty has faded makes more poignant Oscar Wilde's observation that "the tragedy of old age is not that one is old, but that one is young."[3]

The majority of mammalian females maintain reproductive fitness right up to the day they die. A few other large species besides *sapients* experience menopause.* What distinguishes the human one from the others is its prematurity. The chimpanzee and elephant, for example, experience it when they have reached the age at which they could live just long enough to raise their last offspring. A chimpanzee typically breast-feeds each offspring for five years. Female chimpanzees therefore experience menopause five years before the close of their expected life span. A female chimp's life expectancy is forty-five; her menopause begins at forty. Should she become pregnant just prior to its onset, she would live just long enough to wean her last baby.

Consider how different the human female is from her mammalian sisters. *Gyna sapiens* undergoes menopause at a point in her life when she still has many remaining years of health, vitality, and strength. Yet she mysteriously ends her reproductive life at a point between three-fifths to one-half of her entire life span, and at a time when she could perform the duties of motherhood. The puzzling installation of a biological alarm clock set to ring at an ungodly early hour in our species alone is all the more baffling when compared with the absence of a similar device in men.

Most anthropologists posit that the answer can be found in the prolonged length of childhood. The extraordinary amount of care necessary to feed, protect, educate, and socialize children to the point where they can successfully realize their own reproductive potential is greater than that needed by any other animal. Multiple births spaced over years are the norm in our species. A mother caring for more than one child has her proverbial hands full. A male can help in provisioning and protecting, but his contribution is often inadequate. A harried young mother desperately needs a wife.

The Grandmother Theory, also called the "prudent mother hypothesis," proposes that early menopause frees a woman from the responsibility of caring for her own children at a point in her life when she still possesses consid-

---

*An alternate name for us is *Homines sapientes*. I will use *sapients* as the plural form of *Homo sapiens*.

erable vigor and thus can devote her energy to helping her daughter (or her son's wife) raise her grandchildren, whose chromosomes carry one-fourth of her DNA.[4] Mother Nature had to invent grandmothers because childhoods became excessively long. An older, more experienced, middle-aged woman became an indispensable adjunct for burdened ancestral mothers coping with the round-the-clock job of caring for tykes.

Assuming that fifteen is the age when a human becomes marginally self-reliant, and using sixty-five as the age when a human mother's vitality begins to flag significantly, then the cessation of menses in a woman's late forties makes good evolutionary sense. A mother conceiving at this late date can be assured of living to see her last child reach maturity to survive and reproduce. The bonus years beyond sixty-five allow her to spoil her own children's children.

Anthropologist Kristen Hawkes, using sophisticated time-and-motion studies, has demonstrated that grandmothers are more efficient gatherers than younger women and are able to bring their daughters more total calories than the daughter's mate.[5] As Natalie Angier put it, "Mothers bred what grandmothers fed."[6] Older women can also offer younger women assistance in intangible ways. They are repositories of wisdom and can mediate disputes between offspring. They provide an invaluable cushion to help the young to survive. Childhood is a cultural convention made possible by the invention of grandmothers.

The accumulated wisdom of grandmothers was, I shall propose, a key factor behind the origin of the peculiar rite of male circumcision. At present, much thunder and lightning surround this emotional subject. Traditionalists maintain that, because ancient scripture demands it, and God-fearing Jews and Muslims have always practiced it, circumcision must be correct. Another side claims that Biblical people had excellent hygienic reasons for mandating the sacrifice of every male's foreskin. A third group considers the custom a barbaric genital-mutilation ritual imposed on young boys at an age when they cannot protest. When pressed to provide an explanation couched in scientific terms for why this tradition has persisted so widely and for so long, most people fall back on the tautology, "It is because it is." Because we have always done it, therefore we must continue to do it.

Let us examine the practice with new eyes and ponder which group among ancestral people would have benefited the most from male circumcision, and who would have paid the steepest price for the procedure. This risk-benefit analysis assumes that circumcision is a very old and widespread custom. Columbus reported that he encountered many New World indigenous peoples whose males had been circumcised.[7] The earliest recorded example clearly demonstrating circumcision's practice was found in pharaonic Egypt. A graphic depiction of a man clearly circumcising another appears on a mural in the tomb of Ankhmahor, associated with the pyramids of Saqqara dating to 2200 B.C.[8]

Despite claims to the contrary, the lifetime health benefits to the male are paltry. The minimal advantage of a reduction of under-the-foreskin infections, a medical condition known as "balanitis," seems minor compared with the risks and pain of the procedure. Numerous highly successful cultures flourished throughout history, such as the Greeks, Romans, and Chinese, without practicing circumcision. So adamant were the Roman authorities on the subject that, during the reign of the Flavian emperors, circumcision became a capital crime.[9]

Had balanitis been a factor preventing significant numbers of warriors from marching and fighting while living under the unsanitary conditions of a mobile bivouac, then, surely, authorities would have adopted measures to counter the problem. At the very least, they would have mentioned that a problem existed. Hippocrates and Galen, ancient history's two most famous physicians, discoursed on a wide variety of human ailments, but neither listed penile infections caused by an overabundant foreskin as a health issue requiring a culture-wide remedy.

Another health benefit claimed for circumcision is that the incidence of penile cancer is higher in males who have not been circumcised than in those who have undergone the procedure. This condition is so rare it vitiates the argument, and since it has been observed only since records began to accumulate in the modern era, factors such as socioeconomic life-styles have not been adequately taken into account.

Proponents argue that women benefited, because they caught fewer sexually transmitted diseases and cervical infections, which in turn would have lowered their rate of cervical cancer. This defense of circumcision is weak for similar reasons. Cervical cancer as a cause of females' death would be very near the bottom of any list for why women have died throughout history. Other hygienic arguments point to various obscure health benefits for circumcision, but, again, the slight gains do not offset the significant risks to the male who had to undergo surgery in the pre-antisepsis, pre-antibiotic, and pre-anesthesia era.

The most devastating counterargument against any health benefits for circumcision becomes obvious to anyone visiting a zoo. Among the im-

---

*For a more complete historical account of circumcision, see page 374.[10]

mense variety of animals, nonhuman males have evolved some sort of protective sheath to cover the tip of the penis. Surely, humans are not as clever as Natural Selection, and if a covered penis or its retraction within the body was detrimental to the survival of a species, then there should be a whole raft of male animals running around with pale, naked members flapping in the breeze. Several African primates sport short foreskins that permanently expose a small portion of the tip of their penises, but with the exception of these few, the absence of nonhuman males that flout their tips tends to argue against circumcision's health benefits.

Let us reflect on the foreskin and the function it serves. The human male's covers the bulbous glans (tip) and proximal portion of the shaft of the penis when this organ is flaccid. The prepuce's (foreskin's) inner lining, especially its base, is exceedingly rich in nerve endings that greatly enhance male sexual arousal. During intercourse, the prepuce is turned back on itself, much like a rolled sleeve. The inner nerves are now on the external surface, and their stimulation by friction contributes to the degree of penile excitability. When the prepuce is in its resting position, mucous glands lining its inner surface secrete a thick protective substance that keeps the glans moist and slightly lubricated, similar to the resting state of a female's vagina.

A man can easily telescope his loose, elastic foreskin manually when he has to urinate. When the penis is in full erection, the foreskin retracts naturally, exposing the penis's glans and its proximal shaft. Form follows function. This thick cutaneous holster obviously protects a very sensitive and vulnerable part of the male anatomy. Genital injuries are exceedingly common in the many fights observed by primatologists among male chimpanzees. A bipedal mammal would be the most susceptible animal to this trauma, and would require the most protection. No other creature has so defenseless and exposed a penis as does a human, and the function of the foreskin in humans is therefore all the more important.

There are few superfluous parts of the body. Mother Nature has subjected every component to relentless pruning. Resembling an obsessive-compulsive gardener, She continually snips away at deleterious behavioral, physiological, and anatomical adaptations that have edged into a species' genome. At the same time, Mother Nature encourages the retention of mutations that further the survival of the individual by increasing the probability that the holder of such a lucky gene combination produces surviving offspring in greater numbers than those unfortunate enough not to possess it.

Although the human body appears eminently well suited to do what it has to do to compete with other animals, on rare occasions Mother Nature allows a feature, called a "vestige," to persist even though it no longer serves the function for which it was originally designed. For example, resembling a dog, we have a full set of muscles surrounding our ears to twitch them in the direction of the source of sound. But this shadow musculature is so vestigial that most people are unaware they have it. A few people remain capable of an attention-getting party trick and can wiggle their ears on request, but these muscles' original purpose has been lost in time, their function faded with disuse. Many in society have made the assumption that the prepuce is vestigial and of no use. Cutting it away should therefore have no health consequences.*

Besides shielding the penis from injury, the prepuce protects a valuable male sexual asset. The glans has one of the greatest concentrations of nerve endings present in a man's body. The nerve endings connect to microscopic specialized organelles that lie exposed on the surface, scattered among surrounding skin cells. These tiny pleasure centers are extremely sensitive to trauma and can easily be damaged.

An analogous area of a man's body, containing nearly as many nerve endings as his glans, are the pads of his fingers. The sensitivity of his digits allow a man (or a woman) to discern the slightest pressure or the most delicate nuances in the texture of any object touched. But fingers, unlike a penis, are nearly constantly in use. Selecting for practicality, Natural Selection did not protect them with a sheath.

Repeated use of the hands and fingertips, however, subjects their nerve endings to the possibility of chronic trauma, as happens if the fingertip's owner engages in constant heavy labor. Friction and pressure from contact with rough or smooth surfaces will wear down the nerve endings and the upper layer of skin, resulting in a painful blister. To defend against this undesirable outcome, the skin covering the fingertips toughens with use by building a thickened shield of specialized skin known as a "callus." Microscopically, the sensitive nerve organelles appear buried beneath stories upon stories of cornified skin.

The delicate sensation of touch diminishes dramatically because it is filtered through a dense layer of overlying callus that muffles the signal the brain receives from its digital outposts. The ability of a laborer's finger pads to

---

*Occasionally, the appendix, the most infamous vestigial organ, goes beyond what the majority of vestiges do, which is to sit quietly and not interfere with the survival and reproduction functions of the other working parts. This small hollow finger hanging off the right side of the colon can all too easily announce its presence obstreperously by becoming infected with stool, burst, and then deliver its owner to death's doorstep far ahead of the body's scheduled departure. Despite innumerable hours of research on the subject, this fingerling has not been found to perform any useful function. It is the only vestige to pose a serious threat to the viability of its possessor.

The appendix represents a small chink in Darwin's grand work. Evolutionary theory would have predicted that the appendix should be long gone. Appendicitis in ancestral *sapiens* would have most likely eliminated them from the gene pool. Over thousands of years, Natural Selection would have been expected gradually to evolve hominids who did not have this potentially ruinous appendage. And yet there it is, making a classic upside-down obscene gesture. Perhaps the appendix performed a valuable dietary or immunological function long ago that is no longer necessary.

make fine discriminations cannot compare to that of someone whose hands are not subjected to the repeated daily trauma of gripping, squeezing, rubbing, and chafing against rough surfaces.

Carrying this analogy over to the penis, the removal of the foreskin would expose the glans to daily small traumas. Since nearly every man in nearly every culture covers his penis with some kind of protective clothing, an uncircumcised glans will constantly rub against the inner lining of clothes that are far rougher than the moist, snugly fitting scabbard of a lubricated foreskin. The sensitivity of a circumcised glans's delicate nerve organelles lessens with age.

The glans's skin is dryer, more leathery in appearance, and more callused than that of a man who has not been circumcised. One can assume that the sensitivity of the former is less than the latter. Dampening the excitability of slightly buried or traumatized penile nerve endings would mean that a circumcised male requires more friction to reach his climax. He would be less quick to achieve an orgasm than would a man whose foreskin was intact. Although debatable, it is unlikely that any diminution in a man's pleasure occurs, but a circumcised man will take longer to arrive at the same destination than a noncircumcised man, all other factors being equal.

Throughout history, commentators have acknowledged the effect of circumcision on a man's sexuality. Maimonides, in the twelfth century, wrote:

> As regards circumcision, I think that one of its objects is to limit sexual intercourse, and to weaken the organ of generation as far as possible, and thus cause man to be moderate. . . . This commandment has not been enjoined as a complement to a deficient physical creation, but as a means for perfecting man's moral shortcomings. The bodily injury caused to that organ is exactly that which is desired; it does not interrupt any vital function, nor does it destroy the power of generation. Circumcision simply counteracts excessive lust; for there is no doubt that circumcision weakens the power of sexual excitement, and sometimes lessens the natural enjoyment. . . .[11]

Maimonides and many others believed circumcision cooled a man's sexual ardor and lessened his masturbatory impulses, and hailed these changes as laudable within a religious context. Male writers sermonizing on the subject failed to comment on who would have been the true beneficiary of a circumcised male's decreased penile sensitivity.

A woman paired with a circumcised lover would find his climacteric delay extremely salubrious, since premature male ejaculation is the primary source of female dissatisfaction with a male's amatory performance. This is

especially true during the early years of a man's and woman's sex life. Her orgasm occurs considerably later than his, and anything that prolongs his erection would serve to increase her satisfaction. Nothing succeeds like success.
Her satisfaction would ultimately make him believe that he is a great lover,
and incrementally raise his testosterone level, while increasing his pleasure
in sex. Reciprocity would heighten a couple's desire to repeat the act, and the
ultimate result would be more babies deposited in the world. Mother Nature
would smile at these clever creatures for figuring out a way to circumvent
what would appear to these mere mortals as one of her tiny design flaws.

Another benefit of circumcision: A slightly longer time before male ejaculation requires that a man thrust for a longer period. The increased friction on
a woman's vaginal lining stimulates a more copious amount of lubrication.
From an alkaline-loving sperm's perspective, the resting acidic pH of the
woman's vagina is very hostile. Every passing minute of lovemaking raises
her vaginal pH, increasing it to ever-more-hospitable levels. The sperm of circumcised males jump off into a friendlier environment than those of uncircumcised males. The sperm of the former would receive a small boost in viability,
making them more likely to survive before entering the safety of the cervix.

Who, we might ask, among a tribe of children, adolescents, grown-ups,
and elders, would be the group most likely to have ferreted out the above,
most interesting fact of life? Which sex, male or female, would be more likely
to connect the dots between absent foreskin at birth, delayed ejaculation in
youth, more babies in maturity, and more grandchildren to dote over? After a
lifetime of lovemaking, mature women would have spent hours discussing
the sexual idiosyncrasies of their diverse male partners and comparing their
experiences. An older woman is the candidate most likely to have discovered
that *the true beneficiaries of circumcision would be the lovers of the men who
had been converted from hooded cobras to bald eagles.*

Women desiring maximal sexual pleasure seek to delay their lover's orgasm. Contemporary sex therapists recommend thick condoms impreg-
nated with Novocain to desensitize a premature ejaculator's penis and
impede his climax. Removing an infant's prepuce at birth, and letting his
glans desiccate and roughen from constant friction against outer clothing,
serves the same purpose. Having discovered a secret that would increase
their sexual satisfaction, women required only the means to convince men
that they should undergo a little genital tailoring. Every man has caught his
member in his pant's zipper at least once in his life. A Pleistocene equivalent
to this painful experience would most likely also have existed. The benefits of
circumcision are imperceptible to men, and there is not a man alive who does
not wince at imagining placing his best friend and most prized possession on
the chopping block.

The archaic date of the Saqqara circumcision mural adumbrated by a millennium the Hebrews' Declaration of Independence, and it invites the following speculative scenario. Women wielded more power in early societies than they at present do in a world dominated by patriarchy. Helen Fisher, in *The First Sex*, estimates that approximately 15 percent of contemporary cultures throughout the world remain matrilineal—that is, they trace their descent through the female line—an example being the Navajo.[12] Those who control resources exercise political power and can exert considerable influence over a culture's customs. *Perhaps circumcision became a widespread practice not because of health concerns, not because a patriarchal god demanded it, not because it was a superstitious ritual, not to test the mettle of boys in a warrior culture, but because grandmothers recognized that it made men better lovers, women more sexually satisfied, and mothers more fertile.**

*   *   *

Women experience three dramatic markers that trisect their lives. Menarche divides childhood from the period when a woman becomes a *maiden*. Childbirth begins her role as a *mother*. Menopause dramatically initiates the phase of her life ancients called the *crone*.† In classical-Greek mythology, these three phases were represented by the triple goddess: Hebe, the vestal virgin; Hestia, the keeper of the hearth; and Hecate, the crone feared and respected for her sorcery and power. According to lore and custom, each distinct phase carried with it certain pleasures, duties, and onerous consequences.

Before the Trinity became "three-in-one men," the Father, the Son, and the Holy Ghost, it was composed of "three-in-one women." The three Graces, Fates, Furies, and many other triple combinations suggest that trinities once marked the three phases of a woman's life rather than the three spiritual male entities of Christianity.‡

Throughout history and continuing in many contemporary Third World

---

*Reproductive physiologist Mary Jane Sherfey, M.D., agrees that older women were most likely the ones who originally promoted circumcision. "The introduction of the idea into human customs may have come first from the women during Mesolithic times; however, the men must have shown considerable resistance to such a barbaric act of symbolic castration which, as initially practiced, must have taken its toll in actual castration and in lives through infection and blood poisoning (especially since it was first practiced on pubertal boys). It was probably practiced regularly only in the centers where the women wielded unusual power."[13]

†Recently, the word "crone" has been undergoing a rehabilitation, shedding the negative connotations that the patriarchal centuries attached to it. I intend to use "crone" in its ancient and honorable sense.

‡The sex of the Holy Ghost can be determined by the gender article attached to the noun used to describe Him. His first mention is in Aramaic, the ancient language used by both the Jews and the early

cultures, the last phase of a woman's life was and is recognized by both men and women as the one in which the crone attains freedom, power, and wisdom. Typically, a crone is a respected member of the community of elders. In some societies, she is the undisputed leader. Younger members, both men and women, seek her counsel. Others value the crone's forthrightness and valor, and nearly all fear her potency.

The accumulation of wisdom resulting from living a long life is a large part of the reason for the reverential attitude toward the crone. Another major component contributing to her forcefulness, however, relates to the dramatic realignment of the serum concentrations of her estrogen, progesterone, and testosterone caused by her menopause. A woman's change-of-life symptoms result primarily from a steep drop in the ovarian production of estrogen and progesterone. Nearly unaffected by this over-the-cliff drop in her feminine hormones is a menopausal woman's production of testosterone. The jarring recalibration of the percentages of these three crucial arbiters of personality leads to a sudden rise in a woman's testosterone concentrations *relative to* her rapidly declining circulating estrogen and progesterone levels.[14]

Secondary masculine sexual characteristics marking a boy's onset of puberty begin to make their appearance in menopausal women. A woman's voice drops in vocal range, giving it an unmistakably huskier and breathier quality. Scalp hair grows coarser. The peach fuzz marking a boy's first attempt to grow a mustache often appears on a postmenopausal woman's upper lip. Facial hair becomes more visible in other locations.

The relative rise in her testosterone levels also brings about dramatic changes in a woman's psyche. Sleep disturbances and strange sexual dreams often intrude. The indecision, pliability, and relative vagueness of purpose that often marks a woman's youth are replaced by clearheaded assertiveness. Researchers have demonstrated in many different studies that the more testosterone an animal has, the farther away he or she is willing to roam and the more likely he or she will be to challenge and/or dominate a rival.[15] In general, the menopausal jolt of testosterone focuses women on life goals and fills them with a resolve that was often lacking earlier in their lives. (While

---

gospel writers of the first century A.D. The word they employed was *ruach,* a feminine noun that means "breath" or "spirit." Nearly all Hebrew prayers include the word *baruach.* When Greek became the language of the New Testament, gospel writers translated *ruach* into *pneuma,* which is the Greek word for "air" or "spirit." *Pneuma* is a neutral Greek noun without gender.

Then Jerome translated the Greek Bible into the Latin Vulgate version. He translated the word *pneuma* as *spiritus,* which means "spirit" and is a masculine noun in Latin. The Holy Spirit has been male for the last 1,700 years. There was another, more appropriate word for "spirit" in Latin: *anima,* the Latin word for "soul," a feminine noun. Jerome could have chosen this word to describe the third entity. The Trinity would have consisted of a Father, a Mother, and a Son instead of three masculine figures. One can only speculate on how differently the Church might have evolved had Jerome used the feminine *anima* instead of the masculine *spiritus.*

there are surely many cultural factors influencing these trends, I wish to focus attention on the recalibration of a menopausal woman's hormones as one that is rarely discussed.)

Freed from primary child-rearing duties and brimming with testosterone (relatively speaking), mature women re-enter a man's world and begin to exert a wider influence on the welfare of the society. Some older women, as if irrepressible, ascend to the pinnacle of power even in patriarchal societies. Men, in general, recognize that dealing with an older woman is very unlike interacting with a young woman, and not all the differences can be attributed to the acquisition of experience. The Blackfeet of the American West called grandmothers "manly hearted." The force of character and sagacity that have throughout history been attributed to older women are due in no small part to the sudden relative rise in their hormone of aggression and dominance. Postmenopausal women become more virile. Germaine Greer described this stance as "peaceful potency."[16]

A husband, noting that his wife will no longer tolerate behaviors that she may have been willing to overlook in the past, does not always welcome these shifts in attitude. Not uncommonly, some are threatened by their mate's new independence and they leave to seek out the company of younger, more adoring women. Or, conversely, a postmenopausal woman may get up the gumption to end an unhappy marriage, confident that she can manage just fine by herself or seek a new life with someone else. Women who were willing to submit to many onerous restrictions when they were younger suddenly discover that they can no longer endure conventions that restrain them. Relative increases in testosterone boost a woman's spatial awareness and augment her sense of direction. A wanderlust that she never had earlier in her life often seizes her, and she embarks on pilgrimages to distant places or travels to exotic locales, intent on seeing the world and having adventures. Her children often marvel at the change that has taken over her.

If one accepts that the changing relationship between estrogen and progesterone on the one hand, and testosterone on the other, plays a significant role in postmenopausal mental attitudes, then a disturbing question arises: What is the true nature of the bargain a woman makes with Mother Nature when she intervenes to prevent nature from running its course? A woman can choose to roll back the clock by taking exogenous estrogen and progesterone in the form of pills, patches, or injections. This popular but radical subversion of the normal menopausal process, known as hormone replacement therapy, or HRT, can forestall or ameliorate many of the unwanted consequences of a woman's change of life.

The key question, rarely asked: Is HRT a Mephistophelian bargain that trades assertiveness and power for youth and beauty? By artificially maintaining high levels of estrogen and/or progesterone, HRT nearly completely negates the benefits that relatively higher testosterone levels might bestow upon a woman. Not that postmenopausal women on HRT do not become more clearheaded and forceful, but the question is: Would women be even *more* clearheaded and forceful, if the full effect of their relative increase in testosterone was not mitigated by the introduction of HRT?

Current debate aswirl around HRT centers on whether or not it increases the risks of breast cancer or decreases the chances of cardiovascular disease. Long articles in women's magazines discuss the pros and cons of HRT in relation to bone density, libido, wrinkles, and sex appeal.* Often missing from these discussions is the not insubstantial price women pay to acquire what many women consider a near miraculous medical advance. Oscar Wilde's novella *The Picture of Dorian Gray* and the myth surrounding the tragic end of Ponce de León, the Spanish explorer who sought the Fountain of Youth, are cautionary tales reminding us that there are no free lunches in the universe. Rare is the great gift that does not come with a dark price.

I am not suggesting that women should decide not to avail themselves of the advantages of HRT but, rather, I would like to interject into the debate a factor that experts often leave out of the equation. Patriarchy and misogyny overshadow the current structure of human societies and prevent many women from achieving their full potential as leaders. It cannot be known whether or not women would play an even greater role on the world's stage if they were willing to forgo the visible and metabolic benefits they derive from HRT. What would be the result if, instead, they embraced the power Mother Nature intended to give them, which unfortunately comes with what many women believe are certain undesirable side effects. Many men and women recognize the wry humor but deep underlying truth in the T-shirt worn currently as a joke by postmenopausal women that has emblazoned on its front, "I am out of estrogen and I have a gun."

The last several chapters have focused on the dramatic changes that occurred in *Gyna sapiens'* reproductive life cycle. *Homo sapiens* also underwent a makeover, albeit a more subtle one, but in terms of its far-reaching implications on all the other animals and plants, it, too, was radical.

---

*Several months after I finished this chapter, the Natioanl Institutes of Health released a large randomized scientific study on HRT confirming hidden risks I had long suspected. Humans nearly always pay a price for interfering with Mother Nature's plan.

Iron, Sex, and Men

Part II

Human hunting parties were comprised primarily of males. A nine-months'-pregnant woman, a nursing mother, or a woman charged with the safety of small children would not likely be part of a violent struggle such as this one.

Chapter 9

Prey/Predator

A man is the hunter his wife makes him.   —Inuit proverb[1]

"It is a fine day, let us go out and kill something!" cries the typical male instinctively.
"There is a living thing, it will die if it is not cared for," says the average woman almost equally instinctively.   —Olive Schreiner[2]

Women's social standing is roughly equal to men's only when society itself is not formalized around roles for distributing meat.
                                        —Richard Leakey and
                                        Roger Lewin[3]

*H*omo sapiens possesses features that are located so far out at the extremes of the bell-shaped primate curve that they invite the curious to explore them. One of these is the capacity for aggression among humans, particularly the male of the species, and the pleasure most men so obviously derive from hunting and killing other animals.

With the exception of a few small primates that subsist on insects and tiny reptiles, all (save one) are primarily vegetarians. Chimpanzees, baboons, capuchins, and cebus monkeys occasionally hunt, but these quests are sporadic and opportunistic. Less than 2 percent of these few hunting primates' diets consist of meat.[4] From their observable behavior, there does not appear to be a primate species other than *Homo sapiens* that enthusiastically and voluntarily awakens at 3:00 A.M. filled with the happy resolve to spend the day in predatory pursuit.

Accumulating archeological evidence suggests that the verve and skill required to organize highly successful big-game kills consistently did not fully mature in the hominid line until *Homo sapiens*.* And not until about forty thousand years ago did *Homo sapiens* routinely tackle the more extreme big-game adventures, for example, hunting woolly mammoths. Earlier hominids did not pose a significant threat to the survival of other species. Paleontologists have not identified any whose members were driven to extinction because of the predations of earlier hominids.† No doubt exists, however, that, following the arrival of *Homo sapiens*, many unfortunate species have been hunted to the last member.

Explorers reported sighting Steller's sea cows, gentle sea mammals resembling manatees but three times larger, in the waning years of the nineteenth century. Whalers hunted them to extinction. The last aurochs, magnificent large bisonlike creatures, disappeared from the forests of Europe at the end of the eighteenth century. The moa, a large flightless bird resembling the ostrich but triple its size, was once abundant in New Zealand. It vanished from the archeological record coincident with the arrival of the first ancestral Maoris. And so it goes, right up to the bison, condor, and snow leopard in our present day. The one creature that has endangered all the others is *Homo sapiens*. In Genesis, Yahweh encouraged us to "subdue nature" and predicted that "fear" and "dread" of us would be upon every beast. Verily, it has come to pass.

Our aggression has not only been outwardly directed, at other species. Intraspecies mayhem—in the form of wars, duels, fights, persecutions, diabolical tortures, and genocides due to ethnic, racial, and religious hatred—has no parallel in the animal world.‡ The sexual aggression exhibited by some human males toward females is without correspondence in the wild. *No* other male animal would consider murdering the female with whom he has just copulated. And inwardly turned aggression in the form of self-flagellation and suicide is also absent from nature. Are these male behaviors all due to cultural indoctrination, as many believe, or are they hard-wired into the male nervous system? If the latter, to what evolutionary factor do we owe this extraordinary bloodthirstiness, and why is it so skewed toward the male of the human species?

---

*There are a few examples much earlier indicating that the more advanced form of *Homo erectus* known as *Homo heidelbergensis* could organize large kills. The presence of advanced javelinlike, fire-hardened spears and evidence of a mass over-the-cliff horse kill occurring over 300,000 years ago in Terra Amata, near present-day Nice, France, fuel this speculation.
†There might be exceptions to this statement. So many skeletal remains of an extinct giant gelada baboon have been found in the Olargasaille site, in southern Kenya, that some have suggested that the baboon was the favorite prey of *Homo habilis* and this factor contributed to the baboon's extinction.
‡Wrangham and Peterson reported on the warfare they observed between troops of chimpanzees conducted exclusively by the males.[5]

Prior to the agricultural revolution ten thousand years ago, hunting consumed the majority of a male's "working" life. To understand better why
Homo sapiens took so eagerly to his life's chief occupation, we must examine
predatory behaviors in earlier hominid species. That the immediate precursors of modern humans ate meat is not in doubt. Their success at this endeavor, however, is one of nature's improbabilities. Imagine early hominid
hunters' consternation when they realized that they had puny canine teeth
encased in weak jaws, and pathetic fingernails instead of claws. They could
not run fast enough to catch the animals that housed the most meat, nor
were they strong enough to wrestle big prey to the ground barehanded.*
Even if they were lucky enough to come upon a recently deceased animal,
how were they supposed to tear off its hide? What did they have going for
them that would have encouraged them to challenge lions, eagles, tigers,
and crocodiles in the killing department?

The two physical traits that most distinguish a hominid from every other
primate are bipedalism and a big brain, the signature Hominidae traits. They
are the same two that eventually precipitated high maternal-mortality rates
in the human line. How or why we originally acquired either is presently unknown. Bipedalism was, as Stephen Jay Gould claimed, "the greatest single
adaptation in the line of human evolution."[6] Many in differing fields have tried to
understand why we would have adopted such a strange means of locomotion.

Anthropologist Owen Lovejoy commented, "For any quadruped to get
up on its hind legs in order to run is an insane thing to do. It's plain ridicu-
lous."[7] He pointed out that standing up made us more susceptible to being
knocked or tipped over. Our speed was considerably slower and our footing
less sure than that of animals that moved on all fours. We would have had to
practice walking to perfection near the proximity of dense trees, because, as
Lovejoy remarked, "no hominid could have ever ventured out on the savanna
as a stumbling imperfect walker and learned to do it better there. If it had
been unfit for erect strolling on the savanna it would have not gone. If it had
gone, it would not have survived the trip."[8]

Lovejoy believed that hominid males initially reared back on their
haunches so that they could have their hands free to carry food to nursing females. Robert Ardrey proposed that the primary reason early hominids
adopted an upright stance was their need to arm themselves with clubs. Leo
Laporte and Adrienne Zihlman suggested that, in the mosaic habitat in
which the Pleistocene hominids lived, they needed to carry food and water
when trekking across arid grasslands to reach a distant gallery forest.[9]

---

*This is not entirely true. By means of organized and concerted effort, humans can run down a wild
horse or deer, taking turns and keeping pressure constant until the prey is too exhausted to escape.

Many other theories abound. One states that our ancestors needed to run fast in short bursts on the open savanna. Others propose that standing up increased the scope of a vertical primate's vision, a trait called "sentinel behavior." Another theory: Perpendicularity cut down the skin's exposure to the hot sun; bipedalism was therefore driven by a hominid's need to regulate its body temperature.[10] Still another proposes it was a defensive measure, because predators are "triggered" by moving horizontal shapes, not vertical ones.[11] All these variations can be grouped together under what is called the Savanna Theory.

A different premise for why bipedalism evolved is called the Aquatic Ape Theory. In 1920, marine biologist Alister Hardy published an article suggesting that humans diverged from the primate line by wading out into the water and staying there for several million years. Championed doggedly by Elaine Morgan, this theory notes that, somewhere between twelve million and nine million years ago, a great drought settled over Africa. Trees shriveled and plants withered. Deprived of their arboreal habitat, one group of primates returned to coastal and river waters.

Hardy's hypothesis is not as far-fetched as it may seem at first glance. Seals, dolphins, otters, and beavers were all originally land mammals that had made this transition. But after swimming and living in the surf lines and along riverbanks, the ancestors of hominids, unlike the other sea mammals, returned to the land as climatic conditions improved. Millions of years of wallowing in the shallows forced our ancestors to stand up to keep their heads above water.* According to the Aquatic Ape Theory, when we emerged dripping wet from the riverbank we had lost our fur, acquired a thick layer of blubber we call subcutaneous fat, had downward-pointing nostrils streamlined for swimming, and, most significantly, were perpendicular.[12]

The other missing piece of the great puzzle is the evolutionary reason for our massive brain, a subject that will be dealt with in greater detail in a later chapter. Social intelligence, language, accuracy in throwing, and many other reasons have been put forth, yet none of them has been uniformly accepted as the primary one.

Once we had made the transition to bipedalism (for whatever reason) and grew an outlandishly sized brain (for whatever reason), we made an arresting discovery. Free hands could brandish a weapon that a big brain could imagine. Balanced on two feet with two unfettered hands, we were off and running, figuratively and literally. But even these advantages were still insufficient to convert us into the earth's most fearsome hunters.

---

*Japanese primatologists have videotaped proboscis monkeys walking upright in water up to their chins, their long noses (proboscises) held high and dry. These monkeys live in the trees of mangrove swamps and spend considerable time in the water. Humans are the only primate that swims well.

We are the newest predator on the block. Nearly every other major killer was operational by sixty million years ago, with many sporting a longer history. In those sixty million years, very few significant new predators joined the ranks except the Hominidae.* Astonishingly, we made the conversion from prey to predator in a scant three million years. The major rise in our killing prowess occurred primarily in the last hundred thousand years, with the sharpest spike limited to the last forty thousand.

Despite our tardy entry in the predatory sweepstakes, *Homo sapiens* has become the fiercest, bravest, most sadistic, and most successful predator of all. Since large primates can maintain their health quite successfully on a vegetarian diet, why did only one ape take to killing as a way of life? If meat consumption was the path to species longevity, why didn't other ground-dwelling primates, such as gorillas, baboons, chimps, and bonobos, embrace the hunting life as enthusiastically as we did?

An integral part of the Savanna Theory posits that rising temperatures, diminishing rainfall, and thinning forests turned our ancestors' habitat into a patchwork of isolated woodlands and open grasslands. These environmental changes propelled hominids to try their hands at hunting because vegetable foods became scarce.

Given the circumstances that paleontologists surmise existed, why didn't our ancestors evolve a gut that could digest grass and leaves, the most plentiful source of plant protein in drought conditions? Our divergence from *all* other primate species in our lust to kill is still more remarkable because there would have been daunting obstacles to overcome for any primate that aspired to emulate a carnivore.

A courageous hunter requires a killing disposition that revels in danger. When the first hominids contemplated leaping across the high boundary separating prey from predator, they had to override their innate fear reflexes without getting eaten or gored in the process. A primate has finely honed instincts, ingrained over fifty-five million years, that when activated cause its owner to flee from danger. Among primates, only a human is brave enough to stalk an animal that could turn and kill him.† Foraging and gathering are significantly less dangerous than hunting. If no other primates can match *Homo sapiens'* steadfast courage and unquenchable bloodthirstiness, why did we alone acquire such a strong penchant for both? The search for the answer to this question brings us back to the subject of iron.

---

*Wolves evolved around 6 million years ago, polar bears around 10 million years ago. All the precursors to these animals were predators, unlike the vegetarian primate ancestors from which the hominids sprang. Dolphins also fall into this category.

†On rare occasions, some primates (baboons and chimps) will demonstrate courage by mobbing a predator, but this is nearly always a group action.

When women gained the requisite free will to adamantly say *No!* and forestall sex, the specter of extinction began to haunt the human species. A fertile woman could coolly and accurately assess the risks and costs to her of pregnancy, delivery, and child-rearing, and might decide to remain childless. (Throughout history, some women have made, do make, and will continue to make this choice.*)

Without children, *Gyna sapiens* could reasonably conclude that she would not need a male for anything. Rational decisions, however, cannot be made without the input of the older, emotional side of human nature. Maternal instincts, her still incompletely controlled sexual urges, and the pleasure she could, under the right circumstances, derive from male companionship, ensured that ancestral woman would remain engaged with the opposite sex. And there remained another inexplicable and chaotic element—"that old black magic called love."

The male, meanwhile, after much fumbling, rejection, and confusion, gradually realized that a surefire first step toward melting her resistance was to bring her a gift that contained iron—preferably in slab form. Delighted with his perspicacity, he immediately set about planning his quest to satisfy her wish. It was then that he learned, to his eternal dismay, the presence of another very significant obstacle standing in his way.

Food sources such as nuts, roots, berries, and leaves possess a characteristic that makes them very easy to acquire—they can't run away. But an animal must be caught and killed before anyone can sink his or her teeth into its flanks, and cornered prey will always put up a spirited self-defense. The strategy of scavenging carcasses killed by more efficient carnivores has its own risks, not the least of which entailed having to elbow aside bad-tempered competitors such as hyenas, wild dogs, and vultures. Procuring meat was definitely fraught with danger.

Another salutary feature of plant foods is their relatively long shelf life. Most vegetables can be gathered, eaten at leisure, or stored for consumption on another day. Meat will not keep. It decays rapidly, and in a most revolting manner.†

Unfortunately for the eager hominid hunter, the freshest meat happened to be firmly attached to the stout bones of dangerous animals that did not take kindly to being eaten. Early man had to summon up great courage before he could begin his courting routine. Though it was certainly true that he could bring his intended a bunny or a gazelle fawn at no great risk to himself,

---

*The 1909 U.S. Census revealed that 23 percent of women over the age of thirty-five were childless. By 1960, this figure had fallen to 12 percent, but it is now on the rise again.[13]

†Salting, curing, freezing, and drying meat were fairly recent innovations, estimated not to have been practiced habitually until approximately 50,000 years ago.

he knew she would admire him more if he brought her something heftier and possessed of more cachet—like a bear haunch accompanied by a set of its teeth strung as a necklace.

Natural Selection had sixty million years or longer to fashion the reflexes of the mainline predators so that they were fast, cruel, smart, and fierce. To level the playing field, prey animals were equipped with thick hides, tusks, quills, and horns, as well as defense mechanisms to help them escape from, fend off, and confuse predators.

Predators responded to the latter developments by adopting one of two different hunting strategies. The leopard, cobra, or crocodile seeks out prey alone. Social predators like lions, wolves, and hyenas hunt in cooperative groups. Functioning as a unit, social predators can increase the odds of their success and augment the quantity of meat they can garner in a single kill. Ancestral hominids learned to hunt in bands to bring down dangerous quarry an individual hunter would never dare to attack alone.

A common denominator among nonhuman social predators is that the female of the species plays a leading role in both the hunting and killing. Lionesses and wolf and dog bitches frequently bring down the quarry. A dominant female leads a spotted-hyena pack. In the human species, the new transcendent skill called killing evolved into an occupation performed almost exclusively by males.

The reasons for this gender skewing are obvious. Bipedalism makes running an awkward exercise for late-term pregnant women. A menstruating female could signal a wary prey that a group of their dreaded human nemesis was near. A mother cannot take a crying baby along on a hunting expedition; if a nursing mother left her young for long, she would cease lactating. Women responsible for the safety of small children cannot leave them for extended periods. Unlike the females of every other social predator, pregnant women and those burdened by young children are not disposed to hunt large game.*

*   *   *

Footprints fortuitously preserved in volcanic ash at Laetoli in Africa confirm that a bipedal primate walked the land some 3.6 million years ago. Paleontologists have identified these impressions as belonging to a precursor species that led eventually to us. Numerous fossils of *Australopithecus afarensis*—a half-ape, half-hominid—have been excavated; the most familiar and complete

---

*In a large cross-cultural database study of 179 societies, men alone hunt in 166 of them and both men and women hunt in 13. In no society do women do the hunting alone. Among hunting groups of wild chimpanzees, females made fewer than 10 percent of kills.[14]

skeleton found so far is Lucy, named by her discoverer, bone hunter Donald Johanson. Evidence of meat consumption is plentiful among our ancient hominid ancestors. Experts' examination of the teeth of these extinct primates has revealed beyond any doubt that *Australopithecus afarensis'* favorite dinner was a wide variety of other inhabitants of the area, though it is not known for sure whether they were successful scavengers or bold hunters.

Archeologists have also established that both *Homo habilis* and *Homo erectus,* the next two archaic relatives in the line leading to humans, ate meat.* Females of both species may very well have participated in hunting, and males may have provisioned new mothers unable to secure their own share of a kill. Approximately 10 percent of mammalian males exhibit a similar pattern of paternal behavior.

It is highly unlikely, however, that any of these precursor species' males considered meat primarily as a commodity they could barter for sex. Before meat could transmute from comestible to aphrodisiac, females would have had to lose the urgency associated with the estral sexual madness that periodically inflames every other primate female. Not until the obsession associated with estrus drained away could females think clearly enough to understand the connection between sex and pregnancy, which, in turn, precipitated major realignments in relations between the sexes

The males of the human species, similar to males in virtually all other species, compete for dominance, aware that the King of the Hill has the best chance of achieving sexual access. No doubt earlier hominids hunted prey. However, it was not until *Homo sapiens* that the male, prompted by *Gyna sapiens,* finally grasped that his chances of succeeding as a lover were intimately connected to his skills as a hunter.

Anthropologists studying a wide range of hunter-gatherer tribes have amply documented that a man's hunting skills are directly related to his success with women.[16] Kim Hill observed that the men of the Hazda, an African tribe of hunter-gatherers, do not hunt primarily to provision their young ones but, rather, secure meat to trade with women for sex (usually for extramarital affairs).[17] *Homo sapiens* learned that he could satisfy his sexual hunger only after he first satiated *Gyna sapiens'* hunger for iron. "Going downtown to make a killing" and "bringing home the bacon" are modern colloquialisms that tacitly acknowledge the role hunting played long ago, at the dawn of human society.

---

*Electron microscopy of the pattern of wear and tear carved into the fossil teeth of *Homo habilus* does not support the notion that these early hominids were substantially meat-eaters.[15]

Another factor goading *Homo sapiens* to take up hunting as a way of life was the drastic change occurring in his sexual programming. In combination with the immense changes that occurred to *Gyna sapiens'* reproductive cycle, the interlocking needs and desires of men and women became jarringly askew. Whereas she was gaining control over her sex drive, his was increasing its power over him. More and more, his sex urges seized control and dictated his actions. His runaway sexual agitation increased unabated until *Homo sapiens* became the most sex-crazed male of any living species.

With a few rare exceptions, males of other species express *no* sexual interest in females who are not ovulating. In many cases, males and females compete for the same resources, and the male shows precious little chivalry toward a female that a few months hence could be a potential mating prospect.

When the nonhuman female becomes sexually ready, however, the nonhuman male's attention becomes riveted on her erogenous zone. He will engage in deadly combat, put on elaborate displays, perform intricate dance routines, or sing his heart out—all in an attempt to attract a female that, just weeks before, he treated with disregard. The contrast between a male's behavior in the presence of an ovulating female and his behavior in the presence of a nonovulating female is extreme in all of the several million sexually reproducing species—except one. Only *Homo sapiens* has a sexual furnace set at full blast *all the time.**

Doesn't it seem odd that Mother Nature would make *Homo sapiens* obsessed with sex? What could be the benefit to the fitness of the species of having a male lusting for a female during wide swaths of time when there was no chance that engaging in sexual congress would result in a conception? Why was the human male endowed with such a careening, out-of-control sex drive that some would lust to have sex with old women, dead women, menstruating women, other men, little girls, or prepubescent boys? Are there any male animals that impersonate females? Any that find being dominated by a female sexually arousing? Any that routinely fantasize about inserting their penis into the mouth of a female? Any that long to insert it in her anus? In the

---

*In other species, male testosterone levels fluctuate in harmony with the female season of sexual receptivity. Songbirds, for example, experience a sharp rise in their testosterone levels in the spring, just prior to their mating season. The more testosterone they have, the better they can trill, which in turn increases their chances to mate. As soon as the mating season is over, their testosterone levels fall precipitously. This pattern is the norm among most male birds and mammals. Human males, in contrast, maintain a fairly steady concentration of testosterone throughout the entire year.[18]

millions of hours that have been spent observing animals in the wild and captivity, has any ethologist ever observed these desires actually happening?*

Has any ethologist observed other animals engaging in proto-spanking or proto-bondage behavior? Is there a male of another species that appears to be turned on sexually by inflicting pain on a female? Any that experiences heightened sexual arousal from murdering a female with whom he is in the midst of having sex? Have any males of any species been observed masturbating while watching a female's life snuffed out? Are any male mammals other than humans inclined to mate with females of species far removed from their own?

The answer to these questions is a resounding no. The "beasts" do not practice bestiality. Numerous scientific studies have concluded that the male of the human species exhibits a robust capacity for variable sexual behavior, some of which could be considered extreme, some aberrant, and some even deviant.†

What would have been the evolutionary benefit to our species of having the male prone to so many diversionary sexual strategies, none of which result in conception? If it is some sort of spandrel gone awry, what was the original impetus for the adaptation? Are these behaviors simply repressed male sexuality, as the Freudians hypothesize, or is there a component of genetic predisposition to them?

I propose that it is the latter. The Red Queen escalated the male sex drive to such unnatural heights so that he would be willing to take enormous risks to satisfy his urges. To continue to advance the intelligence that the species was increasingly depending upon to survive and prosper, human females needed ever-greater quantities of bioavailable dietary iron (along with a whole slew of hard-to-make animal products). The male, in order to satisfy his sexual hunger, would have to assume new risks and learn new tricks if he was going to find a way to supply women with this most basic element for her and her babies.

The males of nonhuman species appear willing to put their lives on the line for only extreme reasons. However, it is routine for them to answer aggressively the urgent hormonal directive that usurps control of their brain in the presence of an ovulating female. In general, however, dominance fights rarely result in the death of either combatant. *Homo sapiens* does not differ from the males of other species in this regard. But testing his mettle against

---

*Observers have recorded instances of fellatio and cunnilingus among bonobo chimps. No one has observed anal penetration.

†This is not to say that females, too, are not capable of sexual behaviors out at the edge of the bell-shaped curve. "Kinky" is a word that demands cultural context, but its fulsome appearance in both sexes separates us from all other animals. Still, extremes in behavior are more common among men.

potential sources of dietary meat that outweigh him by thousands of pounds and can trample, gore, slash, and bite him to death is a hazard to which other noncarnivorous males have opted not to subject themselves.

Another behavior suggesting that *Homo sapiens* hunters kill more for sex than for food is that all other social nonprimate mammalian predators tend to eat their fallen prey on the spot. A feeding frenzy breaks out among them, and fighting often occurs over the more delectable parts. Rare is the social predator who would think of postponing eating or sharing its meat with a co-predator, much less sharing it with a conspecific* that had not participated at all. The exception occurs when parents bring a portion of a kill back to feed immature young. A male fox will sometimes regurgitate his recently consumed "steak tartare" as an offering to his mate, a nursing vixen that was unable to join him on the hunt.†

Consider, then, the typical aftermath of a *Homo sapiens* big-game kill. After having placed themselves in harm's way, hunters laboriously strip the hides from their prey, butcher the animal, and carry the quartered remnants back to a base camp, leaving a trail of spoor that will inevitably attract predators and scavengers.

Most remarkable of all: The purpose of this unprecedented display of restraint after a kill is to give their hard-earned meat away! Glynn Isaac, the anthropologist who first distinguished the uniqueness of human food-sharing, once remarked, "If you could interview a chimpanzee about the difference between humans and apes including the way we walk, the way we communicate, and our subsistence I think it might say, 'You humans are very odd, when you get food instead of eating it promptly like any sensible ape, you haul it off and share it with others.'"[19]

An extreme form of this behavior is what anthropologists call the "own-kill taboo." Among many extant hunter-gatherer societies, a formidable stricture forbids a hunter from partaking of his kill until he has shared it with everyone else. Frightening myths and potent superstitions ensure that the primary hunter will be the last to eat, if he is allowed to eat at all.

Chris Knight in *Blood Relations* recounts numerous examples collected

---

*A conspecific is another member of the same species.
†After a kill, male chimpanzees and other primates try to gobble down their prize. The other members of the troop—females, elderly, and juveniles—encircle a dining male, gesturing with outstretched hands in a manner that can only be interpreted as pleading. Ethologists call this stance "tolerated scrounging." Not uncommonly, the harassed diner will give away portions of his kill.

by anthropologists from around the world in many different cultures. For example, the Yanomami hunters of the Amazon Basin do not eat the meat of game they have killed themselves, because they believe that if they do they will be deserted by the hawk spirit, which they need to succeed in their future quests.[20] In the case of the Bororo of central Brazil, "A hunter never roasts the meat he has shot himself. Failure to observe this taboo . . . causes a vengeful animal spirit to send sickness and death to the hunter and all who eat its flesh."[21] Among the Juneno of North America, their language even has a special verb meaning "to get sick from eating one's own killing."[22]

What a hunter gains from this combination of self-restraint and generosity is increased stature in the eyes of others. !Kung San tribespeople of the Kalahari, when asked why some men who were poor hunters could not convince a woman to marry, responded, "Women like meat."[23] Each incremental hunting success moves a man up a notch in the male-dominance hierarchy, and each ascending rung brings with it an increasing number of females willing to have sex with him. Kristen Hawkes calls this the "Show-Off Theory" of hunting.[24]

Women of the Aché hunter-gatherer tribe who were interviewed by anthropologist Kim Hill preferred the best hunters for their extramarital affairs.[25] And Hawkes tabulated that the Aché hunters with the most kills to their credit had the most children and that these children had higher survival rates because the other members of the tribe took better care of them than they did of orphans or the children of poor hunters. Anthropologists initially believed that Aché hunters were altruistic grubstakers provisioning mothers and their little ones with food high in iron, protein, and fat, but the reason Hawkes and her co-workers discovered was more self-promoting.

Numerous anthropological studies confirm the connection between hunting skills and opportunities for sex. Landes reports of the Ojibwa of western Ontario:

> A married man who is too lazy to hunt can be supported by his wife for a time. But her tolerance will change to scorn, then to indifference, and finally she will desert him. A man who is unsuccessful on the hunt, and who goes with his wife to her parents' wigwam, can expect to be rejected and left to die of starvation. In one case, the parents' scorn was so great that they took their daughter in to feed and lodge her, but refused their son-in-law. Folk tales are concerned with the same theme.[26]

In other tribes it is the same: Wives refuse sex to men who do not provide meat. Richard Lee, commenting on the customs of the !Kung San, writes, "If a man does not hunt, his wife will make pointed comments about his sexual prowess. And vice versa: if he is no good in bed, he cannot hunt."[27] Anthropologist Janet Siskind, who studied the Sharanhua people of the Peruvian Amazon Basin, summed up the relationship between iron and sex when she observed, "Put at its crudest, the special hunt symbolizes an economic structure in which meat is exchanged for sex."[28]

Primatologists in the field have noted that the cooperative hunting group most closely resembling a human hunting party is that of our nearest relative, the chimpanzee. Males use their hard-won meat from a fresh kill in a similar way. A chimpanzee male will sometimes share his precious meat with close kin or, on rare occasions, use it to obtain a social advantage. But he reveals his understanding that meat can further his opportunities for sex when he tears off a strip of his quarry's muscle and gives it *only* to nonrelated females who just so happen to be exhibiting the florid signs of estrus. Jane Goodall saw these transactions as economic negotiations: "The female's sexual swelling, in a way, serves as a sexual bargaining point."[29] The conclusion anthropologist Helen Fisher derived from this behavior was that only a few of the males had something all the females wanted (meat), and only a few of the females had something that all the males wanted (sex).*[30]

All sexually reproducing species have intricate mechanisms to ensure the "survival of the fittest" genes. When females enter their rutting or breeding season, the males fight among themselves to see who will win the right to mate. For example, male elephant seals engage in fierce combat for the right to copulate with cows, but only 15 percent are dominant enough actually to pass their genes through to the next generation.

When *Gyna sapiens* lost the signs of estrus, Natural Selection had to reprogram the male to ensure that superior male genes would continue to outnumber inferior male genes. Male competition is the great sieve that Natural Selection uses to cull inferior bits of genetic material from the gene pool. *Gyna sapiens* could take her time choosing the perfect male match for her. Infatuation, love, intimacy, and lifelong friendship would all enter the picture.

The man who demonstrated his ability to bring meat home to a woman was the man who would have the most sexual opportunities, because women favored men who exhibited courage, skill, resourcefulness, and strength. They still do. Over time, meat would segue into resources that would, still later, transmute into property and money.

---

*This idea is repugnant to many anthropologists. Richards, for example, disdainfully described it as the "prostitution theory of human origins."[31]

Many contemporary Asian males believe that powdered rhinoceros horn and extract of tiger penis (used to flavor soup) are potent aphrodisiacs. Along with the phallic shape of the former and the obvious sexual connotations of the latter, their sympathetic magic stems from the fact that any man who could stalk and kill such exceedingly dangerous animals and return to tell the tale would be considered by all to be a great hunter.

To prove beyond a doubt that he did what he claimed he did, he would have to provide evidence. Returning with a piece of rhino horn or a tiger's penis would be his proof. He would have only to show his trophy to persuade a woman that he was strong, brave, and resourceful and, by extension, a man possessed of superior genes. She could, with assurance, conclude that he would be an excellent procurer of iron on the hoof. Today, men pay exorbitant sums to obtain small pieces of what they believe are rhino-horn or tiger-penis soup extract in the belief that consuming these substances will grace them with the original hunter's courage and sexual stamina.

Rare furs from animals extremely difficult and dangerous to locate, stalk, and kill also serve this function. Until recently, when such public display became politically incorrect, a man could advertise his skill as a reliable provider, secondhand, by purchasing for his consort, firsthand, an expensive, exotic fur coat for her to wear and all others to see.

As a result of changes in *Gyna sapiens'* reproduction cycle, *Homo sapiens* had to adjust by drastically altering his mating strategy. The most significant factor to emerge was his incredible craving for blood coupled with his willingness to postpone eating what he had just killed. *Homo sapiens* became the first of the predator line to kill for reasons primarily having to do with sex instead of hunger. What he could not know was that his quest was motivated in large part by a mote so small he could not see it—the tiny iron atom.

For the vast majority of the human species' history, a male's primary occupation was hunting.

Chapter 10

Carnivory/Vegetarianism

There are three possible parts to a date, of which at least two must be offered: entertainment, food, and affection. It is customary to begin a series of dates with a great deal of entertainment, a moderate amount of food, and the merest suggestion of affection. As the amount of affection increases, the entertainment can be reduced proportionately. When the affection is the entertainment, we no longer call it dating. Under no circumstances can the food be omitted.

—*Miss Manners' Guide to*
Excruciatingly Correct Behavior[1]

What sort of eaters humans have been has depended very largely on what kind of catchers they have been.　　　　—Jonathan Kingdon[2]

For the human female, monogamy may have guaranteed protection, food, and even love ... for a while. But the price may have been a gradual loss of economic independence and dependence on males from which she is still trying to recover.　　　　　　　　　—Mary Batten[3]

*W*hen philosophers proudly cite the distinguishing attributes that they propose set *Homo sapiens* above and apart from the rest of the animal kingdom, they usually overlook one that, on reflection, is very peculiar. The planet's most relentless predator comes equipped with the animal world's most finicky digestive tract.* The creature that gave new meaning to the "omni" in "omnivore" cannot absorb a whole host of key essential nutrients that other creatures routinely can. No other complex creature possesses so many striking alimentary lapses and quirks as are present in *Homo* and *Gyna sapiens'* dietary regimen. *Sapiens,* it would appear, chew to the taste of a different drumstick.

---

*Savanna baboons can eat just about anything, yet human digestive tracts are highly selective.[4]

Every species of large animal has, at one time or another, been butchered by a *Homo sapiens* somewhere. Entire species have filed two by two into the Ark of Oblivion, dispatched by human hunters who mindlessly engaged in their wholesale slaughter. No matter how quick, how agile, how massive, or how dangerous, each doomed animal was minced into minute pieces to disappear down the small black hole that exists between the tip of a *Homo sapiens'* nose and his chin. In the wild, a small predator does not attack a large predator. Yet the heads and pelts of five-hundred-pound tigers and bears, along with ten-foot-long denizens of the deep, grace the dens of modern *Homo sapiens*. Considering that eagles rarely suffer from indigestion, cheetahs do not require antacids, and hyenas have not evolved discriminating tastes, it is a wonder that a mere human, possessed of such a delicate constitution, could wreak so much mayhem.

Despite its craving for carnage, the world's most efficient carnivore generally disdains eating raw meat. Humans, despite their voracious appetites, partake of daintier bites than any other meat-eater. We are the only predators to waste valuable energy slowly savoring the chewing of our dinner instead of bolting it down, as is the preference of all the others. For a diminutive biped capable of bringing a woolly mammoth thirty times his size to its knees, *Homo sapiens* has a ridiculously small mouth; the dullest, feeblest canines; puniest jaws; weakest chewing muscles; tiniest tongues; and thinnest-enameled, smallest teeth of any other serious meat-eater. Truly, a *Homo sapiens'* eyes are much bigger than his stomach.

The disparities between our culinary ambitions and the reality of our alimentary canal are so great that it would be worthwhile to investigate these nutritional oddities. Let us consider the possibility that Natural Selection may have played a role in advancing many of these eccentric and seemingly detrimental adaptations.

When African Eve forced African Adam to enter into prolonged and tangled negotiations with her over the subject of intercourse, high-quality food became a key arbiter of sexual relations between men and women. The tasty zebra hock spattering juice on the spit began to set the parameters of the health and intelligence of individuals. Ultimately, diet reconfigured the shape and destiny of our species.

Modern science's continuing pronouncements concerning the long-term effects on our well-being of the various foodstuffs that we consume have warped our understanding of how we, in the industrialized world, came to eat the way we do. There can be no doubt that we dine in a manner today very different from our early ancestors. We are sedentary; they were not. We eat processed and fortified foods and have access to supplements, vitamin pills, and mineral capsules; they did not. We can store food for later consumption in refrigerators and supermarkets; they could not. We can pick and choose among seasonal foods all year long; they could not. We suffer primarily from chronic stress; they worried about acute stress. On average, we live long lives; on average, they lived short ones.

So different are the conditions under which they ate, lived, and died from our current ones that we must adhere to a set of dietary blandishments deemed necessary to maintain health. Nearly everyone generally acknowledges that consuming less fat, meat, sugar, cholesterol, and milk (for adults), while at the same time increasing one's intake of fruits, grains, and raw vegetables, constitutes a healthy dietary regimen. Had there been a National Nutritional Advisory Board to which the earliest members of our species could turn to for advice, it would have stood nearly every one of these current recommendations on its head.

Imagine that you are a member of a band of *Homo sapiens* stepping out in the Pleistocene morning 150,000 years ago. Much earlier, more archaic ancestors had radically changed the way they ate, but you will hurry the process along by your new way of hunting, sharing, eating, and making love. It was in the hominid line's formative years of the Pleistocene that the overwhelming majority of the eccentric features of the human digestive tract became the template for everyone living today.

All members of the band to which you belong stay focused on an endless, sometimes desperately fierce, sometimes ridiculously easy struggle to find and stuff into their mouths enough food over a long enough period to live to reproduce. For those adventuresome enough to advance into unknown territory, the search for nutrients of sufficient quantity and quality could occupy the heftiest proportion of each band member's waking hours.* Since every other living thing also strives to attain its ideal daily caloric ration, the stage has been set for a never-ending conflict.

Previously, a nutritional thunderbolt from out of the blue had struck our ancestors. This electrifying event immediately transformed the hominid line and would eventually jolt the entire course of evolution. Humans are beholden to their predecessor species, *Homo erectus,* for adding something so novel to their diet that subsequent hominids bootstrapped themselves into a

---

*One of the remarkable findings to emerge from anthropological study of extant hunter-gatherers was how *little* time they expended actually gathering or hunting. Nevertheless, time-and-motion studies revealed that the majority of activities centered on food. Among the !Kung San, women collected more than 65 percent of the band's daily caloric intake, and they only had to spend about three days of a week gathering.[5] It is dangerous, however, to extrapolate from these studies of contemporary groups and use them to make assumptions about similar ancestral human bands who had to endure the ice ages, periods when miles-deep glaciers covered most of the earth's land mass.

singular category. And this revolutionary ingredient, present in abundance in contemporary human meals, rearranged the enzymatic priorities of the entire hominid digestive system. Not a single one of the other thirty million animal species alive on this planet had ever experimented with what would become a cultivated taste of the striding primate. The new item on the menu was fire.

Somewhere between 1.5 million and 750,000 years ago, a *Homo erectus* decided *not* to do what every other creature instinctively did in the presence of flame. He or she did not flee. Although we can never know whether it was a male or a female who took this first step, the inherent recklessness of the act tends to suggest that only a young male *Homo erectus* would have been foolish enough to risk getting burned.

As his companions cringed in fear, this one intrepid ancestor edged closer to a burning brand at the perimeter of a brushfire that was probably ignited by lightning. In the moment that he vanquished his fear, grasped the lit branch, and then waved it triumphantly over his head, a new chapter in the story of evolution began.*

Homo erectus immediately put fire to a variety of beneficial uses. News of fire's conquest would have spread to nearby bands among a primate species consummately skilled in the art of imitation. Yet, because of poor communication networks and the wide dispersal of *Homo erectus* throughout Africa and Eurasia, evidence of habitual fire use remains scant for hundreds of thousands of years after its discovery. Some groups were undoubtedly too dimwitted to master the techniques of making and tending a fire. Others were perhaps too fearful. Those few members of *Homo erectus* who knew the secret of keeping a flame alive were most likely the ones whose genes we now possess. Archeologists find blackened firepits sporadically among the excavated home bases of *Homo erectus* carbon-dated at 250,000 years ago. Not until modern *Homo sapiens,* however, is there ubiquitous evidence of the construction of hearths within encampments.†

Early on in the evolutionary story of life, plants evolved the complex biochemical pathway to harness the energy of the sun. When photosynthesis became widespread, sunlight, once removed, fueled an explosion of new species of plants and animals. Most flammable substances on the surface of the earth and beneath it have acquired their fuel load because they originally

---

*Naturalists videotaped a species of Australian kite as it intentionally started fires by carrying burning branches from old ones in its talons. These predators cleverly ignite new wildfires as a means to flush prey out of hiding for easier kills.

†When explorers reached the island of Tasmania, off the southern coast of Australia, in the eighteenth century, they found people living there in Stone Age conditions, and some of them were unfamiliar with the taming of fire.[6]

stored up sunlight in the interstices of their substance. When that nameless *Homo erectus* tentatively but valiantly reached for the Burning Bush, the second great evolutionary energy revolution began.

The hominid domestication of fire was a dramatic alternative to photosynthesis. A long fuse was lit that hissed down through all the generations to the present. The leaping flames arising from the first small bundle of super-heated kindling transmuted into hearth, kiln, smelter, steam piston, electrical dynamo, lightbulb, and then spark plug. The stupendous event ignited by that archaic lit brand later detonated a staggering explosion at Alamagordo, New Mexico, in July 1945, followed by the first hydrogen fusion bomb in 1952. A life-form created by the energy of the sun had seen beyond sunlight to reveal the very secret of the sun.

From its earliest beginnings, the taming of fire provided hominids with an evolutionary edge. Warmth, light, parasite-free food, fire-hardened spear points, an opportunity for increased social interactions, a versatile new weapon, increased visibility of facial expressions and body language, the opportunity to make friends with a few other animals such as cats and dogs, and improved security make up the short list of fire's benefits.

In Hebrew Genesis, the first gift that Yahweh conferred on Adam was to teach him how to name. Naming, Yahweh informs Adam, will provide him with the means to gain "dominion . . . over every living thing that moveth upon the earth" (Genesis 1:28). The more practical archaic Greeks believed that a more important boon was fire. In the Greek origin myth, Prometheus, the Titan, stole fire from Mount Olympus and gave it as a gift to mortals. Because the secret of fire belonged to the gods, Zeus punished Prometheus severely. Zeus and the other deities were keenly aware that Prometheus' seditious act forever separated mortals from all other animals.

Fire initiated significant modifications in the dietary habits of the three hominid species to use it: *Homo erectus, Homo neanderthalensis,* and *Homo sapiens.* Cooking flame denatured proteins and magically converted meat that was tough as shoe leather into tender, delicious mouthfuls. Hominids who used fire evolved discriminating tastes and sought flavor. Eventually, recipes became de rigueur.

Cooking food introduced a revolutionary environmental factor in the life cycle of hominids.* Increasing reliance on fire, however, encouraged both beneficial and deleterious mutations in digestion, absorption, and internal biochemistry to replicate exponentially. Grilling meat and cooking tubers

---

*Anthropologist Richard Wrangham believes that cooking food occurred 1.9 million years ago and was the impetus that changed *Homo habilis* into *Homo erectus.* There is little direct evidence to support his hypothesis, but he points to circumstantial evidence to build his case.[7]

made both easier to digest, so the hominid gut began to shrink rapidly in both diameter and length, diminishing the number of the intestine's convolutions. This in turn released a burst of excess metabolic energy that was quickly appropriated to meet the demands of a swiftly enlarging brain.

Homo sapiens' gut handles essential nutrients in a manner that, when taken as a whole, represents a significant departure from the way every other creature digests its meals. All vertebrates require three basic foods to maintain a baseline healthy state: carbohydrates, proteins, and fats.

Digestive enzymes break complex carbohydrates into smaller and less complex molecules. When they have attained a size that is a minute fraction of the original, the cells of the small intestine's lining absorb the byproducts of this reductive process and transfer these simple sugars into the intestinal capillaries on the other side of the gut-blood barrier. These short-chain sugars then can be easily and quickly converted to glucose by the liver.

Glucose is the fodder chiefly used to stoke metabolism to keep the home fires burning. Trillions of mitochondria within trillions of cells simultaneously throw glucose onto tiny cellular bonfires. The primary function of these mini-blazes is to heat and maintain the human body to a core temperature of 98.6°F.* This internal climate control is necessary because virtually all the body's enzyme systems work optimally at this temperature.

The body expends the remaining dietary glucose to fuel enzymatic reactions critical to life processes. If a person ingests carbohydrates in excess of what is needed at the moment, the liver converts this surplus in several stages to fat, in which form, much to the dismay of modern humans, it is amply and conspicuously stored in buttocks, bellies, and breasts. The cells making up adipose tissue (fat) then patiently await orders from on high for their precious fuel, which the body can quickly convert to glucose.

Refined or pure sugar is the fastest absorbable source of glucose. Stripped of any other components, the simple sugars contained in its refined or pure form rapidly enter the bloodstream and can produce a "sugar high." In the context of the Pleistocene, early *Homo sapiens* would have hungrily consumed any source of pure sugar. Honey, one of nature's most concentrated forms of simple sugars, would have been particularly prized.

---

*The heat also comes from millions of mini–muscle contractions called "fasciculations," so tiny that they are nearly invisible. Fasciculations grow in intensity when a person experiences cold. The body's internal climate control orders the force and duration of fasciculations to increase to generate more heat. Then these muscle contractions are very visible. They are called "shivers." Soaking in a hot tub when the water temperature exceeds 98.6°F causes these muscle fasciculations to cease. The sensation of soft-noodle relaxation that comes from sitting in a hot bath is the result of allowing the body's metabolism to shut down the tiny internal furnaces temporarily and give all the muscles a chance to rest.

Proteins are the building blocks of the body. They are necessary to build, repair, and maintain tissues and organs. Pancreatic enzymes break these extremely complex protein molecules present in food down into much smaller components for easier absorption. The smallest absorbable unit is an amino acid. Once proteins are on the other side of the intestinal lining, blood transports these indispensable building blocks to the liver. There they enter various factory assembly lines so that they can be combined and recombined with other amino acids to form the very stuff of life. Albumin, serotonin, and immunoglobulins are just a few of the many intricate molecules fabricated from amino acids like Tinker Toys.

There are twenty distinct amino acids. Most animals can assemble any one of them on the backbone of the carbon, oxygen, and hydrogen present in the fats and carbohydrates in their diets. *Homo sapiens,* along with several other primate species, cannot manufacture eight of the twenty amino acids. These are called the "essential amino acids." They must be ingested. Absence of any one of these eight over any extended period will have a negative impact on health. Besides the eight EAAs, two others are very difficult for the livers of adults to assemble, and children cannot make them at all. These two, arginine and histidine, are called the "semiessential amino acids." Thus, nearly one-half of the protein building blocks critical to a *sapiens'* well-being must come from his or her diet. *Sapients* must find their "missing pieces" in their external environment.*

Now for the strange part: The ten essential and semiessential amino acids only rarely can be found in any single plant food. Generally, some essential amino acids are in one and others are in another. A dedicated forager could get around this by eating a varied vegetarian diet. For example, two essential amino acids are present in high concentrations in grains such as wheat, but one is missing. The one missing in wheat is present in beans, but then the two that were in wheat are in very low concentration in beans. Eating a rich diet of various fruits, nuts, grains, leaves, seeds, and beans can easily prevent a protein deficiency. The critical caveat, of course, is that one must have the opportunity to locate a wide variety of vegetables, or eat the few vegetables that are particularly rich in the EAAs.

In contrast, every single one of the essential amino acids is present in meat, fish, and fowl. Ironically, the most perfect protein food is the much-maligned egg. The yolk and white, intended to make an entire new organism, contain every necessary building block. If an egg is rated 3.92—highest on an overall scale for its food value, especially for the amino acids it contains—then wheat is 1.53 and peas are 1.57. By comparison, beef is 2.32 and fish is 3.55.

---

*Since many large primates share this defect with us, it must have occurred early in primate evolution.

The most complete vegetable food is the soybean, which equals beef in nutritional value.[8]

Vegetarians will eat dairy products, but vegans do not eat anything associated with an animal. Sue Rodwell-Wilton in her authoritative textbook on nutrition and diet comments, "A vegan diet is too poor in required nutrients to sustain childhood growth needs. Vegan children are stunted, and anemic."*[9] They also would not likely attain their full intellectual potential.

When *sapiens* began their journey to lands less hospitable than equatorial Africa, they often found themselves in environments that had seasons during which the local plant stock was poor in the missing eight EAAs. Those ancestral humans who had access to meat protein would have been the most likely ones to make it to the next generation. Those who did not were at a major genetic disadvantage.

Women need protein more than men. Adequate estrogen levels (built up from proteins) are essential for a woman to synthesize properly her other proteins. Progesterone (built up primarily from proteins) plays a crucial role in ensuring that the proteins she manufactures last. Menses causes iron and protein loss. Men do not experience a comparable metabolic event that regularly loses protein.[10]

The challenge to a pregnant or nursing woman would have added an exceptional burden. She would have had not only to fuel her own body's needs but also to participate in building a new one. Every one of the mother's amino acids traversing the placental barrier is lost to her forever. At birth, a newborn plus its attached placenta carries out of the mother a huge amount of what had formerly been the mother's total protein. To create a healthy baby, a woman must procure a diet rich in all the amino acids before, during, and after pregnancy.

Another vital component of the human diet is fat—a food providing the most concentrated energy per ounce. A wild elephant must spend its waking hours locating and eating six hundred pounds of green fodder daily to meet its minimal energy requirements.[11] A wolf, after bolting down a meal of hapless prey, can go for days without eating again. The fat in its meat makes the difference. Fats are large, complex molecules that resist easy digestion and

---

*Present high-tech food-distribution systems in the industrialized nations have made available to vegans an extraordinary cornucopia of plant foods from which they can choose. Benefiting from the knowledge gained in sophisticated research, a vegan can eat a combination diet that will supply all the EAAs. They also have the alternative of purchasing one of the many concentrated amino-acid supplements currently available.

absorption. Enzymes excreted by the pancreas saponify the fats we eat; that is, they break the large fat globules one sees floating in a bowl of chicken soup into smaller and smaller droplets until they are tiny enough to be absorbed by the intestinal lining. The smallest component of dietary fats that can be absorbed is a fatty acid.

Once on the other side of the intestinal lining, fatty acids, like proteins, are transported to predestined storage areas. In a crisis, fats can be thrown into the metabolic furnace for a quick jolt of energy. Some fats that we eat, however, are too valuable to store or burn. These are earmarked for tiny assembly lines scattered in various tissues and organs, principally the liver. There, the special fat is annealed to a protein to become a component of an important class of compound molecules called "lipoproteins."

Lipoproteins are what give life its sponginess. The integrity, shape, and form of every cell membrane depend on lipoproteins. The greatest concentration of lipoproteins exists in the human brain. Sixty percent of brain tissue by weight is fat—but a very interesting, intelligent fat.[12] If neurons can be compared to the transmission wires of an electrical grid, the entire infrastructure—the towers, insulation, and way stations—is constructed primarily of lipoproteins. Biochemist Michael Crawford, who has written extensively on the importance of diet in human evolution, observed, "The real value in animal products may well lie in the fact that they contain a spectrum of structural fats not found in vegetation."[13]

Human metabolism can manufacture all fatty acids the body needs from the materials it has on its internal shelves except one. Linoleic acid is a long-chain fatty acid that many other animals can make with ease. Humans lost the ability to join together the atoms of this one essential fatty acid. Linoleic acid is found in many but not all vegetable oils. Corn, soybean, and canola, to name a few, contain rich lodes of it.* Raw seeds and leaves are another good source.

A much simpler dietary strategy to acquire this essential fatty acid is to eat an herbivore. Leaf- and seed-eating animals expend prodigious amounts of metabolic energy converting linoleic acid into incredibly complex fatty-acid chains. A carnivore can simply bypass all the intermediate steps and devour the finished product. The fat marbling a steak is visible linoleic acid that

---

*All vegetable oils are liquid at room temperature and are called "polyunsaturated fatty acids." Fatty acids that are solid at room temperature are called "saturated fatty acids," and they are found in animal products. There are several other fatty acids that are very difficult to make, but ultimately, in a pinch, human metabolism can construct them using only linoleic acid. Evolutionary nutritionist Michael Crawford has emphasized the importance of the longer-chained, more complex omega-3 and omega-6 fatty acids in building an intelligent human brain. He and many other nutritionists propose that their presence in high concentrations in seafoods makes fish a better brain food than meat.

has been considerably upgraded. Liver and all dairy products are also very rich in this essential fatty acid. This element, so crucial in the construction of a fetal brain, must be in abundant supply during and right after pregnancy in order for a newborn to build out the potential intelligence programmed in its DNA.

The most crucial member of this family of important compounds, the one without which human life is not possible, is cholesterol. The liver and intestinal wall manufacture the necessary minimum requirement of this compound molecule. Most animals can make all the cholesterol they need "in-house." But human metabolism requires so much of the golden substance that it welcomes an outside assist. The liver consumes an enormous amount of energy and resources to build the basic cholesterol molecule. Supplemental dietary cholesterol was an important adjunct to good health for ancestral humans.*

In modern times, people view cholesterol as a villain, because its overabundance has been implicated in atherosclerosis, the leading cause of coronary artery disease, heart failure, and strokes. At the dawn of our species, these modern pathologic conditions competed as a chief cause of death with an attack by a saber-toothed tiger, a goring by an auroch, and starving in the clutch of a relentless ice age. Cardiovascular disease came in last.

Cholesterol plays an indispensable role in everything from the basics of nerve transmission to the parameters of intelligence. It is the precursor molecule to all the steroid hormones: cortisone, estrogen, progesterone, and testosterone are all cholesterol-based. Fleeing, fighting, feeding, and sex are central activities driven by steroid hormones. The brain has more cholesterol in it than any other organ. There is suggestive evidence that high cholesterol levels sustain the mental states of happiness, equanimity, and optimism. Similar studies suggest its deficiency may induce the opposite mood states. Men exhibiting impulsive, antisocial, and violent behavior more commonly have lower cholesterol levels than the population at large, as studies of violent prisoners and inmates of mental institutions have shown.[14] The 25 percent of men with the lowest cholesterol count are four times more likely to commit suicide than the 25 percent of men with the highest cholesterol count.[15]

---

*Underscoring the tremendous amount of metabolic work performed by the liver is the fact that, at any given moment, *40 percent* of a person's circulating blood volume is in his or her liver. In recognition of its role in health, people greeting each other on the streets of old Vienna would inquire, "How is your liver?" rather than the nonspecific English greeting, "How are you?"

Because cholesterol is the precursor molecule for the manufacture of testosterone, estrogen, and progesterone, any process that diminishes serum cholesterol or interferes with the normal conversion of cholesterol to the sex hormones will lessen a man's or woman's libido. Many medical conditions that have low serum cholesterol as part of their clinical picture, such as Addison's disease (a malfunctioning of the adrenal glands), are characterized by a low libido. A woman experiencing low cholesterol will have menstrual irregularities, and these in turn diminish her chances for a successful pregnancy.

There is *zero* cholesterol in 99.9 percent of plant foods. Cholesterol and the precursors necessary for its manufacture are overabundant in animal foods. I speculate that, in the evolution of our species, Natural Selection favored men who ate high-cholesterol foods because they were, on average, slightly more effective hunters and ardent lovers. A female who mated with a man who shared his high-cholesterol bounty with her had a healthier body, a slightly more exuberant sex drive, higher fertility rates, and slightly more intelligent children. Each feedback loop would have had the effect of encouraging humans to stray away from the plant foods that were their primate heritage and develop a hankering for fatty meats.

One need not be a biochemist to assess the importance of cholesterol to the male's sex drive at the moment his testosterone level skyrockets. Simply observe the dietary *volte-face* of boys at the point in their development when they begin to turn into men. Young boys do not particularly like red meat. Ordering a steak for them in a restaurant, as any parent can attest, is a waste of money. Most boys would be perfectly happy subsisting on peanut-butter-and-jelly sandwiches. However, once the pistons of puberty begin to rev, a startling transformation occurs to male taste buds. Peanut butter is relegated to the back of the pantry, and teenage boys hunger for red meat—the greasier and bloodier the better. A significant component of this change in culinary tastes is due, I believe, to the male interest in sex. Men seem instinctively to understand that red meat increases their sexual stamina. And the cholesterol in his hamburger may be one among a series of vital components a young man wolfs down because he is unconsciously driven to satisfy his restless, ever-pacing libido.

Here, then, is the puzzle. Why did the human species lose the precious knowledge of how to manufacture the ten amino acids and one essential fatty acid critical to building youthful smart brains? Why do we thrive better if our diet contains supplemental cholesterol? The human brain's colossal size commands that its owner resupply this voracious organ with gargantuan quantities of these vital components. Natural Selection favored a Rabelaisian appetite over a dainty one.

Many other animals, including primates, have varying degrees of the same biochemical lapses, but the major difference between all the others and us is that they do not have an impatient brain tapping its toe waiting for its daily shipments. Any early humans who botched his or her access to animal foodstuffs were greatly disadvantaged in the genetic contest to survive and leave offspring. This was especially true for those adventurers who left the bounties of equatorial Africa and inadvertently walked right into the teeth of ice ages that lasted fifty thousand years each. On average, children of these individuals would not have been as strong or as smart as those whose parents supplied them with generous portions of cholesterol, essential amino acids, and the one fatty acid we cannot make. The red spice sprinkling this mix of delicacies would be the metallic presence of iron.* Once Mother Nature had begun to shove this one reluctant prey primate toward becoming a fearsome hunter, She made sure that his retreat would be difficult. Whereas the other primates are primarily vegetarians, with a handful eating occasional mouthfuls of meat, *Homo sapiens* would not go exploring without some beef jerky in his pouch.†

One reason *Homo sapiens* became a ruthless hunter was that, as he became more successful in his endeavor, Natural Selection closed off the avenue for him to remain a pure vegetarian. Because we are a transitional creature, we can, if we make a concerted effort, overcome our biochemical lapses. But a vegetarian must spend more time gathering a larger quantity of

---

*The human need for meat does not diminish the importance of vegetables in a healthy diet. A pregnant woman who eats only meat will deliver an underweight infant with a tendency to fail to thrive. Nevertheless, diets rich in meat are suspected of increasing intelligence. The Flynn Effect, named after political scientist James Flynn, tracks the worldwide rise in IQ points occurring at the rate of three points per decade over the last seventy years.[16] The cause of the Flynn Effect remains a mystery. Many scientists believe it is the result of improvements in the diets of children. When two Guatemalan villages were given protein supplements for several years, the IQ of the children rose significantly when measured ten years later.[17]

Many researchers believe that the abundant bioavailable iron that can be consumed by eating meat but that is not present in the majority of vegetables plays an important role in establishing intelligence. Dr. Jill Halterman reported on the results of giving girls with low-normal iron levels iron supplements in over 5,398 cases over an eight-year period. The result was a six-to-eight-point rise in test scores in math. Dr. Halterman comments: "Past studies have shown a superiority of females in math achievement during elementary and middle schools and a reversal of these trends in high school and college years. This study suggests that iron deficiency may contribute to this gender discrepancy."[18]

Ann Bruner of Johns Hopkins also conducted a similar randomized double-blind study giving Baltimore adolescent nonanemic girls iron pills versus placebos. The group receiving the iron supplements had a statistically significant rise in test scores.[19]

The World Health Organization, attempting to address the immense problem of iron deficiency in developing countries, began a pilot program among poor villagers in Ethiopia. They distributed iron pots for cooking. Some of the iron mixed with the food on heating, and the results were impressive. Health and cognitive skills among children rapidly improved as their internal iron stores increased.[20]

†Insects are rich sources of proteins, cholesterol, and essential fatty acids. Anthropoids regularly consume large quantities of them. Chimps and bonobos regularly eat termites, and orangutans and gorillas savor ants. These primarily vegetarian apes avert many major dietary deficiencies by eating bugs.

plant food to equal the same amount of calories available in meat. Vegetarians must spend more time and energy chewing their meals. And they must allot more time to rest until the massive diversion of the body's blood supply to the gut passes. Attempting to digest a large meal containing hard-to-absorb nutrients mixed with cellulose takes its toll on the brain, as anyone attending a one o'clock lecture after a heavy noontime lunch knows. * This waiting period also occurs in carnivores, but the total time spent resting while digesting is less.

Sexual selection was at work in the human species in the same way it was in all other species. Women chose men who were the best hunters. A man who could provide a woman with a diet containing the fattiest meat won her heart over someone bearing a sling filled with kumquats and papaya.

An animal that abandons critical enzymatic pathways to seek finished products in its immediate environment can be deemed to be using a clever strategy. It has been argued that an organism's fitness can be measured by how many metabolic processes it can "out-source." An organism that locates a rich lode of hard-to-manufacture molecules in its habitat has freed up surplus energy that can better be allocated to the business of survival and reproduction. For instance, a carnivore is a more "efficient" eating machine than an herbivore, because rather than expending energy maintaining all the digestive machinery necessary to break down plant food into readily usable constituents, it overcomes its enzymatic deficiencies by simply letting the herbivore fatten up. Then, in a few gulps, the carnivore quickly devours the finished product, overarching the intermediate steps that the herbivore so painstakingly accomplished.

In this sense, it is a dog-eat-dog world, but in another, we are all interwoven in what the Disney classic *The Lion King* called the "Circle of Life." A popular rendering of the science of ecology, the Circle of Life emphasizes the interconnectedness of all organisms. Biologists use the concept of symbiosis to describe how two different species coexist to assist each other. The unlikely plover bird that sits in a crocodile's mouth cleaning the reptile's jagged teeth busies itself, unconcerned that the crocodile could eat it with a snap of its jaws. The crocodile opens wide for its avian dentist because it has evolved a behavior pattern that allows the plover bird to perform a function beneficial to its survival. The bird symbiotically benefits, dining on tasty leftovers

---

*Humans cannot digest cellulose, so it passes through our gut. In those animals that do digest this plant substance, an enormous amount of time and energy must be expended to digest it.

caught in the crevices between the croc's teeth, and does not have to spend a morning searching for breakfast. The symbiotic relationship that exists between prey and predator is less obvious, but in the ecological overview, each aids the other. Lions that eat the old and the weak wildebeest serve the function of culling the herd and preserving fit genes, so that in the long run the herd prospers.

As hominid predators became bolder and more successful, they began to discard essential enzymatic pathways because they could eat what they needed to maintain their health. At some point in the recent past, *Homo sapiens* ceased being a highly successful predator living symbiotically within its ecological niche and instead became a *parasite*—a very large parasite, but a parasite nonetheless.

Parasites derive so many essential nutrients from their hosts that they cannot survive independent of them. Ten percent of all living species are classified as parasites. One familiar example is the tapeworm.

The key to performing as a successful parasite is to siphon just enough nutrients away from the parasite's unlucky victim so that the host's ability to continue to provision the pesky hanger-on remains unimpaired. A tapeworm luxuriating in the intestinal tract of a human can achieve lengths of over thirty feet. The person plagued by such a hitchhiker must eat to satisfy his or her needs and feed the stringy, unwelcome guest as well. The tapeworm has cleverly adapted in such a way that it does not have to lift a finger to find dinner, because it can depend on its host to bring it food. And the tapeworm is not such a glutton that it eats more than the host needs to maintain a minimum of health, so the host can continue to forage for the lazy tapeworm.

A parasite can be considered stupid when it makes such excessive demands on its host that it kills it. Then the parasite has sealed its fate, because not only does it lose its meal ticket but it, too, dies. An organism that kills its host ceases to be a parasite and is reclassified as a "pathogen." It could be argued that *Homo sapiens* has degenerated from its beginnings as a symbiotic prey to a symbiotic predator to a parasite and has now transformed into a planet-devouring pathogen.

Think of the entire planet, with its blue oceans and pristine mountains, as a host. The roll call of species that humans have dispatched to the Land of the Extinct, when combined with deforestation, pollution, strip mining, overgrazing, or overfarming, has distinguished the bipedal primate as the planet's most exasperating parasite—all in the space of 150,000 years. We have arrogated many of the earth's resources simply to satisfy our craving for material comfort. While we have been congratulating ourselves on our species' unrivaled domination, alarm bells are beginning to sound in all regions of the planet. From the perspective of other life-forms, we have transmogrified into the planet's most virulent pathogen, and our frenzied degradation of our host, Earth, signals that we may be just another stupid parasite too feeble-minded to realize that one should never bite the hand that feeds one.

Another extremely odd feature of the human digestive system pertaining to iron is vitamin C, otherwise known as ascorbic acid and present in abundance in fresh fruit. Linus Pauling, the quirky Nobel Prize laureate biochemist, alerted people to their bizarre vitamin C metabolism by lecturing on the subject with theatrical flair.* Pauling would ascend the stage wearing a rumpled suit whose inner coat pockets fairly bulged. He began his presentation by informing the audience that a healthy 180-pound pig made a certain quantity of grams of ascorbic acid per week. With a flourish, he would reach into his pocket and produce a test tube brimming with a white powder Pauling identified as swine ascorbic acid. He then claimed that a 180-pound goat produced a certain amount of vitamin C in the same time frame, and then retrieved from his voluminous coat pocket another test tube, filled with goat ascorbic acid. After running through nine familiar 180-pound mammals and revealing to the audience how each one was capable of producing its required vitamin C entirely by itself, Pauling then produced one last test tube. Pausing for emphasis, he displayed the empty tube and proclaimed, in a stentorian voice, "And this, ladies and gentlemen, is all the ascorbic acid a healthy 180-pound human can produce in the same time period."

By his dramatic style and the prestige of his reputation as a renowned chemist, Pauling initiated the vitamin-C craze. Anyone sitting in the audience would have to wonder why humans had lost such a vital adaptation as the biochemical pathway to make a substance that is so indispensable to the maintenance of health. Why did the human liver "forget" the secret formula for the ingredient without which gums, bones, blood clots, cell membranes, and myriad other vital structures cannot properly perform their function? Perhaps the answer has to do once again with the quirks of human iron metabolism.

Earlier, I noted that the abundant iron contained in leafy vegetables is very poorly absorbed because a human's digestive machinery has great difficulty freeing the vital atom from the clutch of compounds binding it. In the presence of vitamin C, however, the chelator reluctantly releases some of its iron atoms, allowing the human digestive tract to more readily absorb plant-

---

*Pauling did not win his Nobel Prize for his work on vitamin C. Many of his recommendations concerning taking megadoses of vitamin C and other constituents have since been discredited by scientific trials.

based iron. If Popeye ate an orange along with his spinach, he would receive more of an iron jolt. Unable to manufacture sufficient ascorbic acid internally, the human body must locate a source in its diet.

Humans are apes. Other apes, such as chimpanzees and orangutans, principally eat a diet of fresh fruit. They never migrate to environments that lack plentiful sources of this staple of their diet. Scientists estimate that the primate ancestor that evolved into the anthropoid line lost the enzyme to make ascorbic acid twenty million years ago. For this ape or any of its descendants, the inability to internally manufacture vitamin C posed no problem because it was superabundantly present in their diets.

Fresh fruit, however, is not always available in temperate zones, especially during winter. One successful strategy to stave off iron-deficiency anemia secondary to ascorbic-acid scarcity would be to switch to a diet high in heme iron. Although one would still be susceptible to the other protean manifestations of vitamin-C deficiency, at least a lack of iron stores would not be among the most important ones.*

The lack of vitamin C can create unusual side effects. Aldous Huxley in his book *Heaven and Hell* posed the question why so few people in modern culture see visions or experience the kind of religious ecstasy that was so commonly recorded back in the Middle Ages. The literature of that time is replete with reports of these phenomena, occurring on a regular basis to ordinary people. When someone in contemporary society claims "to talk to God, hear voices, or see visions," we often admit them to the psychiatric ward to cure them of what the rest of us would consider deranged behavior.

Huxley attributed the relative disappearance of trance states in contemporary society to improvements in diet, and he proposed this as a major factor in lessening the influence of the Church. Prior to these developments, Europeans endured long, cold winters during which there was an absence of fresh fruit. Scurvy, the medical name for vitamin C deficiency, was pandemic.† To make things worse, in the dying days of winter, the Church mandated that everyone fast for Lent. Body reserves of proteins and fats already

---

*Carnivores must eat their veggies, too. The favored item on the menu of lions, tigers, and other meat-eaters is offal. By first devouring the intestines of the herbivores they have just killed, carnivores ensure that they will have consumed (albeit secondhand) the crucial constituents of vegetables, vitamins and minerals that are not present in meat, because these missing ingredients are soaking in the herbivores' digestive enzymes inside their intestines—the very enzymes that the carnivore doesn't have. Only after dining on their "salad" will carnivores turn their attention to the entrée, the prey's marrow, brains, and muscles. When zookeepers attempted to feed lions *only* meat, they became ill, because they lacked the vegetables present in offal that had been an indispensable component of their diet in the wild.

†Vitamin C shaped world history. Long sea voyages meant scant fresh fruit to distribute, and, inevitably, scurvy would lay the crews low. In 1753, James Lind, a Scottish naval surgeon in the Royal Navy, discovered that if sailors were given fresh limes at intervals, then the symptoms of the scourge would not appear. Britannia went on to rule the seas, and her sailors are still nicknamed "Limeys."

depleted by poor winter diets would have created the conditions of border-line starvation. Nerve transmissions in the brain, verging on serious disruptions, would begin to falter.

Adding to the complications of near starvation was the religious practice of self-flagellation, which the Church tacitly encouraged. A whip applied to the thick skin of the back excoriated it, leaving long superficial lacerations, not deep enough to cause death but severe enough to cause bleeding and anemia. These wounds, superimposed on a body reeling from the effects of incipient starvation, would then begin to suppurate. A low-grade bacterial infection would incrementally add to the debility of one who had scurvy, anemia, and starvation. These conditions would coalesce just as Easter approached. The mental states of many people were so affected by these dietary deficits and systemic toxicities that whole cities had mass hallucinations. Clerics in an age of extreme religiosity assumed that they were bona-fide revelations.

One wonders how many "visions" were due to cholesterol-starved, ascorbic-acid-deficient, linoleic-acid-deprived, amino-acid-depleted neurons trying to cope with the toxicity associated with a subclinical streptococcal and/or staphylococcal infection. The predictable result: the overt syndrome of delirium. Such are the hidden factors that may influence history.

*   *   *

Let us step back and reassess the peculiarities we have just discussed. We apparently lost the vital means to make eight amino acids (ten in youth), one vital fatty acid, and ascorbic acid long ago, when primates differentiated away from other mammals. As long as primates remained in an environment in which there were plentiful dietary sources for these nutritional components, and their brains had not yet embarked on its hyperinflation routine, all was well.

Our metabolism was lazy when trying to make these vital substances internally because they were readily available and we were eating them. A major problem arose when we moved away from the lush vegetable sources containing these components and found ourselves encased in the ice that characterized most of the Pleistocene age. Having lost the ability to manufacture them, and deprived of easy pickings, we had all the more reason to search out and kill the animals that contained these substances. Like the Greek army under Agamemnon attacking Troy, we had burned many of our ships on the beach, so that retreat would not be an option. We had to increase our ruthless ways or die.

You are what you eat. The introduction of fire made meat much easier to chew and digest. A dietary strategy that relied primarily on plants segued into one increasingly dependent on meat. Fire also greatly improved our ancestors' choices among previously inedible vegetables. A human who attempts to eat raw rice, wheat, or potatoes will experience significant digestive uproar. Tamed by fire, these dietary staples became "staffs of life."

Keeping in mind the many ancillary nutritional human needs discussed above, let us refocus the discussion on *Gyna sapiens'* need for iron. As the key characteristics of her reproductive life history veered away from what had served the multitude of other sexually reproducing females so well for so long, she could only dimly surmise that she was trapped in a quandary. For her and her offspring to survive and thrive, she would have to evolve new adaptations that would balance the ones that caused her to lose iron persistently. Like two runners yoked together in a three-legged sack race, these adaptations had to advance in precise synchrony or the human species would likely be deleted from the taxonomic catalogues. Natural Selection had to furnish *Gyna sapiens'* metabolism with a credible strategy.

In a seemingly self-defeating countermaneuver, Natural Selection encouraged adjustments in the enzymes in *Gyna sapiens'* digestive tract that would make her goal very difficult to achieve. Despite her frequent forays for food, and despite the variety of comestibles she collected, she was unable to establish a stable iron source by herself. The toddler tugging persistently at the hem of whatever ancestral mothers wore, the infant at her breast, or the fetus in her womb did not make her quest any easier.

Mother Nature had played a cruel trick on human mothers. In every direction a woman looked from her position at the home base, there was an invisible sea brimming with iron atoms. Nuts, roots, fruits, shoots, and leaves were abundant with iron. But, to paraphrase the lament of Coleridge's Ancient Mariner, there was "Iron, iron everywhere, nor any filing to absorb." The iron, so close, was present in its inaccessible form. The iron running around on the hoof was better, but how was she to secure it? Menses and the five other major factors that cause a female to lose iron made her search urgent. If meat was the most reliable source of easily absorbed iron, and she, burdened with small children, was unable to attain it by herself, she would have to deploy another strategy. And then she hit upon the solution: There, lumbering around in the underbrush, was the key to her success and her species' continued existence—*Homo sapiens.**

During the long hominid evolution, *Homo sapiens* segued from a frightened vegetarian ancestor to a tentative scavenger to a skilled hunter to a

---

*A woman was perfectly capable of setting snares and hunting small animals. Her successes contributed substantially to the overall diet and health of herself, her family, and her tribe. Nevertheless, large-game meat is, in all cultures, more highly prized than small-game meat both for the courage a hunter displayed in bringing it home and the concentrated amount of minerals, fats, and proteins it contained.

fearsome predator. Our closest primate relatives—chimpanzees, gorillas, orangutans, and bonobos—thrive on nuts, fruits, shoots, tubers, and insects. So can we, if we have to. But deciding to hunt big, dangerous animals that would just as soon kill him wasn't something *Homo sapiens* was likely to embrace enthusiastically just for the thrill. Considering the vegetable bounty that must have existed in the home of our species' origin, it is doubtful that hunger was the primary motive. I nominate two other prime candidates—sex and money. Since money had not yet been invented, by elimination that leaves only sex.

*   *   *

One can witness in any romantic restaurant the political and economic ramifications of women's gaining veto power over sex but losing iron. Despite the dietary-cholesterol awareness presently widely disseminated in the popular press and the recent shift to vegetarianism, the ritual persists. More often than not, a young man can be observed wooing a young woman over dinner by lavishing her with food purchased with his hard-earned hunting money. The table often is lit by candlelight, just as it was by firelight at the dawn of this exchange. Whether he hopes to gain exclusive, permanent sexual access (marriage), temporary sexual access (an affair), or fleeting sexual access (a one-night stand), his ultimate goal is to persuade her to say *Yes!*

In most settings (more so in the past then the present), he understands that tofu is not a proper offering; neither is a solitary salad or a bowl of soup. An appetizer or dessert alone will rarely advance his suit. His "nuptial gift" most often must be red meat, and good red meat at that.

Filet mignon is harvested from an animal's *psoas major* muscle. Positioned alongside the vertebral column, deep in the most inaccessible interior region of the slaughtered animal's abdomen, this muscle is a difficult one to retrieve. But more than its inaccessibility, its high preference value lies in its extremely low collagen content. Collagen is the substance that holds everything in the body together. It is tough stuff. Scar tissue, tendons, joint capsules, and adhesions consist mainly of collagen. The gristle in a steak is collagen. Its relative absence in filet mignon makes this the tenderest cut.

The young woman sitting across from him, in her turn, appears to understand the terms of this transaction and will often order her filet medium rare. In the candlelight, one can discern iron's presence glistening on the filet's cut red surface. This is a very old deal.

A young girl's awakening to the power of her fertility marks a major milestone in her life.

Chapter 11

Menarche/Mustaches

Sex is an antisocial force in evolution. Bonds are formed between individuals in spite of sex and not because of it. —E. O. Wilson[1]

Put simply, in our evolutionary history, it seems likely that a woman's value was usually her reproductive value, and a man's value was his resource value.
—Bobbi Low[2]

Everywhere sex is understood to be something females have that males want.
—Donald Symons[3]

The differences and correlations between the puberty rites of boys and girls further forge the evolutionary link between sex and iron. Menarche, a girl's first menses, typically begins around twelve years of age in industrialized societies. This life-changing event occurs several years later in agricultural and hunter-gatherer societies. Historical cultures celebrated menarche with elaborate rituals, and many contemporary ones still do. Though these rituals varied from place to place, two features remained constant. Elder women isolated the initiate and instilled in her that she was not to dispense future sexual favors easily, because they were extremely valuable.* The key purpose of the menarche ritual was to impress upon the girl that she was now in possession of both a great power and an enormous responsibility.

While many menarche rituals are celebratory and loving, the ethnographic record drawn from a wide variety of cultures is replete with numerous exam-

---

*Even in those few cultures in which postpubescent girls were encouraged to engage in unrestricted sex, the message conveyed remained essentially the same.

ples of draconian ones. Among the Loango of East Africa, menstruants are confined to isolated dark huts and are prohibited from setting their foot on the ground or looking at the sun for two years. Far away in New Ireland, in the South Pacific, girls are similarly confined for four years in small darkened cages and also not allowed to touch the ground. When the symptoms of a girl's first bleed occur among the Guaranis of South America, she is sewn into a cocoon hammock strung between two trees, with only a small slit through which she can breathe. Shrouded as if she were a corpse, the girl has to maintain a vigorous fast until her bleeding is over.[4]

A postmenarche girl will, by fits and starts, begin to settle into regular menstrual cycles, which will continue, punctuated by monthly bleeding, for the next forty-odd years (interrupted only by stress, illness, starvation, pregnancy, or nursing).* Sexual intercourse during the first five years, however, seldom results in conception, because of another unusual feature of *Gyna sapiens'* reproductive life history—for the first few years after menarche, anovulatory cycles are the norm.

An anovulatory cycle is one in which the complex hormonal algorithm that drives a woman's monthly menstrual cycle occurs on time and in the proper sequence and concentrations, the uterus sheds its lining, and external bleeding appears; indeed, every aspect of the cycle proceeds on schedule except for the one that is the cycle's raison d'être—the ovary fails to release an egg. A few other mammalian females, notably the chimpanzee, experience anovulatory cycles during a span known as "adolescent subfertility."† The frequency and duration in years of a young *Gyna sapiens'* anovulatory cycles, however, exceed that of any other mammalian female.

What would have been the advantage to the individuals of our species of this adaptation? Young, nubile females exhibit all the signs of potential fecundity, yet it is a sham, invisible even to the girl herself. An evolved trait enticing males to mate with what ostensibly appears to be a sexually mature female, even though her ovaries are only sputteringly active, would not seem to benefit the fitness of the species. Mother Nature, however, often works in circuitous ways. Anovulatory cycles are a boon to young women.

A postpubescent male does not shoot blanks. His testes, unlike her ovaries, are not shy. They prodigiously pump out sperm that are more potent, swim faster, and leave the starting line in greater numbers than they

---

*Writer Eugene Weinstock chillingly documented the influence of stress and starvation on menses. One hundred percent of inmates who survived the Nazi concentration camps reported that their menses ceased during incarceration. While much of this phenomenon can be attributed to starvation, a goodly proportion of it was due to stress.[5]
†For chimpanzees this period can last from six months to three years.[6]

will at any other time in his life. In the event that a young Lothario seduces an adolescent girl (or the more unlikely converse), or in the circumstance of rape, anovulatory cycles act as a safety mechanism to prevent the death and, ultimately, protect the health and welfare of both the girl and her potential future offspring.

The dimensions of a female's pelvic channel are critical to the success of childbirth. An increase in diameter of even a few millimeters can make the difference between life and death. A young woman's pelvic circle of bone does not finish growing until she is in her late teens—four to seven years after menarche. Death in childbirth for both mother and infant is higher among adolescents than it is for mature women.* Also, postpubescent girls tend toward emotional immaturity and generally do not mother offspring well. Moreover, the likely father of her child is rarely in a position to provision her adequately, much less accept the emotional responsibility of fatherhood.

If anovulatory cycles ensure that sexual intercourse will fail to result in a pregnancy, then one might ask why Natural Selection didn't opt for delayed menarche. *Gyna sapiens'* pelvic bones would encircle a larger channel, and her increased emotional maturity would make it more likely that her baby would survive to reach its own reproductive age. Perhaps anovulatory cycles evolved to serve another important function. I propose that they allow girls as much as six extra years to learn how boys think.†

An adolescent male's sexual thermostat is set on high from the first moment his puberty switch is thrown, hitting its peak in early adolescence and then gradually waning over his lifetime. Testosterone concentrates a man's mind singularly. *With the onset of puberty, male testosterone readings skyrocket to twenty to forty times their prepubescent levels.* A postpubescent boy's sexual desires nearly drive him mad as his brain soaks in a cranial tub laced with testosterone. Most men can recall with crystalline clarity the precise moment in their transition through puberty when it suddenly dawned on them that girls were not just soft boys. During an adolescent male's waking moments, sexual fantasies are omnipresent, and nights spill forth highly exciting and disturbing sexual dreams.[7]

In contrast, a female's libido builds slowly during adolescence and doesn't

---

*This factor is somewhat mitigated by a young girl's more flexible pelvic joints in contrast to an older woman's more brittle ones.

†Anovulatory cycles are shorter in the females of extant hunter-gatherer societies because the age of onset of their menarche is later than it is for young women in contemporary technologically advanced societies. Although no one knows what the age of onset of menarche would have been for young females living in a principally meat-eating ice-age culture, I hypothesize that this feature of the human female's reproductive life-cycle has been present since ancestral times and has played an important role in human sexual relations.

reach its peak until her mid-thirties.* A pubescent girl's estrogen level increases to about three times what it had been prior to her menarche. Though this rise is not strictly analogous to the testosterone spike in boys, the 20–40:3 ratio gives a fair indication of how much more boys, in general, are addled by sex than girls are. In both men and women, testosterone levels have a profound effect on libido. An adolescent girl's testosterone level rises minimally. The average male has ten times more circulating testosterone than the average female.[8]

The phenomena discussed below do not apply to all boys and all girls but occur often enough that a broad generalization can be made. Whereas pubescent boys are "sex-crazed," adolescent girls are frequently afflicted by a condition familiar to parents as "boy-crazy," an interest not necessarily related to fantasies of sexual intercourse. Instead, teenage girls become obsessed with studying boys' behavior and seek to penetrate their psychological makeup.

An anovulatory girl learns how to be alluring as she hones strategies to entice boys. Her intense fascination extends to carefully observing the behavior of other girls and young women as it relates to the opposite sex. All her concentration has but one purpose: to understand how the man-woman thing works. It is a fine line she must walk. She desires to flummox all young males within range with her flirtations, but if she excites them too much, she might lose control of the situation. If she falls—emotionally or sexually—waiting to catch her is the safety net of her anovulatory cycles. Boys "think" with their penises, girls "think" with their hearts. Neither gender thinks with their brains.

Anovulatory cycles amount to a free pass for several years, during which a woman-in-training can expend considerable time and energy experimenting. Mother Nature cleverly evolved a temporary tough skin to cover the surface of a female's ovary, one so thick that it overrides the authority of her hormones, thus retarding the release of ova. This gift gives her a grace period to study the responses of males (and other females) to her dress, walk, speech, repartee, gestures, cosmetics, hairstyles, adornments, figure, and facial expressions. As she gradually gains an appreciation of the power she has over a male, early on, she also grasps the leverage inherent in the word *No!*

Another safety factor operational at this period of development I shall call "pubescent reverse sexual dimorphism." Sexual dimorphism, remember, is the disparity between the size of the male and the female of a species. Hu-

---

*As reported in numerous large scientific surveys of the sexual behavior of women, cited in previous references. Adolescent girls rarely report experiencing the graphic sexual dreams that roil the sleep of adolescent boys.

mans fall somewhere in between the most sexually dimorphic species and the least. Men are about 8 percent taller and 20 percent heavier than women—except for one brief span during which girls are larger, taller, heavier, and in many cases as strong as or stronger than boys of the same age. As any adult chaperoning a middle-school dance can attest, eleven-to-fourteen-year-old girls are, on average, a head taller than their male peers. This period, coincidentally, circumscribes the years when anovulatory cycles are most frequent. No other species exhibits the typical human adolescent growth spurt.[9] And no other species features so pronounced a reversal of sexual dimorphism.

The physical advantage men have over women, enabling them to impose their will through superior strength, is absent during the one period when they are least able to control their urges. Should a twelve-year-old boy get carried away and try to satiate his sexual urge by force, Mother Nature has conveniently equipped the girl with the physical means to defend herself.

Anovulatory cycles, differential sizes, and the gaping disparity in libidos are factors that allow an adolescent female to "get up close and personal" without risking pregnancy. They allow her to hone what men, in their befuddlement, refer to as "women's intuition." This "sixth sense" will be of paramount importance to her. First, it will help her determine which future suitor might be the most steadfast and successful in supplying her with iron and other important resources. And, second, it helps her learn how to influence males to do her bidding. A woman who fails to learn how to choose wisely and influence men endangers both herself and her unborn children.

Aware of the quirk of anovulatory cycles, many ancient cultures held a very liberal attitude toward sex among their young people. Accounts from Mesopotamia, ancient Egypt, and classical Greece record young people indulging in unbridled sexuality during fertility rites and certain religious holidays. Herodotus tells of young girls serving as sexual temple priestesses. After several years of duty, they would leave to marry and begin families. There apparently was no social stigma attached to this service.[10]

A number of isolated indigenous cultures throughout the world (for example, the Trobrianders, in the South Pacific, and the Mangaians, in the southern Cook Islands in central Polynesia) continue to sanction, and even encourage, matings among their unmarried youth.[11] If pregnancy were a common consequence of these activities, no doubt elders would discourage the practice. The current rise of teenage pregnancies may not be due to the obvious. Perhaps modern girls are having sex as frequently as did their predecessors in older cultures, but the rise in their pregnancy rates may be due to a sharp reduction in anovulatory cycles.

Some unexplained factor in modern societies has greatly reduced the incidence of cycles without an egg in teenage girls, and as a result, they become pregnant at an earlier age and more frequently than their sisters in earlier historical periods. Some scientists have speculated that high-fat diets have brought about this change. Others claim it is the increased estrogenic compounds in our environment, such as pesticides, other chemicals, or the high estrogen levels in commercially processed cow's milk. Another intriguing hypothesis fingers the massive increase in artificial lights, which has upset the natural timing of sunlight and moonlight. No one knows with certainty.

I suspect that another key factor stimulating earlier menarche and diminishing anovulatory cycles has been the authorities' recognition of the importance of iron in the diet of females. Iron-fortified foods consumed by children and teenagers, along with a consistently high meat diet, have increased their young bodies' iron stores. This increase, combined with adequate stores of fat, has signaled the feedback controls in the brains of these young girls that it is now permissible for pregnancy to proceed.

*   *   *

Males do not experience anything so dramatic as menarche. Yet, prior to the advent of modern society, diverse cultures commonly created an initiation rite for boys crossing the threshold to manhood. The fundamental difference between the male and female transition into puberty is that a man must "prove" himself worthy of manhood, whereas a woman has only to begin menstruating to be accepted as a woman. Although some societies have rigorous menarche rituals, none exist in which a young girl can "fail." By contrast, among the Plains Indians male-initiation rites were so challenging that boys were stratified according to how well they passed the tests; the highest achievers were elevated to the warrior caste.

Uniformly, male initiations, designed by male elders, tested a young man's mettle. Hunting and warrior societies especially prized fearlessness in their young men. They inculcated values important for a future hunter who must work in concert with others. Trials of skill, strength, stamina, courage, obedience, resourcefulness, cold-bloodedness, and the ability to withstand pain and privation were common initiation elements. As if to mimic menarche, many hunter-gatherer and agricultural societies incorporated bloodletting into the ritual.* The ultimate goal of ancient male-initiation rites was to encourage initiates to be brave and willing to take risks.

---

*Subincision is the practice of slitting a boy's urethral opening from the tip of his penis along the underside of its shaft, splaying the glans to make it resemble the vaginal orifice and labia minora. This initiation rite is also attendant with a prodigious amount of bleeding. Many anthropologists interpret this

Fortuitously, Natural Selection created ideal conditions to ensure that adolescent boys would seek thrills by flirting with danger. The reason parents often have occasion to screech at their young sons, "What *were* you thinking?" is the acute imbalance that exists in pubescent boys' brains—insufficient myelin combined with exceedingly high levels of testosterone.

Myelin is a gigantic molecule that binds fat globules within a lattice of protein. Once formed, it serves to sheathe individual neurons, the information-transmitting cells of the nervous system. Myelination is the process by which a human brain's nerves receive their myelin coatings, the function of which is similar to the insulation used on copper wires. Disparate areas of the brain and the peripheral nervous system myelinate at different ages during growth.

Both nerves and wires conduct electrical currents that generate electromagnetic fields extending into the surrounding space. These fields will, in turn, activate currents in otherwise dormant wires or nerves nearby. This is the principle behind radio and television transmitters and receivers, dynamos, and transformers, and as with these man-made devices, "interference" (or static) in nerves or wires is a potential problem. To protect signals from corruption by neighboring electromagnetic fields, each individual nerve is encased in insulating material.

In the electrical industry, this substance is the familiar plastic or rubber that sheathes the wires of appliances. In the brain and peripheral nervous system, the insulating substance is myelin. The fetal brain contains very little myelin. Evidence of a newborn's lack of it is apparent in the Moro reflex.* Hands clapped loudly near a newborn will startle it. The sharp sound sets off a chain reaction, so that the infant responds as if every nerve in its body had been activated. The newborn adopts a signature open-armed grasping manner, as if reaching out to embrace its mother for protection.

As an infant's brain grows, it steadily lays down myelin around its nerves. In general, myelination proceeds from bottom up, back to front, and right to left, and takes some twenty-odd years to complete.

If we are to understand better the consequences of this pattern for relations between the sexes, a brief detour into brain development is in order. Neuroscientist Paul MacLean introduced the concept of the triune brain in 1973.[13]

---

peculiar "feminization" ritual as a strong indication of male "vagina and menstruation envy." Psychoanalyst Bruno Bettelheim wrote extensively on this subject.[12]
*First identified by German pediatrician Ernst Moro in 1918.

According to MacLean, the human brain consists of three components derived from different evolutionary periods. Arranged in a vertical hierarchy, each stratum sits atop the older one.

The oldest brain component—the brain stem—was inherited from our dinosaurean ancestors. This reptilian brain responds to stimuli in a rigid, programmed fashion and contains the primitive instincts behind basic survival behaviors. Above the reptilian brain sits what MacLean calls the "paleomammalian brain," the one humans inherited from our mammalian ancestors. Comprised primarily of the deeper layers of the occipital, parietal, and temporal lobes, this stratum (more commonly called the limbic system) enlists emotions to evaluate current sensory input and will often cross-check against past emotionally charged situations.

The most recently evolved part of our brain, and the last layer of MacLean's triune brain, is the neomammalian brain. Composed of the neocortex, it is the thin rind covering the brain's outermost wrinkled layer. Two features that distinguish *Homo sapiens* from other animals are the extent of the neocortex and the brain's outsized frontal lobes. Together, they are primarily responsible for language, foresight, reason, judgment, and delayed gratification.

Again, myelination begins from the bottom up, coating first the neurons of the reptilian brain, essentially completing this phase within the first months of life. The spinal cord and brain stem are the first parts of the brain to receive their coats, beginning at twenty-two weeks in utero. The paleomammalian brain is next, as myelination proceeds from back to front, a process mostly completed by the end of the first five years. The neurons in the visual cortex, residing in the occipital lobes at the rear of the brain, receive their insulation first; then the process moves forward to the temporal and parietal lobes. The frontal lobes, along with a few other, highly specialized neocortical patches, are the last to be completely myelinated.[14]

A subtle differential exists between the myelination of the right and left sides of the brain. In utero, the visio-spatial right hemisphere develops earlier than the logical-language left one.[15] A more recent evolutionary addition, the highly specialized left hemisphere undergoes its major growth spurt beginning in the ninth month of pregnancy and continues this accelerated phase through the first four years.[16]

The most highly specialized and dichotomous brain belongs to a right-handed heterosexual male. Over 97 percent of the language centers of this subset are located in these men's left hemispheres. This extreme skewing of brain function is not present in the brains of either right- or left-handed females, gays and lesbians, or left-handers of either sex. All three of these categories have their language centers (and other cortical functions) more evenly distributed between the two hemispheres.

In right-handed people in general and right-handed heterosexual males in particular, the left frontal lobe is the seat of the Executive Function. Accepted as a useful concept by many neuroscientists, the Executor is the final assembly location that assimilates information from all other areas of the brain, and then evaluates risks, judges outcomes, overrides urges, delays gratification, and makes decisions.* After contemplating multiple risk-benefit scenarios, it chooses among several courses of action. Then the Executor exerts its "will," instructing the motor centers to carry out its command. Having decided to drink a beer that is located in the refrigerator, for example, the Executor will order certain muscles to twitch in an orderly prescribed sequence, propelling the body toward the kitchen.†

Natural Selection found it advantageous to the species to place an incompletely myelinated left frontal lobe's Executor in charge of the rapidly growing, gangly body of a teenage boy. Add to this "accident waiting to happen" the fact that the Executor, like the rest of the adolescent boy's brain, is pickling in a potent brew of testosterone brine.

This sets up a unique situation. Postpubescent boys will assume extraordinarily dangerous risks for the sheer thrill of it. After their left frontal lobes have been completely myelinated in their mid-twenties, they will look back and shake their heads in disbelief, reminiscing about the impulsively foolish, even life-endangering stunts they performed on instructions from their faulty Executors.

From an evolutionary standpoint, however, it was a pubescent boy's willingness to engage in risky behavior, combined with a young woman's unique requirement for a constant supply of iron, that played a paramount role in the development of our species. The proximity of a young woman can make a young man behave stupidly. He will try to impress her with his prowess or reckless behavior—attributes that fortuitously put him in harm's way in the pursuit of iron on-the-hoof.

A girl's left frontal lobe is also the last part of her brain to myelinate, but

---

*Recent research on brain imaging has surprised neuroscientists with the large number of far-flung areas in the brain that light up when the subject in the scanner is given instructions to carry out even the simplest actions.

†The concept of an "Executor" is a hotly debated issue. The weight of current neuroscientific research points to the left frontal lobe (in right-handed people) as the location within the brain that integrates all competing emotions, sensations, and memories with evidence of what is happening at the moment in the real world and knowledge from past experience. The Executor evaluates complex whorls of weltering information and, if necessary, makes snap judgments that precipitate action. The Executive Function also resides in the right hemisphere's frontal lobe. Neurologists Antonio Damasio and Joseph LeDoux have demonstrated that input from the right frontal lobe is crucial for sound decision making. The input from the right, nondominant hemisphere helps the left, dominant hemisphere exercise what we call "good judgment." Because the left hemisphere controls both speech and the preferred right hand, however, it plays a more commanding role in initiating action.[17]

she does not depend on it as much as boys do. Nor is she being driven as mad by hormones. A female's greatest risk revolves around getting pregnant. If her Executor fails, there remains the safety cushion of her anovulatory cycles. The greatest threat to a young ancestral woman's life was a premature pregnancy; the greatest threat to an ancestral adolescent boy was premature injury or death when hunting. The two great themes of the earliest art preserved on cave walls and chiseled in stone are hunting and fertility. Iron and sex bind the two tightly together.

❦

Once puberty begins, male testosterone levels soar. Thoughts of sex are never far from a young male's mind.

Chapter 12

Premenstrual Tension/
Masturbatory Tension

Males act as kind of a genetic sieve: Only the best males get to breed and the reproductive extinction of bad males constantly purges bad genes from the population. From time to time it has been suggested that that this is the purpose of males, but that commits the fallacy of assuming evolution designs what is best for the species. —Matt Ridley[1]

We come to blows, I think, because sex is the only behavior that to be fully satisfying requires another person. You can find food and eat alone, you can sleep, think, and play alone. But to have a decent sexual interaction you need the enthusiastic participation of another person. And this is the problem.
—Meredith Small[2]

The two traits paired in the chapter title appear to be uniquely human. No other animal manifests either of them to the degree to which they crop up in our species. From their observable behavior, the other thirty menstruating primate species do not appear to experience the symptoms women have come to associate with the onset of their periods. Nor does any other male animal masturbate as often or as intensely as a human. To appreciate why these two rooted and then flourished in our genome, an understanding of three physiological concepts—internal milieu, homeostasis, and stress—is necessary.

In the late nineteenth century, French physician Claude Bernard proposed the existence of an invisible field within the body that he called the "internal milieu." Influenced by physicist Michael Faraday's earlier discovery of the insensate electromagnetic field, Bernard broke with previous thinkers who had compared the human body to a machine. He proposed that within all the tissues of each organism there exists an optimal state the organism strives to maintain. Throughout the human body, all systems tend toward stability.

For example, the concentration of sodium ions (determining the blood's saltiness) circulating in the big toe is essentially the same as that making the loop around the skull. The body seems to know what is going on in its far-flung corners and deep internal recesses, and it manages to titrate the levels of its vital components, keeping them all within a narrow optimal range. Homeostasis is the process by which the body regulates and maintains its delicate internal equilibrium through a continual series of feedback loops. An organism experiences stress whenever some factor disturbs its internal milieu's homeostasis. Stress comes in two varieties: acute and chronic. Acute stress is sudden and precipitous. Chronic stress intrudes over an indefinite period of time.

The human response to acute stress is mediated by the reptilian part of the brain. It is involuntary, immediate, and instinctual, and in all but the most extreme cases, homeostasis is quickly restored. Chronic stress results when the factor in question disturbs homeostasis for a longer period and cannot be ameliorated by reflexive instinctual measures. The higher brain centers must respond by developing a strategy to thwart its source.

Seemingly minor changes in the internal milieu can ruffle homeostasis. Sensors scattered throughout the body identify which particular component of the internal milieu is disturbing the peace, and they, in turn, send out a cascade of nerve impulses and hormones, forcing the organism to take action. The example of water regulation is both apt and familiar, since water is the most common constituent of the human body.

Within the sealed waterproof bag we call our skin, feedback sensors positioned strategically around the body monitor water levels in the blood as it filters past. Whenever water loss exceeds intake—for example, through excessive perspiring—water levels in the blood fall, the blood becomes thicker, and the symptoms of dehydration appear.

Warned by the sensors of the problem, homeostatic regulators immediately institute safety measures: The kidneys begin to conserve water, causing concentrated urine to appear darker; sweating diminishes; and the brain secretes hormones that parch the throat, creating the sensation of thirst. Overseeing the movements of the body from its perch in the frontal lobes, the Executor advances thirst to the top priority slot on its to-do list.

A thirsty person will then interrupt whatever he or she is doing and take the necessary steps to locate a water source. Continued concentration on the task at hand becomes increasingly difficult until it approaches impossibility. If something prevents the person from drinking, a condition of acute stress is activated. Internally, red lights flash, sirens sound, and every bodily system is marshaled toward achieving a single goal.

Once water has been imbibed, the stomach absorbs it and transfers it into the bloodstream. The water level in the internal milieu begins to rise, resembling a filling reservoir. When fluid levels return to normal, the water sensors deactivate.

Much like the shipping lanes of a vital port, the internal milieu is a busy place. An organism must consume a constant stream of raw materials to build, repair, and maintain the integrity of its tissues. Every twenty-four hours, the pancreatic cells that secrete the highly corrosive digestive enzymes die and must be replaced. Essentially, the body must manufacture a new secretory pancreas every day. The cells lining the stomach last but three days, and most white blood cells live for only five days.* Homeostasis is an incredibly dynamic process.

Defending the integrity of the internal milieu is an organism's highest priority. The process called life is a ceaseless struggle against the dour second law of thermodynamics. Also known as the "principle of entropy," this adamantine rule states that all things in the universe tend to run down and decay. In our world, the arrow of time points in only one direction, and a system that begins with a high degree of order, such as a brand-new car, will, with the passage of time, pass to a condition of less order. New cars inevitably become clunkers, demonstrating the harsh truth of the second law.

Just as there are mechanisms that drive an animal to quench its thirst and satisfy hunger, there are those that drive a creature to correct more subtle imbalances. Iron loss will result in a desire for foodstuffs high in iron. When a battered Spartan warrior returned home after battle, friends and family greeted him with a goblet of red wine containing a rusting iron nail. Apparently they understood, as did many ancient peoples, the relationship between blood, iron, and vitality.

The most extreme example of the homeostatic mechanism controlling iron is the strange medical syndrome of "pica," the overwhelming craving to eat dirt. In many malnourished cultures, physicians have observed people, mostly children and women, eating handfuls of soil. The most plausible explanation for this anomaly is a diet deficient in the elements commonly found in soil, notably iron. It shows how important the body considers iron

---

*Fortunately, not every cell of these organs dies simultaneously. Their out-of-phase rotational death ensures that the organ keeps perking along without any discernible evidence of the massive destruction ongoing within it.

homeostasis that, in a desperate attempt to correct a deficiency, humans will resort to such a drastic measure.*

Though the components of the internal milieu are basically the same for men and women, several are more specific to one sex than the other. Women need significantly more iron than men do; men have incredibly high testosterone levels, which create intolerable sexual tensions demanding release. Although at first glance seemingly unrelated, the disequilibrium of these two features of human sexuality conveniently complement each other.

Many peoples across the globe have a distinct word in their vocabularies for "meat hunger."[3] (Other components of animal fats and proteins are beneficial to humans, but for the moment I wish to concentrate on the iron in meat.) A pregnant or lactating woman raising small children has evolved instincts that use male desire so that she can be assured of a reliable iron supply during her reproductive years. These instincts are activated at the onset of menarche, long before her health and the well-being of her offspring depend on that supply. Many other examples exist in nature of creatures, from ants to squirrels, that lay away supplies in good times to tide them through the lean ones.

This coincides with the period in a woman's life prior to motherhood, when she is the most sexually desirable to most males. And so we return to another evolutionary benefit of the adolescent girl's anovulatory cycle. The slow but steady blood loss caused by early menstrual cycles stimulates in her a desire to eat meat. She now has an incentive that did not exist before menarche to set about learning how to acquire iron-rich foods at the least risk to her and her future children's safety. This knowledge will serve her well when pregnancy, delivery, and lactation create enormous demands on her internal mineral reserves.

By the time a woman manifests an iron-deficiency anemia, her iron stores are already exhausted. Maintaining sufficient reserves is critical to keeping her vital organs, especially her brain, fueled with an unvarying oxygen supply. Individual iron atoms come and go to the warehouses in the liver and bone marrow far more often in a woman than in a man.

*   *   *

Though *Gyna sapiens* does not experience estrus in the nonhuman-primate sense of the word, a faint imprint of her once exuberant sexual heritage per-

---

*Ethologists postulate that chimpanzees fulfill their iron requirements by eating roots and tubers that they do not typically clean. The dirt clinging to these foodstuffs contains sufficient iron to prevent iron-deficiency anemia. Both gorillas and chimpanzees have been observed to eat handfuls of dirt, presumably for their mineral content. In the modern human quest for cleanliness, we have scrubbed away one of nature's subtler methods of replenishing the bone marrow with the critical minerals existing in soil.

sists. The symphony of hormones secreted during her monthly ovulatory cycles reaches a crescendo just prior to the moment of her ovulation, increasing her libido to its maximum. Other physiological changes serve to increase the likelihood that she will engage in sexual intercourse. Her body temperature rises by one degree, causing her to experience excessive warmth and remove more clothing, thus exposing more skin. In an attempt to reduce her body's temperature further, her skin becomes ever so slightly flushed.[4] These extremely subtle physiological changes tend to attract all males within visual range, even though they may not be consciously aware of what attracted them.

A woman becomes more restless during the time surrounding her ovulation. She travels farther and to more unfamiliar places. She is more willing to take risks, and to engage strange males in conversation.[5] In one study, ovulating females' taste in males tended more to the rugged, jut-jawed he-man; during the rest of their cycle, they gravitated toward softer, less masculine-looking men.[6] This change in preferences, however, can also be seen as a paradox, because it has been abundantly demonstrated in numerous studies that high testosterone levels are antithetical to parenting behavior: Macho males are less likely to remain faithful or help with children.[7] In another study, American women at midcycle entering a discothèque wore more jewelry and makeup and were more often touched by men than those at other points in their cycle.[8] These behaviors make good evolutionary sense and are echoes of the archaic hormonal tug of primate estrus.

But many women experience a second libidinous peak just before and during the height of their period. Penelope Shuttle and Peter Redgrove in their book *The Wise Wound* refer to this phase of a woman's cycle as the *paramenstrum*. This second lusty spike is more difficult to explain in terms of evolutionary theory, since sexual desire in other animals coincides with ovulation. In *no* other animal have females been observed to manifest a surge of sexual desire during a time of nonovulation. This feature of female human sexuality would most likely not have advanced unless it possessed an adaptive advantage. Why would a woman desire sex when she is least likely to conceive? What is the evolutionary payoff?

Therese Benedek, a Jungian analyst, teamed up with clinician Boris Rubinstein to study the differences between the dreams of women during these two antipodal spikes in monthly libido.[9] On the nights on either side of ovulation, women reported sexual dreams in which they felt open and receptive. Romantic feelings welled up, and, in general, the subjects commonly reported a sense of goodwill toward males. A few women prone to masochistic fantasies reported their increased incidence at midcycle.

In contrast, during the paramenstrum, the subjects described very different libidinal emotions. They felt more dominant and often initiated the sexual act. Many tended to regard men in a more negative light. Their dreams and fantasies, if not overtly sadistic, were more controlling, and were, in rare cases, tinged with cruelty.* Why would a woman be dangerous and antagonistic to a man, yet sexually aggressive at the same time? The answer, I believe, lies in her homeostatic response to her need for iron.

A major component of a woman's libido is the relative concentrations of several quite different hormones, one of which is testosterone, a molecule that induces aggressive and libidinous behavior. A woman produces this androgenic substance in her ovaries, adrenals, and fat cells. Her testosterone peaks at ovulation at midcycle, and its level is about the same as those of both estrogen and progesterone. The precipitous fall in estrogen and progesterone, however, during a woman's period allows the level of testosterone to *exceed* the levels of the other two hormones significantly for a few days. This changing balance serves a woman well, for she is about to lose a small but incrementally important amount of blood.

Approaching a man provocatively during her menses, a time when sexual intercourse is least appealing to a male, reverses for a few days each month the natural relationship between *Homo* and *Gyna sapiens*. Woman assumes the persona of the fierce hunter, and man the submissive gatherer.

The human male's sexual drive is—for most men, particularly young men—stuck in permanent overdrive. Mr. Ever-Ready is not known to be particularly discriminating in his constant quest to alleviate his lust. For a young man, the prospect of engaging in sex that is ragged with risk appeals to his incompletely myelinated left frontal lobe. For him, sex with a menstruating woman is dangerous.

In his limited experience, women so rarely play the role of sexual aggressor that when a woman does, he would be unlikely to turn her down. If she asked for something in return, such as a demonstration of his vaunted hunting skills, he would be eager to comply even if her request entailed considerable hazard. *The first libidinous peak in a woman's monthly cycle, coincident with ovulation, is for the purpose of conception; the second libidinous peak, coincident with menses, is to restore her iron homeostasis.*

Another aspect of the human female's monthly sexual cycle stands out when compared with those of other female mammals. In order for her uterus to shed its lining, a very dramatic sea-change must occur within her hor-

---

*In one study, headmistresses of schools meted out stricter punishments during their paramenstrum than during other phases in their cycle.[10]

monal system just prior to the onset of menses, and this ebb tide washes away in a relative instant.

Every major hormone has an effect on both body and the mind; sudden disruptions in the serum levels of any hormone produce profound physiological and psychic phenomena. Anyone who has taken steroidal drugs, such as cortisone, or who knows someone who has taken them, can attest to the profound changes in both appearance and personality that manifest. An examination of the physiological and emotional changes associated with a woman's period will provide a clearer picture of why human menses was such an extraordinary adaptation.

At the beginning of a new cycle, the interior uterine landscape is sere and appears dead. The remains of a few unshed necrotic cells lie heaped here and there on what has become a vast raw plain following the expulsion of the last menstrual effluvia. Then, with each passing day as ovulation approaches, new cells, like sprouts of fresh grass, begin to appear, forming a resurgent healthy nap. These cells steadily grow taller, forming stalks, and at their extremities, tendrils called "villae" begin to wave gently. What at first resembled new-grown prairie grass quickly becomes a luxuriant rain forest. An exquisitely sequenced orchestration of hormones stimulates this lush uterine growth. The levels of estrogen, the sex hormone, and progesterone, the pregnancy hormone, both ascend a steep curve as the egg's launch site on the ovary pumps them out in prodigious quantities (see the figure on page 61).

These, in turn, stimulate the growth of the uterine lining. Blood vessels, resembling the root systems of trees, corkscrew and pulse with energy. They grow fat and full as they transport increasing amounts of structural material for the uterine mucosal buildup, which has but one purpose—to be ready if conception occurs. Should a fertilized ovum wander in from either of the two narrow fallopian tubes, it will find itself in a dark wonderland, enveloped by a sumptuous jungle of villae. This nutrient-rich tangle is waiting with inviting tentacles outstretched, eager to embrace a new life.

But Mother Nature is no sentimentalist. Should conception fail to occur, She will survey the now unnecessary field of waving villae with the keen eye of a farmer about to torch a field that had failed to yield its expected harvest. Estrogen and progesterone levels begin to retreat—at first gently, but as the paramenstrum approaches, with increasing acceleration. Hormonal support for the thickened uterine lining withdraws. The once tall, mighty columnar cells that line the surface wilt. And then, just before the onset of menses, a most dramatic event occurs. As if the rug were being pulled out from under the uterine lining, hormonal levels crucial to the whole enterprise plummet precipitously.

This causes the root-system arteries to constrict in a death spasm. What had been gradually decaying now becomes, within minutes, a holocaust for all the cells lining the uterine surface. The uterus becomes a living shell encasing a dead inner core and serves momentarily as its tomb. Then the arteries that were in constriction suddenly go limp, and blood floods into the space directly under the lining. A hematoma, a lake of blood, forms in the space neatly separating the now dead uterine lining from the dense undergrowth of blood vessels. Resembling a dam that can no longer hold back the swollen river behind it, the necrotic uterine lining ruptures; blood and the remains of formerly lush uterine tissue cascade down toward the cervical opening.

Whenever any tissue of any organ dies and remains within the body, toxic substances released by the dead tissue threaten the existence of the entire organism. In other female mammals that do not menstruate, the uterine lining dies gradually in the absence of conception. Toxic substances are released at a rate that the host's defenses can easily handle. The uterus's dead lining is reabsorbed into the mammal's interior, to be broken down into its valuable constituents and recycled.

Too much of the human uterine lining dies too fast for this process to be a viable alternative for *Gyna sapiens*. The human female (and the handful of other mammals that menstruate) had to evolve an efficient way to rid her body of what will rapidly become a highly toxic mix of pathogenic bacteria and the poisonous byproducts of cellular death. The solution: The cervix, gatekeeper of the uterus, opens; the uterus contracts rhythmically in what are known as menstrual cramps; and the discarded lining, mixed with about four to eight tablespoons of blood, is expelled down toward the vagina. A woman's period has begun.*

*   *   *

Besides menses' physical manifestations, a potent psychological component accompanies it.† In her 1997 book, *Molecules of Emotion,* neuroscientist Candace Pert argued that what we experience as an "emotion" is, in fact, due to the rise or fall in serum concentrations of a particular polypeptide molecule or combination of molecules. Many of these complex molecules are neurotransmitters that carry or inhibit signals across synapses, the short gaps between neurons—serotonin, norepinephrine, and dopamine being among the more familiar neurotransmitters. Those that carry their messages over longer distances are called hormones.

---

*In 1940, J. E. Markee, in an ingenious study, discovered the phases of menses by transplanting a wedge of a monkey's uterine lining into the anterior chamber of the monkey's eye. There he was able to observe the visible effects of a female's cycle on this extraordinarily volatile tissue.[11]

†We can only surmise about the psychic effects of menstruation in other primates. However, from their observable behavior, the few nonhuman primates that menstruate appear to be unaffected psychologically.

The word "hormone" means "messenger" in Greek. Released into the bloodstream at one site, a hormone carries information to a distant tissue, causing the target tissue to change its state. For example, the hormone thyroxin, produced by the thyroid gland in the neck, circulates throughout the body and regulates the rate of metabolism for nearly every tissue. Too little thyroxin, and cellular functions gradually slow. Fatigue, weight gain, slow speech, and inability to burn glucose properly with insulin produce the clinical symptoms of hypothyroidism. Too much thyroxin, and eyes bug out, body temperature heats up, and the affected person is so agitated he or she can't sit still.

In the seventeenth century, René Descartes posed a conundrum that was to intrigue generations of philosophers. How, he asked, does matter affect mind, and how does mind affect matter? Where is the interface between ephemeral mind and solid matter that allows thinking ultimately to change matter? Conversely, by what mechanism does matter, impinging on consciousness, change thinking? Dr. Pert, building on the earlier work of Walter Cannon and Hans Selye, answers these questions by bridging the gap between hormones (matter) and emotions (mind), claiming that the one is indistinguishable from the other. A caveat: Any particular individual may manifest a variable response to different hormone levels.

Estrogen, testosterone, and progesterone profoundly affect mood, especially sexual mood. Men and women treated for medical conditions that require the administration of one or the other of these three potent hormones routinely report psychic fluctuations, increased libidinous thoughts, and vivid sexual dreams. If we accept Pert's premise that, in one sense, emotions *are* hormones, what, then, is the effect on a woman's mental disposition of the sudden withdrawal of estrogen and progesterone that occurs premenstrually? When estrogen and progesterone levels perform their monthly sky dive, a woman's testosterone level rises dramatically, relatively speaking, even though its serum concentration barely changes.*

Relative testosterone rise is the proximate reason some women experience dramatic mood swings, a sharp rise in libido, and increased aggression during their periods. The ultimate evolutionary reason for this protean display is to entice (or coerce) the male to restock *Gyna sapiens'* soon-to-be-tapped iron reserves.†

A few women become combative, argumentative, and, in some extreme

---

*At ovulation, estradiol (estrogen) levels average 300 units and testosterone is at 350 units. During menses, estradiol plummets to 25 units, compared with a minimal decline in testosterone to 275 units.
†Evolutionists make a distinction between a "proximate cause" and an "ultimate cause." The proximate cause affects the organism in a very direct way; the ultimate cause may be the slight selective advantage the trait conferred on the fitness of the individual within a species.

cases, dangerous at this time in their cycle. Studies reveal that women who have committed violent crimes are statistically more likely to have done the deed during the paramenstrum. In some legal codes, a crime committed by a paramenstrual female is considered to have an extenuating circumstance.[12] Examining at autopsy the ovaries of women who had died by their own hand, pathologists ascertained that slightly over half of these suicides occurred during the paramenstrum.[13]

A man paired with a woman who experiences paramenstrual mood changes soon learns that for a few days of every month he would be prudent to hunker down and batten the hatches. As her testosterone surges (relatively speaking), his levels will drop. A man learns (and if he is young and doesn't know, he will learn soon enough) that this is the monthly moment when it would be best for him to adopt a conciliatory posture known among men as the "Jell-O Defense."

Even if he were not motivated to leave his cozy rock shelter, the behavioral manifestations of the paramenstrum would, in some cases, give him an excellent reason to disappear for a few days. "Uh, you know, hon, I think I'll just gather up my spears, and me and the boys will go see if we can get us some deer, dear." If he were to calculate it precisely, he would return just before her ovulation, when his mate was in a more loving and receptive mood. If he returned lugging a huge haunch of venison, his mate's homeostatic mechanism, which urged her to replenish her iron depots by any means available, would have been well served.

*   *   *

Although the male of the human species doesn't teeter-totter through monthly cycles, his persistently high levels of circulating testosterone constantly roil the waters of his internal milieu, particularly in adolescence. Since there is little opportunity, especially in his early years of manhood, for the sexual release he craves, Natural Selection provided him with a surrogate safety valve—masturbation. Human males masturbate to ejaculation more often, and more intensely, than any other animal.* I shall propose that the male's masturbatory proclivity is complementary to the female's anovulatory cycles. Both appear at first glance to be counterevolutionary, sham sexual activities because they never result in offspring.

Besides the futile loss of considerable quantities of perfectly usable protein-

---

*Some male animals in captivity—for example, lions and monkeys—frequently masturbate, but it is a practice rarely observed among most animals in the wild. Observers have recorded that some species of primates, such as male baboons, do masturbate in the wild. But none approach the frequency of a young human male.

rich fluid and sperm, masturbation during ancestral times would likely have been very dangerous. Most animals eat privately and engage in sex publicly. Humans, in contrast, eat publicly and engage in sex privately. It would be fair to assume that young ancestral adolescents whose urges demanded a lowering of their unbearable sexual tension would have sought privacy, too. This meant leaving the safety of an encampment and wandering off to be alone. Lost in the focused attention of the act, the onanist's normally alert defenses would be markedly diminished, making him a tasty target for any number of predators. There could have been no more humiliating way for a hunter to die than to be devoured by a beast while caught in the act of masturbating. What would have been the evolutionary advantage of having the male engage in this prodigious sexual charade?

Many researchers posit that masturbation is necessary because a male must constantly refresh his supply of sperm in case he gets lucky and finds a receptive female. His sperm, upon entering the Great Hall, might encounter another male's sperm that had reached the much-desired anteroom first. Since to the fastest and feistiest go the spoils in the race to the ovum, this theory suggests that masturbation serves to maintain a fresh supply of vigorous sperm capable of outswimming the competition—a theory recently bolstered by the discovery that many sperm (called "kamikaze sperm") function to deter a competitor's swimmers from reaching the prize.[14] By sacrificing themselves so others can outpace the competition, they selflessly forfeit the chance to be the One.

Each month, a woman releases one egg. During that period, a *Homo sapiens* will produce approximately 3.6 billion sperm. This number is enough to impregnate every single woman of reproductive age alive at present. Recent research has revealed that the ovum does not just sit passively by while all this commotion is going on around her but rather plays an important role in determining which one among her busy crowd of suitors she will permit to merge into her interior.

The Ever-Ready Sperm Theory of masturbation arose from observations of chimpanzee mating. Female chimps in estrus actively solicit males, who patiently queue up, waiting their turn to copulate. Often, the female will mate with a wide assortment of males in a very short time. Lower-ranking males have a chance on both sides of her ovulation, but it is the alpha male who most often occupies her time at the optimal moment.[15] Many males copulate, but it is primarily the dominant who propagate.

Field observers clocked one female chimp having sex with eight different males within the space of an hour. Another female holds the record of fifty copulations in one day.[16] A female chimp mates on average 138 times with thirteen different males for every infant she births.[17] Male chimpanzees have considerably larger testicles relative to their size than do humans. Making great batches of fresh sperm hurriedly is a critical factor in the intense competition.

This commonly accepted account for *Homo sapiens'* prodigious propensity to masturbate, however, cannot be the entire story. The fresh-sperm-at-the-ready hypothesis would predict that male chimps would spend a large part of their day masturbating, but they do not. Human males masturbate the most when their prospects of securing sexual intercourse are the slimmest. Even if they convince an adolescent girl to have sex, it's unlikely that any of their sperm would need to fend off another male's sperm. Further, if an adolescent's sperm were to swim upstream all the way to the fallopian tubes, the odds are against its encountering a viable ovum, given the frequency of young females' anovulatory cycles.

If the selective pressure for frequent masturbation was for the purpose of enhancing sperm competition with other males, then the practice could be expected to be more prevalent in the years when men are *really* competing for women. Instead, masturbation peaks in the years when access to females is the least likely, and the practice is in decline during the years when male competition for females is at its maximum.

The extraordinary masturbatory frequency of young human males has little to do with sperm competition but much to do with extremely high testosterone levels. It serves the species well to have its young males in a constant state of homeostatic disequilibrium, doggedly trying to deduce what they must do to obtain "the real deal." The male experiences chronic stress because he cannot satisfy his sexual urges. Masturbation alleviates some, but not all, of that stress.

In the vulgar wordplay of the locker room, young men refer to their penises as their "meat." A common remark made between high-school boys on a Friday afternoon, usually delivered with a punch to the shoulder and a knowing smirk, is, "Have a good weekend, and remember, don't let your 'meat' loaf." There are many other slang expressions that link a male's penis, a woman, and meat. "The meat rack," the "meat market," and other terms referring to human mating frequently appear in the vernacular. One might wonder why many a male has conflated in his mind the object of a hunt, his member, a woman, and what he will offer her to entice her to say *Yes!*

*   *   *

Women acquired an important bargaining chip when they gained partial control over their sexual urges and began to demand that sexual intercourse take place on their terms. Unfortunately, their advantage came with a catch.

Women can refuse sex, but they must have iron. Men do not need iron, but they do crave sex. Herein lies the evolutionary secret behind the sudden, dramatic appearance of the bold *Homo sapiens* hunter.

One major impediment, however, remained to bar the road to success for this redoubtable primate adventurer. The basic primate design is an awkward starting point for building an efficient killing machine. Compared with predators like hawks, lions, jaguars, and crocodiles, *Homo sapiens* was not a very sleek, swift, or imposing beast. Moreover, he lacked the most basic killer accouterments. Mother Nature had to jury-rig the body plan of a vegetarian, converting it in a jiffy into a carnivore.

This, then, was the problem. *Gyna sapiens* had dispatched her man to "bring home the bacon," but *Homo sapiens* was woefully ill-equipped for the task. To ensure that he would triumph, she would have to come to his rescue. It is at this important juncture that *Gyna sapiens* became the heroine in our evolutionary story. She accepted a series of interlocking adaptations that, taken as a whole, represented a nettlesome burden. She had to run this evolutionary gauntlet in order to acquire an incredible gift. And no sooner did she acquire it than she bestowed it on her sons and lovers. *Homo sapiens* would transform the astounding bounty that *Gyna sapiens* gave him into the deadliest weapon ever hefted. Her boon allowed him to become the most fearsome, efficient predator the planet had ever known.

Sex and
Time

Part III

The similarity between the periodicity of the moon and the length of human menstrual cycles, in combination with the correlation of the length of human pregnancy as a precise multiple of lunar cycles, seems to be too rich to be mere co-incidence.

Chapter 13

Moon/Menses

A good model of human origins should provide a number of hypotheses about our ancestors' behavior and anatomy that interweave in a sensible way. That is, the model must be both internally and externally consistent. It must explain the origins of those traits that are uniquely human above all else, since other traits that we share with the great apes are likely primitive ones that we possess simply due to a common ancestry.

—Craig Stanford[1]

Egyptian priests venerate the moon as the Mother of the Universe, having the light that makes plants and animals moist and pregnant.

—Plutarch

But the menstruant, having the most direct connection with the lunar cycle, would surely have been the first to know; she had motive, method, and opportunity to be the originator of lunar notation.

—Judy Grahn[2]

I magine a group of goggle-eyed, newly arrived intergalactic ethologists sitting in a concealed blind somewhere in the Great Rift Valley in Pleistocene Africa. Their assignment: Observe and record the habits of the diverse animals roaming around all about them. At the moment, all the ethologists have trained their binoculars on the edge of a clearing. One of them has spotted a band of stealthy *Homo sapiens* huddling low, trying not to attract the attention of unwanted carnivores. The odd-looking predators appear to be stalking a large herd of wildebeest.

The visitors could not be faulted for bursting out laughing at these puny, presumptuous big-game hunters. The alien observers could clearly see that the arsenal with which each *Homo sapiens* was naturally endowed—speed, strength, and killing appendages—was ludicrously inadequate to his mission.

But the ethologists' laughter soon fades, and they begin exchanging quizzical looks. These resolute hunters exhibit surprising pluck. The naked bipeds obviously have a trick up their metaphorical sleeve.

The visitors could not have known that bolstering the confidence of these early *Homo sapiens* was an invisible weapon so deadly that it would make the vaunted predations of *Tyrannosaurus rex* pale in comparison. The new armament: a consciously constructed, expansive sense of time hereafter referred to as "deeptime." A human hunter's ability to learn from the past and mentally pole-vault himself into the future allowed him to sharpen his sense of deeptime into a formidable piercing stiletto called "foresight." From the moment he hafted this deadly lance-tip to the shaft of his crude spear, a collective shudder rippled through the animal kingdom. No flora or fauna inhabiting the forest, savanna, or sea would ever be safe again. Learning to fear and avoid him became a life-and-death imperative for all the other animals whenever he appeared in their neighborhood.

Only human hunters could predict that elephants would migrate through a specific pass two months hence. Only *Homo sapiens* could methodically prepare an ambush to attack an animal that none of them had yet seen, heard, or smelled. Some creatures, ranging from spiders to beavers, also exhibit similar types of behavior, but they, with very rare exceptions, depend on instinct, whereas human actions are almost always based on deliberate plans rooted in experience.*

But here was the strange part. *Homo sapiens* was the first predator who did not kill primarily to feed himself. He risked life and limb for *Gyna sapiens*—first for sex, then for love, and still later for intimacy. And then he killed for his children—first because he was afraid to die, and then because he loved them for who they were. But that's getting ahead of the story. Before he usurped the crown of King of the Beasts, and before these changes in his psychic programming fully matured, *Gyna sapiens* had to rush to assist him.

To obtain the iron she needed to maintain her vigor and increase the intelligence of her children, *Gyna sapiens* had sent an ill-prepared *Homo sapiens* on an errand fraught with danger. It would be up to her to armor him with a wonder weapon more potent than slashing claws, more destructive than snap-crackle jaws, and far more powerful than leonine strength. What *Homo sapiens* failed to appreciate then, and has refused to acknowledge since, is that his fearsome hunting skill was a gift from *Gyna sapiens*. He never

---

*At the Kruger National Park in Africa, game wardens installed a chain-link fence at one boundary. Lions at the reserve learned, over time, that wildebeest kills would be easier if they worked in concert to drive the prey against the fence. Other examples in nature also indicate that chimps and a few higher carnivores are capable of limited foresight. The human facility with foresight is distinguished from other animals in both its extent and degree.[3]

thanked her for the considerable sacrifice she made so that her species could acquire its precious expanded sense of time. As has happened so often throughout our history, *Homo sapiens* would later brag that foresight was entirely his idea.*

Gyna sapiens, however, has nursed little outrage regarding this slight. The radical reprogramming of her reproductive cycle was the key that led her to an increased sense of time. Once she made the leap into the future, she would teach this invaluable lesson to her men, so that they had a better chance of procuring for her what she found difficult to obtain. Her gift, for which she paid dearly, returned to her and her offspring incalculable dividends in the long run. Both sexes benefited, and the knowledge of deeptime made the difference. As the novelist Samuel Butler once remarked, "A hen is an egg's clever strategy to create another egg."

Leaping out of the zoological record is the inexplicable fact that, among the myriad species that have inhabited the earth, we are the only one that has ever acquired a sense of deeptime. Why? If in the contest for survival the acquisition of foresight by humans became Nature's most effective weapon of mass destruction, why didn't this trait erupt all across the mammalian spectrum? Where are the horse seers? Why can't one make a fortnight appointment with a dog? Why don't lions plan what they will do next Tuesday? No other creatures wear watches because, for even a single animal species, to learn how to tell time was an exceptional evolutionary development.

The trajectory of evolution appears to travel from the simple to the complex. The arrival of each new phylum introduced creatures sporting a more intricate brain, eye, or mode of movement than the earlier versions. Yet, looked at in another way, the entire history of life on this planet could be conceived as a striving by life-forms to attain an ever-greater appreciation of the vectors of space and time.

It must be a given that, to appreciate a dimension of either space or time, an organism must have the requisite nervous apparatus with which to recognize it. Plants, among evolution's earliest experiments, do not possess a nervous system, so it would be fair to argue that plants are not sentient and do not appreciate their extension in space or time.† They could be said

---

*Considerable circumstantial evidence for this misappropriation of the credit for foresight can be found in ancients myths. See the pages 378–80 for a more thorough discussion of the issue.[4]

†Sentience is a very difficult concept to define! Lynn Margulis, Elizabet Sahtouris, James Lovelock, and others have argued that simpler life-forms have a form of consciousness beyond the comprehension of humans. For the sake of the narrative flow I will restrict the term to its narrower classical sense.

to represent the "point" of Euclidean space—a point having no extensions in space.

Simple one-celled animalcules move toward nutrient stimuli and away from noxious ones. Although they live in three-dimensional worlds, it is as if they lived their existence on a clothesline. A paramecium or an amoeba resembles the line of Euclidean geometry, representing only the vector of length.

Planaria, belonging to the phylum of flatworms, is the first creature manifesting a bulbous bifurcated area in the forward part of its primitive nervous system. The presence of this exaggerated right-left symmetry most likely signifies that a planaria is aware that it moves from side to side as well as to and fro. Its existence can be likened to the square, a geometrical figure possessed of height and length but no depth.

With the evolution of vertebrates, the first cerebellum appeared. And this component of increasingly complex nervous systems provided its owner with an awareness that it existed in a three-dimensional world. Any animal with a cerebellum "knows" that space extends in the three directions of height, length, and depth. Early vertebrates can be thought of as resembling the Euclidean solid form, the sphere.

What is missing from the existence of these life-forms is any appreciation that they exist in any moment of time other than the present. Frogs, fish, and geckos cannot easily hold past events in their memory and have virtually no ability to predict the future. They are condemned to live out their existence trapped in the thin slice of the ever-present *now.*

With the arrival of the first mammals, a new element entered the animal world. Mammalian nervous systems contained a specialized receptacle called a memory capable of recalling past events that no longer existed.* For the first time, a living form could re-create what *had* happened *within its mind.* The heavy hand of brute instinct began to lift. The arrival of a primitive memory allowed the first mammals to learn from past mistakes and react to their environment in a flexible way that depended on their remembering who their friends and enemies were.

The addition of the duration of the past into the mental functioning of animals paved the way for mind to enter the world. An animal that can remember the past is "mindful." Lizards may be alert, but they do not possess the qualia of mindfulness.

With the evolution of animals that had the biggest and most complicated brains, such as elephants, dolphins, and higher primates, a new concept

---

*A rudimentary memory exists in extremely simple organisms, such as sea slugs. Not until the emergence of mammals, however, did memory evolve to become a sophisticated factor in the contest for survival.

known as self-awareness appeared in the world. Self-awareness means that an animal is aware it is aware; and that it has a distinct sense of self. A monkey placed in front of a mirror never grasps that the monkey it sees on the other side of the glass is none other than *me*, because it doesn't possess a sense of "self." A chimpanzee, however, will, after a time, realize that its mirror image is actually a mere representation of itself and will begin to make faces and examine parts of its body.*

Primates can use the past to predict what may happen in the future, but their ability to "plan"—that is, to choose a course of action in a duration of time that hasn't happened yet—is extremely limited. Humans, on the other hand, are the only animal that knows that the sun will extinguish in ten billion years. Elephants may never forget, but no other creature is as easily capable of making forays into the dimension of the future as a human. We *know* that we exist in a reality bounded by the three vectors of space (length, height, and depth) and the three durations of time (past, present, and future).

Our solitary occupation of the land of the future is due to the formidable obstacles Natural Selection would have had to overcome before a life-form could appreciate time in all its fullness. To have created even one deeptime animal was a near miracle. To comprehend why we are the only animal to have crossed this critical threshold, let me conjure a fantasy. Disclaimer: I do not intend to impute purposeful intent to the natural processes of evolution but, rather, will use the artifice of poetic license to simplify and enliven the story of how humans became astute at telling time.

Let us imagine that, on one fine morning somewhere around five million years ago, Mother Nature stretched, yawned, and began Her morning perambulations. Feeling in a particularly artistic mood, She padded about drinking Her morning coffee. Suddenly, a novel notion struck Her. She wondered whether She could create a creature that could anticipate the future.

Ever since She introduced mammals, She had been occupied primarily with perfecting a more specialized paw or a tougher hide. Having discarded the ponderous dinosaurs as too unwieldy, She had introduced a whole new class of birds and mammals, about which She was quite proud. Still, despite Her obvious successes, She itched to pull off one more feat of evolutionary virtuosity.

She began to toy with an outrageous idea—a new "software" program vastly increasing a mammal's memory capacity, along with new neurocircuits that could recall past experiences and use them to solve present problems by anticipating events that might transpire in the future. She sat down

---

*The mirror test is only one criterion cognitive neuroscientists use to define the slippery concept of self-awareness.

at her drawing board and began to doodle on a scratch pad the minimum re-quirements in order to begin. Let us put ourselves in Her place and imagine that we are Mother Nature in Her workshop. Which animal would you choose to begin Her daring experiment? What intermediate steps would you envision necessary to execute such an audacious plan?

The first stumbling block would be the necessity to build an outlandishly massive brain, and brains, you were aware, are exceedingly expensive organs to maintain. Also, a complex reorganization of the brain would have to occur while the putative creature you were trying to design continued to compete with the others—an extremely delicate and dangerous operation. Reconfigur-ing the brain would require setting aside existing areas upon which to exper-iment, and older areas would have to be hastily rewired to perform double duty.

Then there would remain the seemingly insurmountable problem of coaxing your new changeling into leaving the safety and familiarity of the present to venture forth and explore purely mental locales that did not really exist, except in an ephemeral past and an even more tentative future. No other animal had ever been similarly rewired, because the job had been too complex.

You are convinced, however, that a creature capable of planning would swiftly gain dominion over all the others. So you press ahead. Let us assume you begin your survey in the continent of Africa and its surrounding seas. You scan the waters, rain forests, savannas, and mountainsides on the look-out for a suitable animal with which to begin. Since rapidly enlarging a brain will be the most fundamental aspect of your plan, you would prefer an animal already equipped with a large brain. You quickly narrow your choice to the four most intelligent mammals: dolphins, whales, elephants, and primates. You are tempted to use one of the first two, but eliminate them because they lack an indispensable appendage.

Imagine a fishing boat out at sea, a tackle box perched precariously on its transom. The boat pitches, the box falls overboard, and, weighed down by an assortment of lead sinkers, it descends slowly toward the bottom. The falling object attracts the attention of a pod of nearby dolphins (or whales) that cir-cle it excitedly. Possessed of a sophisticated means of communication, the dolphins whistle and click back and forth. Able to understand dolphinese, you eavesdrop on their speculations as to what this strange object in their midst is all about. After the container settles on the sea floor, they inspect it more carefully, nudging it repeatedly with their bottle-nosed snouts. But no matter how curious they are or how clever might be their guesses, dolphins and whales can *never* discover the nature of the tackle box's contents. With flippers for forelimbs, they lack the means to open the box.

After consideration, you also eliminate elephants. They are very smart and possess capacious memories. Their six-foot-long trunks contain over one hundred thousand muscles, enabling them to perform intricate tasks.

You reject pachyderms because you favor the animal you have endowed with an even more versatile appendage than the elephant's trunk: Primates possess two hands, and each one sports an opposable thumb accompanied by four delicate fingers that allows them to grasp and pick objects apart effectively. You surmise that the constant feedback between hand and brain, and brain and hand, will greatly facilitate your task of reconfiguring the brain's circuits.

Once having settled on the order of primates, you begin gradually to increase the size of the brain of one line that someone else will later name "hominids." At first you inflate the brain of these primates relatively slowly, but as you grow more confident, you rapidly enlarge it over a short period in one species of the hominid family. At the same time, you have encouraged this one primate to walk upright so that its hands are free for other uses. You lavish your attention over the one that walks proudly upright. About 150,000 years ago, you zero in on one small band of these big-brained bipedal primates who are living along the shores of Lake Victoria in Eastern Africa.

Now you begin to plan your experiment in earnest, anticipating the problems to come. As you rapidly increase the size of your chosen primate's brain, you must also identify a periodic feature of the primate's life cycle and greatly exaggerate it. This modification will serve to attract the hominid's attention to its recurring physiological clock. You must make the event dramatic enough so that the primate could not possibly fail to notice it—only something terrifying, yet magical, will do.

Then, somewhere in the landscape, you must identify an event that repeats at precise regular intervals. The external event's periodicity would have to be spaced just right. If the episodes are too close together, it would be useless to teach an animal the majestic secret of time; too far apart, and it would tax your experimental subject's newly evolving memory. The final steps of your experiment will be most challenging: entraining the inner, physiological timer you created with the outer, naturally occurring clock you selected. All the while, you must energetically twiddle the dials on the immensely complicated new neurocircuitry you had to install hastily in the brain of this creature.

In the past when you created something new, you had cleverly grafted it onto something old. For example, rather than make a bird's wing from scratch, you started with a reptile's forelimb and modified it. You planned to construct deeptime upon the foundation of an earlier time-related instinct of primates. Circadian rhythms are involuntary timepieces ticking away deep in the neuro-innards of virtually all complex species. The majority of these timers regulate breeding.

For instance, once a year, gray whales swim nearly half the circumference of the earth to congregate and breed in one small lagoon in the Sea of Cortés. Millions of delicate monarch butterflies cover thousands of miles, and then camouflage a preferred grove of trees in northern Mexico for the same purpose. Leks, rookeries, and otherwise isolated, barren, windswept beaches periodically teem with life, lust, and conflict driven by circadian rhythms' uncompromising commands. Carefully timed to maximize a species' survival, the congruent peaking of male rut and female lust at the moment of the female's ovulation ensures that conception has the best chance to occur. Depending on the species' average length of gestation, these seasonal corybantic convocations occur so that subsequent births arrive in the month optimal for mothers to nurture their young—usually, when the bounties of summer allow for easy pickings.

Unfortunately, you observe with consternation that the primate you selected does not come equipped with a circadian rhythm that you deem suitable for your intricate experiment. Primates never migrate vast distances to mate, nor do they have preferred breeding grounds conducive to mass assemblies. Instead of a sexual cycle coordinated with the planetary seasons, as is the case with many birds and large mammals, female primates' estrus cycles vary considerably from species to species, and rarely do the females instinctively coordinate their estrus.

Plagued with self-doubts, you reconsider your choice of the bipedal primate and resurvey the entire primate order. You briefly consider choosing the ring-tailed lemur of Madagascar because the females all ovulate in synchrony. But you immediately reject the idea, forced to admit to yourself that the lemurs' lack of intelligence, stemming from their relatively small brain, would be a disadvantageous starting point for you to try to achieve your ultimate goal of a really smart, time-aware primate.

You note that, each day, primates awaken, stretch, yawn, and begin their daily search for food. Intermittently, they eat, rest, nap, groom, fight, play, search for more food, and then sleep again. It would appear, observing this pattern, that little else occupies their brain—except when they periodically engage in breeding behaviors. In the majority of species, courting and mating are punctuated highlights that break the monotony of an animal's otherwise repetitive daily grind.

Meticulously, you inventory the primates' other circadian rhythms. Primates sleep at night and are active during the day. Their metabolic needs vary slightly during this diurnal cycle, causing their temperature to fall one degree in the middle of the night and rise one degree in the late afternoon. Nocturnal animals (among which are a few species of primates) have the reverse circadian rhythms. *Homo sapiens'* testosterone levels rise in the morning and dip in the evening. His levels are higher in the fall than in the spring. None of these slight variations translate into obvious observable changes in behavior. You pass over each of these primate circadian rhythms because they are too subtle to be of much use for teaching an animal about deeptime.

Running out of options, you repeatedly return to the idea of using a primate's striking estrus cycle as the key internal timer, and you seek a way around the problem of its staggered estrus. During your initial survey, you had noted that a few primates featured a unique component of estrus missing from the majority of mammals and absent from all other sexually reproducing species. Your interest piqued, you eventually focus on the small cohort of thirty-one primate species that menstruate. "Hmmm," you muse, "menses is a reliable circadian rhythm—a most unusual clock, but a clock nonetheless."

The germ of an idea begins to take form within your mind. You note that menses marks the exact counterpoint of these few primates' estrus cycles with an event—external bleeding—that just so happens to be both dramatic and episodic. The interval of approximately a month is also appealing to you, for its length fits in with your ambitious plan. The problem: This subset of female primates experiences so minor a menses (some barely bleed) that they hardly notice the event.

In a thunderbolt of inspiration, you hit upon the solution for how to teach this particular animal the meaning of deeptime: *Abandon the idea of using estrus as the marker and substitute menses as the key element instead.* Like Mary Shelley's Dr. Frankenstein, you lock the door to your laboratory, roll up your sleeves, and begin to fashion a creature the likes of which no one—on earth or above it—has ever seen.

The Russian physiologist Ivan Pavlov performed the classic experiment psychologists have since named "Pavlovian conditioning." Pavlov trained a dog to salivate at the sound of a ringing bell. Normally, a dog begins to salivate at the sight and smell of his owner bringing him food. Pavlov structured his experiments so that he consistently rang a bell whenever he fed the dog. Even though the sound of a bell had nothing to do with eating, the dog's brain slowly built new neuronal connections that caused it to link ringing with the arrival of its dog dish. This conditioning required that Pavlov repeat the drill over and over again. Eventually, however, his dog learned to associate the arrival of food with the sound of the bell. One day, Pavlov clanged the bell's clapper but did not deliver the food. The dog salivated, even though no food was forthcoming. Until the dog learned this new, disappointing lesson, it would continue to salivate at the ringing of the bell.

You decide to use a similar form of Pavlovian conditioning as the primary teaching tool to instruct *Gyna sapiens* how to recognize deeptime. Instead of a bell and food, you begin redesigning the estral and menstrual cycles of this singular bipedal primate. You set to work tinkering with three reproductive features in six interrelated steps to pull off what you are sure will be your most sensational feat of evolutionary showmanship.

Step One: You begin by synchronizing each *Gyna sapiens'* estrus and menses with those of the other females whenever they congregate in tight communities over extended periods of time. This alignment means that both their signs of estrus and the subsequent sloughing of their uterine linings occur in unison. (Although we cannot know which of the six steps to be outlined occurred first—most likely they occurred nearly simultaneously—let us arbitrarily designate the harmonizing of women's cycles as Step One in Mother Nature's radical makeover.) Over thousands and thousands of years, the coincidence of women in close proximity bleeding together would begin to impinge upon the awareness of a few of the more intelligent and observant *Gyna sapiens* of the tribe.

Step Two: You begin to suppress all external signals of ovulation. You had briefly considered using these well-demarcated intervals as the key learning marker, but quickly discarded the idea. With hormones raging, animals in heat can think of nothing but sex. The mating instinct is so powerful that it would interfere with any attempt to teach an animal a new trick. Dog trainers are reluctant to introduce a new routine to a female in heat when nearby males are urgently barking. Conversely, the scent of an ovulatory female distracts a male dog, making him mad with desire. You reach the conclusion that you must select a less discombobulating periodic event than estrus.

Gradually, you suppress the smells, sights, calls, and gestures associated with ovulation. Your final goal is to make ovulation so opaque that even the female in which it is occurring would have difficulty recognizing it. Finally, external ovulatory signals that mark a most awesome moment in the life of virtually all other female animals disappear entirely. By eliminating a significant diversion from the newly emerging consciousness of *Gyna sapiens*, you refocus the female's attention on the one remaining periodic event in her life—menses.

Step Three: In a stroke of genius, you sharply escalate the experimental primate's menstrual symptoms. During the time that ovulatory signals were fading, *Gyna sapiens'* menstrual bleeding became more copious, more odiferous, and an increasing nuisance. During her period, she experienced psychic and physical changes, many of which she came to dread, such as migraines, depression, anxiety, irritability, cramping, bloating, metallic tastes, and tender breasts.

Gradually, over numerous generations, you caused the problems surrounding menses to mount. The physical, psychological, and physiological events associated with menses increasingly intruded on *Gyna sapiens'* attention. Finally, they reached a tipping point at which the females could no longer continue to ignore them. Momentous changes had overtaken the hominid female destined to become the ancestor of modern women. She began cycling in harmony with other women, she lost estrus signaling, and she gained a bothersome monthly period. But all the elements of your complicated plan were not yet in place. Three more steps would be required.

Step Four: Intently, you searched for a recurrent event in nature that marked off an appropriate interval with which you could begin to entrain your experimental subject's menstrual cycle. You knew you could not start your Pavlovian conditioning experiment until the rapidly evolving sections of *Gyna sapiens'* brain, grappling with the novel concept of linear time, had two events, one internal and one external, to guide it. Another requirement: The recurrent event had to possess some sort of a spectacular indicator. The choices were limited: a day, a month, a season, or a year. A day was too brief, seasons too variable and their transitions too subtle, and a year was too long. The 29.5-day interval between two new moons, however, was just right.

Your original decision to use a primate for your evolutionary experiment now proved extremely fortunate, for a primate is among the few animals that can actually *see* the moon. Excepting birds and several other mammals, the majority have poor eyesight. Many have an eye positioned on either side of the head. At any given moment, they are looking at two different scenes. If one eye could see the moon, the other would be peering at the ground. Getting an animal to pay attention to lunar phases, you realize, will be much easier if both eyes happen to be looking at it.

The vast majority of animals habitually align their vertebral column parallel to the earth.* Few ever direct their gaze into the sky, preferring instead to keep their noses pointed at the ground, since for the vast majority of large animals the sky is boringly uninteresting. No food comes from that direction; neither do threats. Many live under cover of an impenetrable forest canopy and could not track the moon even if they tried. Animals that live in the open during daylight often seek dense brush at night to hide from predators.

---

*Some animals—giraffes, ostriches, and flamingos, for example—appear to be upright, but none has a vertebral column that is actually vertical throughout its length. Some animals, such as meercats and chimpanzees, occasionally adopt a vertical stance, but it is not their normal posture. The tiny seahorse, and penguins when they are not swimming, are examples of the few other perpendicular vertebrates.

Predators that hunt out in the open during the day often spend the night pursuing prey in dense foliage.

The moon is 248,000 miles distant. Vision is a primate's supreme sense. The function of seeing co-opts one-third of all the neurons in a monkey's cerebral cortex. All primates are blessed with two of the most forward-facing eyes in animalkind. Both eyes look at the same scene, but view it from an ever-so-slightly different angle. This quirk of the primate's visual apparatus allows it to knit these two overlapping scenes magically together.

Animals with forward-facing eyes converge the information garnered from their two eyes and use it to construct the coordinate of depth somewhere in their brains. Close one eye, and look at the world, and it appears flat. Open the second, and instantly the landscape transforms from a flat, two-dimensional world into a rich, three-dimensional one. Stereoscopic vision, a primate's forte, along with other visual clues, allows it to see very distant objects and instantly estimate their spatial relationship to nearer objects with a degree of precision unsurpassed by any other animal, including eagles.*

Another feature solidly connects an upward-glancing bipedal primate to the moon. A magical moment occurs in every child's life when he or she realizes that the moon is the child's personal companion! As we move through the nighttime landscape, approaching objects glide by and then recede into the distance behind us. Not so the distant moon, which always keeps pace right alongside us. We stop, and the moon stops. We move, and the moon moves. The darkness of night accentuates this observation, because the moon is usually the brightest object. During the day, the same phenomenon applies to the sun, but because there is so much more to see in daylight we rarely if ever notice it. Moreover, looking directly into the sun is extremely blinding and hazardous.

There is something vaguely comforting, especially to a small child who has a natural fear of the dark, in knowing that the moon is a reliable and faithful companion that will not only light the child's way but also be a steadfast companion during nighttime excursions. This simple but gripping observation establishes an early personal lunar connection that predisposes many people to feel kindly toward the moon throughout the rest of their lives.

When night falls, there is no other object in the visual landscape so distinctive as the moon. The appearance and disappearance of this silvery orb passing overhead has always enchanted humans. She leads a stately procession of her retinue of stars and planets in a celestial pageant. To prevent monotony, the moonshow changes its routine nightly and repeats its run every month. Before the invention of electric lights, the moon's performance was a most closely

---

*The nearest competitor to a primate in this attribute is an owl.

watched show that held ancestral people enthralled. To a significant number of the population today, the moon continues to exert a mystical attraction.

Step Five: Having assembled all the key components, you begin the penultimate, crucial maneuver. You must coordinate *Gyna sapiens'* menstrual cycle so that its average length is *exactly* aligned with the moon's periodicity. The thirty-one primate species that exhibit menses have cycles ranging from seven to forty-two days. For example, the lion-tailed macaque's cycle recurs every forty days on average; the tiny tarsier averages twenty-four days. The length of a lunar month is 29.5 days.* Many nonhuman primates have cycles falling close on both sides of this magic number. But, for your purposes, *Gyna sapiens'* menses has to hover consistently around 29.5 days. A half-day less or more would undermine your entire enterprise.

The Barbary macaque's menstrual cycle is thirty-one days. If a Barbary macaque menstruated on the night of a full moon, in the next cycle she would bleed when the moon had been waning for one and a half days. In the following month, the moon would have been waning for three days, and by the fourth cycle menses would occur a day short of a quarter moon. Within ten cycles, the Barbary macaque's menstrual cycle would be completely out of phase with the moon. The low-wattage brain of this monkey, attending to a seemingly insignificant lunar synchronization with her nearly inconsequential menses, would make it extremely unlikely that she would recognize a correlation between the moon's cycles and her menstrual cycles.

The opposite problem occurs for the Japanese macaque, whose menstrual rotation is twenty-eight days in duration. If she begins to bleed on the full moon in one cycle, she will bleed again during the waxing phase of the moon, one and a half days earlier, the next time. Thus, within ten cycles she, too, will be completely out of phase with the lunar cycle.

You realize with a sigh of relief that not every ancestral *Gyna sapiens* within any tribe will have to have a cycle that matches the exactitude of soldiers marching in lockstep. Many modern women's cycles vary considerably from the ideal 29.5 days. And not every woman who experiences a 29.5-day cycle does so for every cycle throughout her life. You understand that such perfection will be impossible to achieve. But you appreciate that, *if the majority of Gyna sapiens' cycles average 29.5 days,* then sufficient numbers of them will entrain with the moon to shake out a few of them who will recognize this critical pattern.† And with the parallel development of an exciting

---

*A mean synodic month, the time between one new moon and the next, is 29.53 days in duration.
†There are a number of large studies that both confirm and dispute the correlation of the moon's cycle and a woman's. The majority of them, however, confirm a statistically significant relationship. The largest, conducted by Walter and Abraham Menaker, collected data on over 250,000 menstrual cycles in over 2,700 women. Their findings revealed that the average length was 29.5 days! Virtually all the stud-

new form of communication, you gleefully anticipate that all it would take for your experiment to succeed would be for a handful of smart, observant *Gyna sapiens* to begin pointing out the connection to the others.

Diffusing out into the population, this wondrous discovery would rapidly become a crucial component of the culture's lore, to be continually renewed down through the generations by the most inventive myths, stories, taboos, and superstitions. At the same time, women shamans would mark the connection by creating a wide assortment of notational calendars, amulets, talismans, and other abstract representations. So vital is the knowledge of the link between the moon and menses that it would become embedded in the very core of a wide variety of languages. Along with their mother's milk, children would imbibe it with their native tongue.

Step Six: Having accomplished the main part of the experiment, you hasten to add one final exclamation point to the link between lunar and menstrual cycles. You delicately calibrate *Gyna sapiens'* onset of bleeding to coincide with the moon's ides. Within every lunar cycle, there is a dramatic three-day period when the earth interposes between the moon and the sun. To an observer looking to the sky, the moon mysteriously disappears. The effect is to plunge the side of the earth facing the hidden moon into near-total darkness. These three nights, in some cultures called the ides, seem longer and more ominous than others in the cycle.*

Moonless nights would have been a time of rest and introspection for ancestral humans. With only dim starshine to guide their way, few would dare venture forth. Because of the sudden withdrawal of a woman's primary sex hormones, menstruation also tends to induce an introspective state of mind. Darkness, repose, dread, alertness, and introspection are a fortuitous mixture, out of which often emerges a deepened receptivity to new insights. If you could arrange to have *Gyna sapiens* begin to bleed during the dark of the moon, the increased premonitory mood experienced by a primate that depends on light to see would significantly heighten its alertness.

Distinctive mood changes, unmistakable physical symptoms, and the dramatic external sign of blood loss occurring coincident with a fear-inducing event happening in the sky at a time when all the other females in the band were similarly indisposed could not fail to attract the attention of at least one clever *Gyna sapiens.* You now feel confident enough to begin to exult. You are

---

ies have been conducted after the invention of artificial light, a fact that could influence each study's outcome. Sung Ping Law studied women in rural China, a group minimally affected by artificial light. Her published data strongly suggest that the link between the moon and menses was statistically significant.[5]
*Shakespeare emphasized the dread associated with the dark of the moon by placing the date of Caesar's assassination in the Senate precisely in the ides. A seer warned him prophetically that morning, "Beware the ides of March."

sure that it will not be long before you witness positive results from all your complex and careful time-related preparations.

To reinforce *Gyna sapiens'* awareness of the intertwining of the moon's cycle and her menstrual cycle still further, just for insurance, you add another, related lunar-arithmetical coincidence. At present, obstetricians predict the date of a woman's delivery by counting the onset of her last menstrual period as Day One, reckoning a pregnancy's length by this method to be nine calendar months, or 280 days. A calendar month, however, is a day or two longer than a lunar month. If the count begins from the date of the last full moon prior to a pregnant woman's first missed menstrual period, then the length of a human gestation can be more accurately calculated. The average length of a human pregnancy using these parameters is 265.80 days. Nine lunar months equal 265.77 days. The statistical odds against this correlation's being just a coincidence are astronomically high.

More often than not, conception occurs on the night of a full moon, with most babies born nine lunar months later, on the night of another full moon.* We still tacitly acknowledge the moon-sex-birth connection when we refer to newly married couples about to embark on their culture-sanctioned sexual marathon as going on their "honeymoon." Anthropologist Chris Knight remarked, "No other primate shows so close a correlation between menstrual cycle length and the lunar month, nor between any whole number multiple of the menstrual rhythm and the length of gestation."[7]

Before Western culture converted to solar calendars, midwives used the lunar method to calculate delivery dates. It is a common practice even today in many less developed countries and rural areas of industrialized societies. Skilled midwives kept careful track of the moon's orbital periodicity to guarantee the accuracy of their predictions. Full moons also gained significance, because for many women it was a reliable external clue that they were ovulating.

Numerous reports in the ethnographic literature confirm that women the world over understand the connection between the moon and the length of pregnancy. Among the Tiwi in Australia, as soon as a woman knows she is pregnant she starts to "follow the moon." Anthropologist Franz Boas, reporting on the Kwakiutl women of North America, comments, "The child is generally born on the night of the full moon." In Vietnam, first-time mothers are advised by midwives, "For your first pains wait until the moon is full."[8]

You appreciate that a human pregnancy equaling nine lunar months might not be noted immediately, but you decide it is worth putting in place as one more factor tying the moon to menses. You have created conditions so

---

*The existence of small groups of women living together in close proximity, the norm in ancestral cultures, has been undermined by the invention of cities. The introduction of artificial light over a century ago is another factor that has greatly attenuated the connection between birth and full moons.[6]

that a majority of women bleed on the dark night of an absent moon and birth their babies in the light of a full moon. Your intense labor done, you can now sit back and wait. Over the course of many generations, a few of your chosen group of females begins to perceive a pattern roughly twenty-nine days in length. Since a month is an exceptionally long time for an animal to hold steady in its consciousness, you clap your hands in glee when first one, then another, and then another finally gets it.

Then you stand back and watch in awe. Once the first group of *Gyna sapiens* have breached this critical time barrier, the rest pour through the gap. On the other side lie the wide expanses of the future, a dimension uninhabited by any other life-form.

Freedom of movement back and forth along the salient of time gave first women, and then men, the means to control their personal destinies better. A short time later, this skill would present *sapiens* with control over the earth's destiny.

*　　*　　*

Menstrual cycles do not fossilize. We may never know what exactly was the relationship between ancestral *Gyna sapiens'* periods and the periodicity of the moon. There is, however, a rich historical and ethnographic record linking the moon to menses. Historian and anthropologist Robert Briffault accumulated a vast number of such connections in diverse cultures. Many of his methods, from another era, might not be up to current standards of research; nevertheless, the sheer volume of his findings makes his work difficult to dismiss. Briffault records that German peasants called a woman's periods "the moons." The French term is *le moment de la lune.* Mandango, Susu, and Congo tribes in Africa call menstruation "the moon." Maori in New Zealand refer to the same event as "the moon." The Fuegians, living at the tip of South America, call the moon "The Lord of the Women." In rural India, the moon is believed to be "the cause of time," just as it is the cause of menstruation.[9]

Close links among the words for "moon," "menses," and "time" are present in every language. The English "menses," "month," "moon," and "measurement" have their roots in the Latin words *mens* ("mind") and *mensis* ("menses") and the Greek word *menos* ("menses"). So, too, do "mental," "meter," "metric," "mentor," "diameter," "commensurate," "immensity," "parameter," "perimeter," and "dimension." "Calendar" comes from the Latin *calendare,* which means "to proclaim." In ancient times, it was the task of a priestess to scan the sky searching for the arrival of a new moon. All anticipated her public pronouncement, as it was an event laden with significance. The timing of the Islamic holy month of Ramadan and the Jewish High Holy Days depends on a priest "proclaiming" his sighting of a new moon. Easter falls on the first Sunday following the month's first full moon after the vernal equinox. The Arab peoples once worshipped a moon goddess, and Islamists have retained the crescent moon as a symbol of their culture, prominently displaying it on their flags and emblems.

The same correlation exists in foreign languages. In German the word for menses is *Regel*, in French *règles*, and in Spanish *reglas*. All of these words are synonyms for "measure" or "rule." "Regulation" is our English word derived from the same root. The ultimate measure of value, money, is also linked in this ancient numbers game. The relationship between silver and gold throughout classical times up to and through the Middle Ages intrigued economist John Maynard Keynes. Silver maintained a constant 13:1 ratio to gold, regardless of marked fluctuations of market forces. He speculated that, since value is usually determined by supply and demand, the constancy of the 13:1 ratio had more to do with magico-religious considerations than with the actual value of these two metals. The ratio of thirteen and a half silver to one gold approximates the ratio of the number of lunar orbits to the number of Earth's orbits in a solar year.[10] Reminder: There are twelve calendar months but closer to thirteen lunar cycles in a year.

Barbara Walker, an encyclopedist of women's issues, has collected considerable evidence from a wide variety of cultures that express poetically the notion that men acquired an immense gift of knowledge from women, and it had something to do with their menstrual blood. The Norse god Thor owed his enlightenment to bathing in a river of menstrual blood. Odin was similarly gifted with shrewdness because he stole and drank the wise blood from the Mother Goddess, a myth quite similar to the Hindu god Indra's theft of knowledge from the Primal Matriarchs via their menstrual blood. *Soma*, the Sanskrit word for "body," was also the name of the mystic drink of the Hindu gods. A synonym for *soma* was "wise blood." The Hindus particularly revered *soma* on Monday, the day of the moon. Celtic kings acquired their right to rule by drinking the "red mead of the Fairy Queen Mab." Egyptian pharaohs, in a holy ceremony, ingested an ambrosia called *sa* that was called "the blood of Isis."* Perhaps all these references allude metaphorically to a profound truth:

---

*This connection persists right up to the present. Christianity bears many similarities to the ancient moon-menses-goddess religions. Jesus dies, disappears for *three* days, and then reappears to his disciples as the resurrected God. Blood plays a central role in Christian ritual. After Paul had his revelation on the road to Damascus, he did not begin his ministry for three years. Historians report that he traveled extensively in the southern deserts on the outskirts of the Roman Empire before returning to its cosmopolitan centers to the north. During his wanderings he would have had to be familiar with the reigning religion in those regions. Attis, the beloved shepherd of the Asiatic fertility goddess Cybele, died at the foot of a pine tree. His death and resurrection were celebrated each spring by tying an effigy of a young man to a tree to symbolize Attis' death and resurrection after a three-night interval. The dying-resurrected son story was originally based on a moon religion.

Women's menstruation was the key to men's learning how to measure time. Men acquired this knowledge from women and used it to catapult themselves to become, first, formidable hunters and then, masters of the universe.[11]

Only since the 1950s has the scientific community begun to examine the link between women's menses and the moon, and many continue to insist it is an artifact. Penelope Shuttle and Peter Redgrove admonish the naysayers in their 1999 book, *The Wise Wound:* "In view of this 'constellation of coincidences' it seems churlish not to postulate some connection between these astronomical and biological phenomena."[12] I agree. The consonance between moon, month, menses, and measurement is too rich, too universal across geographic regions, involves too many diverse language groups, and spans too many historical periods to be a mere accident. Something of immense significance occurred to the human species long ago that intertwined menses, the moon, and the duration of a month.

*   *   *

In 1964, American anthropologist Alexander Marshack proposed that a puzzling group of hand-crafted objects excavated from Ice Age Europe were lunar calendars. Widely distributed through space and time beginning some thirty-five thousand years ago, they were originally thought to be ceremonial objects. Some continue to insist that this is the case. But Marshack believed otherwise. He presented compelling evidence that these objects were lunar calendars. One of the oldest, dated at thirty-two thousand years ago, was a segment of antler containing an elaborate notching pattern. Examining the notch marks under a microscope, Marshack noted that they were not uniform but seemed to have been fashioned by an assortment of instruments, and he posited that a human hand had methodically picked at the bone at different intervals. A periodicity seemed to be involved in these objects, for the marks were placed in a sinuous configuration with nodal breakpoints at twenty-nine days in some of them. Some ancestral individual (or group), Marshack hypothesized, was keeping track of the moon.*[13] To reiterate the words from the epigraph that opened this chapter, of the two human sexes, a woman, far more than a man, "had motive, method, and opportunity to be the originator of lunar notation."

---

*Some archeologists dispute Marshack's interpretation. Randall White, for example, thinks these designs are imitations of natural seashells.[14]

A drawing of an etched bone that Alexander Marshack proposed was a lunar calendar fashioned by an Upper Paleolithic man or woman.

Few other subjects fascinate as thoroughly as do the sex lives of Nature's creatures. Despite the originality exhibited by many unusual specimens, not a single one diverges so far from the average as do humans, particularly the female of our species. A yawning chasm separates the reproductive life history of *Gyna sapiens* from that of the females of the other three million sexually reproducing species. Features such as harmonized menses, loss of estrus, cryptic ovulation, lunar synchrony, year-round receptivity, early menopause, the need for delivery assistance, fulsome orgasms, dramatic periods, and iron loss are far, far away from the standard boilerplate designs of nonhuman females. In the Great Game of Sex played by the forces of the environment on one side and genetic mutations on the other, fielding so unusual a female as *Gyna sapiens* marked a notable event in the history of the game. Although numerous commentators have proposed theories to explain one or a combination of two of her unusual features in isolation, perhaps it

The prehistoric Venus of Laussel holds in her right hand what appears to be a crescent moon (or an animal horn). The thirteen notches on it approximate the number of lunar cycles in one year. A coincidence?

would be more productive to examine *all of them together*, as if they were part of a single magnificent huge adaptation.

I propose that *Gyna sapiens'* reproductive novelties were *not* for the primary purpose of immediately advancing the fitness of the human species. Too many of them either had a neutral effect or were counterproductive to survival and reproduction. The reason for their persistence in the human genome is that, when linked together, they were the agents that taught our species how to tell time.

Foresight has proved to be the sexiest idea that Mother Nature came up with since Her clever invention of the penis two hundred million years earlier. Whereas the penis significantly advanced the fortunes of every reptile and mammal species that acquired one, foresight dramatically increased the fortunes of *only* humans, at the expense of all other species.

Here, then, is the answer to the key question I posed in the preface. *The reason women bleed so copiously every month is so that humans could anticipate the future.* Gaining the ability to maneuver conceptually in the dimension of time was so powerful an adaptation that whatever price the human species would have to pay would be worth it, because it guaranteed that they would exercise dominion over all the other animals. Unfortunately, one sex was more disadvantaged than the other. The Faustian bargain *Gyna sapiens* unwittingly and involuntarily entered into was an awesome tradeoff. Iron-deficiency anemia, loss of estrus, and potentially debilitating menses were the tolls she paid to do something no other animal had ever done before—see beyond the moon to the next month.

Speech is a human's signature feature.

Chapter 14

Woo/I Do

Plenty of animals can express the fact that they are hungry, but none except man can ask for an egg or a banana.   —Julian Huxley[1]

What kind of beast would turn its life into words?
                                                     —Adrienne Rich[2]

"No, no Bessy ... I meant what I said to stand for summat else; but never mind—it's puzzling work, talking is."   —George Eliot character
                                                     Mr. Tulliver, in conversation
                                                     with his wife[3]

*A*mong the many innovations that issued from an enhanced sense of time, none is more spectacular than the human capacity for language. Linguists, anthropologists, neuroscientists, and archeologists have engaged in a lively debate as to how and when this wondrous invention came into being. The preponderance of circumstantial evidence suggests that human language is a relatively recent development. Most experts in the field believe that *Homo sapiens* forded a linguistic Rubicon about a hundred thousand years ago.*

A question haunting the discourse: What evolutionary factor, either pos-

---

*However, many reputable authorities believe that the development of human language was more gradual and archaic. Ralph Holloway, like Norman Geschwind and Alan Galabruda, noted that the endocasts of the brain cases of a *Homo* species that lived 400,000 years ago indicate that an asymmetry of the brain existed at this early date. The areas critical to speech, both located in the left hemisphere, were larger than comparable areas on the right of the few skull remnants they were able to examine. Based on this evidence, they propose that language began with the species some call archaic *Homo sapiens* and others call *Homo ergaster* or *Homo heidelbergensis*. Other evidence suggests that earlier hominids possessed a rudimentary form of language. Most researchers, however, place the Big Bang of speech, as it is called, approximately 100,000 years ago.[4]

itive or negative, could have transformed a grunting, relatively inarticulate *Homo erectus* into an eloquent Demosthenes, spellbinding his fellow Athenians in the agora? Why did humans, alone among the multitudes of life's creatures, evolve so towering a form of communication? There remain enormous swaths of ignorance waiting to be filled in by scientists concerning the complexity of other animals' systems of communication. Still, not another creature seems able to employ the pluperfect subjunctive tense. Philosopher Ludwig Wittgenstein mused, "A dog cannot have the thought, 'Perhaps, tomorrow it will rain.'"

And how to explain the fantastic abruptness with which this all came about? One moment, our distant ancestors were probably lip-smacking and pant-hooting at each other, and the next, two of their descendants were sipping espresso at the Deux Magots, engaged in sophisticated discussions about the meaning of Jean-Paul Sartre's existentialism and Jacques Derrida's deconstructism—all in a wink of evolutionary time.

Most believe that language was the premier behavioral innovation that gave the *sapiens* the critical edge enabling them to cage all the other species in zoos. Why, then, did it develop only once, rather than erupting all across the evolutionary spectrum? When Natural Selection finally smoothed out the aerodynamics of the wing, it released into the air an immense variety of swooping, gliding, and thrumming creatures. The need to see was so pressing that an adaptation as indispensable as an eye appeared in many different configurations and evolved independently forty different times. One would certainly anticipate that an innovation as superior as human communication might manifest in loquacious llamas, articulate hippos, and chatty tabbies.

To understand better just how complex human language is when compared with all other forms of animal communication, consider the following. Having discovered a food source, bees use a distinctive waggle dance to inform other bees as to its distance, nature, and direction.[5] Their repertoire contains twenty-two different routines, each one conveying a slightly different message. The bees' waggle dance is all the more remarkable when one considers that this industrious insect *doesn't even have a brain*. Clumps of nerve cells, called "ganglia," are the meager stagecraft behind the curtain. Adding an arabesque, recent research indicates that the bee's ability to remember details about former treasures is more extensive than we had previously imagined.

Now fast-forward the evolutionary talent show several hundred million years. Chimpanzees are our closest kin. They sport a brain of immense complexity when compared with the nervous system of a bee. Chimps are capable of deceit, empathy, morality, self-awareness, toolmaking, and complex

problem solving. Yet, after logging hundreds of thousands of hours of observations in the wild, ethologists have been able to identify only a measly
thirty-five distinctive chimpanzee calls and gestures. Despite the sophistication evident in chimpanzee neuronal brain wiring, communication between members of a troop has not advanced all that much beyond that of
the busy bee.

It is entirely possible that chimps possess many subtle signals that human observers cannot discern. But even if we grant chimpanzees a thousand
additional signals, their calls and gestures would fade to faint echoes and
pale shadows when placed alongside the full-throated three-dimensionality
of human speech.* Compared with the bee's twenty-two and the chimp's
thirty-five variations, the English language alone contains a half-million
words that can be arranged in an infinite variety of sentences to convey the
most delicate distinctions. Insects can signal and some complex creatures
can inform, but only a human can ask a compound question and, further, dispute the answer. Which brings us again to the overriding question: Why did
we *need* to evolve such a robust, high-speed, and exquisitely intricate ability
to communicate?

Over the centuries, numerous theories have been posited. So many, in
fact, that in 1866 the Linguistic Society of Paris banned the subject altogether.
Some of the society's members had bloviated so much about the origins of
human speech that those with more sensible dispositions declared the topic
unfit for consideration by such an august body. Their proclamation, however, did little to still speculation.

Darwin, in his 1871 book, *The Descent of Man*, proposed that speech, similar to the function of birdsong, evolved gradually through sexual selection
as a means to display verbal skills to impress prospective mates. Early on,
one school of thought suggested that human speech evolved primarily so
that cooperative hunters could better coordinate their movements while
stalking large prey. This idea has steadily lost favor because of emerging contradictory evidence from anthropological studies of the habits of extant
hunter-gatherer tribes. Stealth and silence are more typical of a group hunt
than is running commentary.

Watch a pride of lions spread out to bring down a wary gazelle coopera-
tively. Observe the sophisticated division of labor deployed by chimpanzees
intent on snaring a colobus monkey. Nonhuman social predators proceed
without resorting to the most basic subjects and verbs. (The planet contains

---

*Apes can be taught simple syntax, but it is not something that they do naturally. Cetaceans—for example,
dolphins—appear to be the only animals whose language is complex enough to possess a recognizable
syntax. Dolphins use a two-level (word-phrase) language structure.

over six thousand tongues, but one universal gesture, acknowledged by all humans, is a forefinger placed upright against pursed lips, meaning, "Shhh— no talking." Finger-to-lips is commonly encountered during a hunt.) The hunting hypothesis, therefore, seems inadequate to explain completely the exuberant robustness of human language.

Another popular theory is that the social complexity of human groupings required a better way to communicate the needs and desires of their expanding membership. Kinship, friends, cheaters, enemies, and reciprocal relationships created a pressing selective pressure to track the complexities of who-is-doing-what-with-whom-when-where-and-why. Anthropological linguist Robin Dunbar is one of the most articulate proponents of this school. Gossip, according to Dunbar, has replaced grooming as the social glue used by primates of the human persuasion.[6] Machiavellian intelligence, the ability to deceive others while remaining undeceived, demands that human language possess the nuances necessary to implant the mendacious messages of the kind that a scheming Iago whispered into the ear of a gullible Othello.[7] Leda Cosmides proposed that the human brain evolved a special language module designed to detect cheaters.[8]

Psychologist Nick Emler proposes that the majority of human speech is for the purpose of what he calls "reputation management."[9] Anthropologist Glynn Isaac suggested that language was the strategy an individual used to play the complex game of "social chess."[10] Chris Knight and Camilla Power hypothesize that language arose from humans' first ritualistic practice of adorning their bodies with red ochre. Ritualistic signaling laid the foundation for mutual trust between humans. This led to the fundamental prerequisite for language—agreement among individuals that a particular sign or sound stands for a specific thing.[11]

The influential linguist Noam Chomsky advanced the idea that language may not be due to natural selection at all, and gained an important ally in Stephen Jay Gould, who admonished evolutionary biologists for jumping to the conclusion that language evolved to solve an environmental challenge. Gould believed that language might be the end result of a long ago process far removed from its present use. Steven Pinker and Paul Bloom have attempted to bridge the gap between the Chomskyites and Darwinists by proposing a middle ground. Perhaps, they posit, language developed as an offshoot of a form of mental thinking they call mentalese that in turn was in response to environmental stresses. The brain's ability to think in these abstract terms was necessary to reconstruct a true picture of the world.[12] Owen Jesperson posits that language grew out of a form of play.[13]

William Calvin proposed that language evolved out of our need to master complex serial maneuvers.[14] As do Doreen Kimura and Philip Lieberman,[15]

Calvin argues that the complex serial gesture of accurately throwing a stone can be likened to the motor skill necessary to assemble sequentially a syntactically and grammatically correct sentence. Merlin Donald proposes that both language and art evolved out of a process of mimesis, a word he defines as the human ability to reconstruct three-dimensional space using spoken words or graphic images.[16]

Derek Bickerton set forth the theory that language was largely the result of a magnificent beneficial chance mutation, and that its primary function is to represent abstract thought internally. Only later did language serve as a communication vehicle to transfer information between individuals.[17] Terence Deacon attributes the astonishing rise of language to the human ability to create symbols.[18] Michael Corballis argues that the development of the spoken word arose from human gesture.[19] Morton Christiansen proposes that language is a parasite that is using humans as its host, and that we have become dependent on each other to propagate. Language needs humans, and humans need language. We are in a weird kind of symbiotic relationship with a new kind of incorporeal organism called language.[20]

These are a sampling of the many inventive theories proposed to explain the origins of human language. But among the welter of competing claims, no one explanation has yet emerged triumphant. Something seems missing, something that could help to explain our relatively recent logarithmic leap to expressiveness.

When an adaptation as intricately filigreed as human speech bursts upon the evolutionary scene, it is usually in the wake of one or more dreadful environmental pressures bearing down, forcing a species to choose between evolving or exiting the scene. An alternative explanation: A grand new opportunity for a species makes a fortuitous abrupt appearance. To take advantage of this potential bounty, the species must hastily throw together a novel adaptation.

What, then, was the looming environmental disaster that, if not adequately and urgently addressed, would cause the human species to "go missing," as the English are fond of saying? Alternatively, what was the manna-from-heaven opportunity that required speech?

I propose that the threat was maternal mortality, and the reward, the opportunity to be the first and only animal to roam freely in the lush uplands of the future while all the others remained individually marooned on isolated islands in the immense archipelago of the present. Suppose human language evolved so that men and women could negotiate the terms of when, where, and how they would have sex. Suppose, once they began to talk, they discovered that other creatures were at a significant disadvantage in trying to compete with them.

Psychologist Geoffrey Miller, in his recent book *The Mating Mind,* puts forth an intriguing theory. Miller proposes that language, instead of being prodded into existence by the exigencies of survival, was a function of sexual-selection pressures. Females chose men based on their ability to communicate effectively, and "runaway sexual selection" did the rest. Runaway selection, remember, is the evolutionary process that caused the male peacock to have such large, spectacularly gorgeous, but useless tail feathers. Females noted that any peacock that could survive with such an outlandish ensemble sported an honest handicap. Once the peahens began to select for males with ever-larger useless tail feathers, the race was on! Miller proposes that women selected men who were smooth talkers over inarticulate ones: "Language puts minds on public display, where sexual choice could see them clearly for the first time in evolutionary history." [21]

His theory would predict that males would be far more facile with words than women. Yet numerous studies of gender-related speech patterns indicate a slight skewing in the opposite direction.* He explains this feature of human language by asserting that men, too, began to use language to gauge the intelligence of women. One feedback loop reinforced the other, and in a remarkably short time— *Voilà!*—human language achieved its present state of suppleness.†

Human language, like a fine Swiss watch, has many moving parts. All of them must be present and mesh with one another nearly simultaneously for the whole enterprise to generate intelligible sounds. The larynx (voice box) had to sink deep into the neck, far below the opening to the pharynx (throat). This descent endowed humans with the tallest larynx in all of animaldom (relative to size). Possessed of such a splendid resonant chamber, humans can articu-

---

*Girls speak earlier, make fewer grammatical mistakes, and use a larger vocabulary than boys, who also experience more speech and reading problems than girls. [22]

†An intriguing piece of supporting evidence Miller provides is the extraordinary number of synonyms present in languages. If one word adequately describes something, why, Miller asks, is it necessary to have so many words that mean the same thing that most people rarely, if ever, use? For example, "azure" is another way of saying "blue." How often have you ever incorporated "azure" into a sentence? Miller contends that using an extensive vocabulary alerts a prospective mate that you are a cut above the average. By monitoring the conversation of a new prospect, both men and women can evaluate the relative intelligence, social class, and resource potential of the person very quickly. [23]

late a wide range of vocal sounds—one of the prerequisites for nuanced messaging.

The mouth opens into the oral cavity (lips, tongue, gums, teeth, and palate), which leads to the pharynx, which becomes the esophagus (gullet), the long cylindrical tube that serves as conduit between pharynx and stomach. Food travels this gustatory highway on its way down the alimentary tract.

When the opening (glottis) to the human larynx drifted past the vestibule of the pharynx, a major design flaw became evident. The entrance leading to the delicately lined tracheal (windpipe) and bronchial (lungs) airways became exposed on a routine basis to the likes of beer and peanuts every time swallowing occurred. A trap door (epiglottis), resembling a hatch cover on the deck of a boat, evolved at the top of the larynx, so that it could slam shut the instant a warning arrived, via nerve reflexes, that food was on its way past the tongue.

This Rube Goldberg* arrangement was necessary in order for a human to alternate between eating and breathing through the same tube (pharynx). We are among a handful of land animals that must suspend breathing in order to eat. If one makes the mistake of inhaling at the moment of swallowing, an energetic paroxysm of coughing will ensue. The simple act of swallowing does not pose this hazard for the vast majority of other animals.

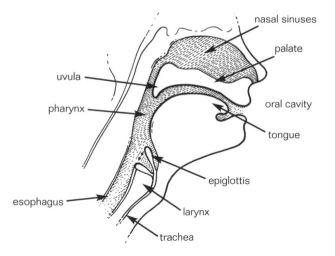

Schematic of the human vocal tract.

---

*Rube Goldberg was a famous cartoonist from the early to mid-1900s who designed humorous, inordinately complicated contraptions whose purpose was to achieve simple ends.

An extremely powerful incentive must have propelled us to talk to each other to justify fouling up such a vital and basic feature as the ability to breathe and eat simultaneously. That we would deviate from the standard anatomical alignment, present in virtually every other oxygen-dependent creature, points to the urgency associated with the development of human speech.*

Given the potentially fatal misalignment of pipes and tubes in their neck, it might have made more sense for *Homo sapiens* to heed their bizarre throat plumbing and sensibly eat in silence. But human speech is irrepressible. It is what we do best. Despite the dangers inherent in talking and eating simultaneously, people the world over consider mealtime to be the most propitious opportunity for social interaction. "Let's do lunch" is really an invitation to shmooze; the food is secondary. Communal eating and talking over food is the *sapients'* second-most-popular form of entertainment. The word com-*pan*-ion derives from the Latin for "to share bread."

The number-one favorite first date between a man and a woman is a dinner for two by candlelight. Typically, he does most of the talking. Typically, she listens carefully, alert to the meta-messages of his conversation as she begins to gauge his worthiness. Typically, his primary purpose in all this folderol is to convince her to accept the advances he will likely make at her door at evening's end.

Of course, there might be many other currents ebbing and flowing over that dark table in the corner, but this basic scenario represents the crest of the bell-shaped curve. One can easily imagine this scene being repeated by firelight throughout innumerable Pleistocene dinner dates.

Along with the gumming up of *Homo sapiens'* neck-works, there were additional prerequisite adaptations for fluent speech. A crucial one was the need to control respiration voluntarily. Breathing in and breathing out (or its equivalent) is *the* most fundamental exercise that fish, amphibians, reptiles, birds, and mammals do. A prolonged stoppage of breathing stills the heart, and all cellular life downstream abruptly dies. Very early on in the course of evolution, Mother Nature decided that the most efficient way to deal with something as basic as breathing was to place it under the control of the auto-nomic branch of the nervous system. "Autonomic" means "automatic." One wag once characterized the autonomic nervous system as being so far from the head it doesn't even know there is a head! Breathing occurs involuntarily

---

*Among the few animals with this anatomical arrangement, sea mammals are disproportionately repre-sented. Elaine Morgan has made the case that this is one more feature that should alert archeologists and neuroscientists that humans underwent a period 9 million to 12 million years ago when we began to adapt to living in the water. A freelance writer, Morgan has become a knowledgeable physiologist and stalwart champion of Hardy's Aquatic Ape Theory of human origins.

with boring regularity until the day of death. Respiration is the sine qua non of life. We pronounce someone who has stopped res*piri*ng as ex*pired*.

Ponder, then, the human. While breathing remained, for the most part, under autonomic control, something remarkable occurred. Exceptionally large nerves peeled off from the spinal cord and exited the upper-thoracic vertebrae on their way to innervate the respiratory muscles. Resembling rodeo roughriders grasping the reins of bucking broncos, these robust, jumbo-sized nerves wrangled control of the muscles of respiration away from the autonomic centers. Neurons within these upper-thoracic bundles report to the highest cortical centers in the new left brain. They take their orders directly from here, rather than from the more primordial autonomic breathing centers lower down in the brain stem, the part of the brain analogous to a basement.*

A moment before the left hemisphere's Executor decides it has something important to say, it orders an interruption of the cycle of breathing to store up air, which increases pressure in the lungs. The assembly line producing talk originates deep in the chest, at the level of the diaphragm. Broca's area, in the left hemisphere, directs the rib and thoracic muscles, through messages carried by these specialized nerves, to release the trapped air in controlled, short spurts.† These puffs provide the raw material that will soon be tooled into sharply chiseled words farther up the line.

Next, our incredibly complicated vocal apparatus (larynx, twin vocal cords, pharynx, tongue, nose, sinuses, teeth, and lips) springs into action. As the unprocessed sausage tube of air comes hurtling out of the trachea, each of these workers on the assembly line of speech must snip, mold, sculpt, and package the exhalation streaming by into a distinct, discernible word, in milliseconds.

Paleoanatomists offer a prime piece of fossil evidence refuting those who argue for an earlier date for human speech. Compared with the generous diameters in *sapiens*, the upper-thoracic vertebral openings for nerves exiting the spinal cord for other hominids are unimpressively stingy.[24] This would seem to indicate that antecedent hominids lacked the fine motor control over respiration that human speech requires. The shape of their pharynx,

---

*Contrary to what is generally believed, it is not the gasp for oxygen that drives respiration. The more critical motivation is the organism's need to exhale carbon dioxide. The buildup of this gas in the bloodstream is the result of the accumulating waste products of cellular metabolism. CO_2 must be ejected from the body or a dangerous change in the acidity of the blood will occur that will ultimately interfere with the function of every enzyme system. Thus, the internal milieu places greater emphasis on taking out the garbage than receiving fresh deliveries of the fuel necessary for thinking.

†Broca's and Wernicke's areas in the left hemisphere's neocortex (of right-handed people) are the two primary speech centers. They are named in honor of the two nineteenth-century neurologists who first described them.

larynx, and oral cavity also seems to militate against the possibility of silver-tongued orators holding forth atop a rock early in the Pleistocene Serengeti.

As wondrous as the mechanics of speech are, it staggers the imagination to contemplate how the brain reorganized itself to learn, generate, receive, and interpret the high-pitched squeakings of others. And at such an early age. Anyone who has witnessed the rapidity with which a toddler acquires speech can only marvel at the efficiency inherent in brain modules that convert babble into syntactically correct sentences between a couple of birthdays. The speed with which children acquire speech is in marked contrast to other, far less complicated motor skills, such as mastering a musical instrument, or accurately throwing a football, or building a model airplane, all of which take years of practice, diligence, and patience. Two-and-a-half-year-olds begin to master the categories of subject, verb, and object—when they haven't yet completely mastered their bowel function.

Every day, we speak sentences that we never rehearsed. Out they come, tumbling from our mouths with little (if any) editorial interference. Where were these words just before they were spoken? What gremlin wordsmith hurriedly assembled them into a grammatically near-perfect ensemble just prior to their tripping off our tongue?

And then there is the receiver. A listener does not hear distinctly enunciated words interrupted by pauses, as if spoken words were bullets fired by a slow-action machine gun. Instead, a more-or-less continuous sound slurry issues from the speaker's mouth. Typical conversation does not set individual words apart between empty spaces (as you see while reading this sentence).

A listener must have first learned how to establish what linguists call "auditory constancy." Certain sounds are treated as discrete "objects," just as a person learns to recognize things in the visual field as distinct visual objects. The smallest auditory object is a word. The phonetic sound "tree," for example, can appear in a sentence with an immense variety of phonemes hemming it in on either side. Yet the word "tree" can be easily identified by the listener's ear, regardless of its surrounding context.

Once each word has been identified, the listener must swiftly collate the sentence's syntax (the linear arrangement of each individual word or phrase and the relationships among words and clauses that determine how speech is organized and interpreted). After specialized regions in the left hemisphere accomplish this feat, another part of the brain has to fish out from speech's bouillabaisse the meaning without which a listener cannot grasp the connection of this particular sentence to the preceding or link it to the one that will follow. The brain then accomplishes something utterly magical. From mere sounds, *meaning* emerges. Somehow, two English-speakers accept that the uttered sound "tree" represents the tall, lush Douglas fir they both saw yesterday in a distant location. Even though the tree is no longer within view, they blithely discuss the auditory object this puff of air represents *as if* the actual tree stood before them. Human speech is a truly virtuoso performance!

Dramatic changes in the anatomy of *Homo sapiens'* vocal apparatus, breath control, and hemispheric lateralization are the reasons that speech became possible. But the question remains: What compelled Mother Nature to slap together so rapidly so many diverse components, creating one of Her most extraordinary and singular adaptations? Most likely, there would have been a multifactorial mix of whirring feedback loops reinforcing speech, including social interactions, toolmaking, child-rearing, and hunting.

I believe that the primary impetus was sex—sex in the context of an increased awareness on the part of the female of the inherent danger of childbearing. Speech was hurried along because *men and women had to talk to each other.* And the most important subject they had to negotiate was sex— with whom, when, where, under what circumstances, what must be exchanged before, and at what cost to each party.

The complex chain of events and adaptations that led to this spectacular development began around the time of the tragic death of the unheralded Unknown Mother. Given the combination of bipedalism and their fetuses' rapidly enlarging brains, mothers began to suffer a high rate of mortality.

A factor making maternal mortality unique was that it affected only one sex. No young male ever had to confront a death threat that regularly originated from within his body. Combined with a few other deleterious factors, some isolated *Homo erectus* groups found their continuation endangered. Most likely, quite a few of them disappeared into extinction's maw. One, however, survived by adapting. The key: Women acquired a rapid increase in mental endowment, enabling them to exercise control over the sexual act.

For this to come about, a critical prerequisite concerning time had to be in place. Women had to make the causal connection between two events that were separated by months—sex and signs of pregnancy. Having learned the extent and duration of the time frame of a month, some woman, somewhere, strung together a handful of months and had a momentous insight. She realized that her morning sickness, cessation of menses, or the quickening she felt in her womb was the direct result of the transient moment of sexual pleasure she had experienced with a particular man months earlier.

Having made this backward-looking link between sex and pregnancy, she peered more months into the future and realized that she *had risked her life* by engaging in sex. Soon, many women grasped the sex/pregnancy/labor-fraught-with-danger connection and its implications. Lengthier excursions into the future soberly impressed upon women the staggering output of energy required of them to raise their offspring to maturity. The ability to leap from a past event and extrapolate it into the future is the hallmark of foresight. *Gyna sapiens* modified her behavior. Distinct from all of her mammalian sisters, she would no longer enter into sexual relations without considerable forethought.

Meanwhile, *Homo sapiens'* nervous system was undergoing a physiological change that made him interested in having sex *all* the time, a sharp deviation from what had been the norm.* *Homo sapiens*, driven by his constantly high testosterone levels, craved sex *with all women*, with a few exceptions, *all year long*. There was little to disincline his lust at the outset of the human experiment. An archaic human male found sex to be both pleasurable and entertaining, and, best of all, it initially cost him almost nothing in the long run. Only one obstacle stood in his way: He needed the cooperation of a woman.

As the sexual thermostat rose in men, it lowered in women. Men became more eager as women became more coy. An untenable situation was building that could threaten the continuation of the species. Nothing like this dissonance of desire had ever occurred before to any of the millions of other species that reproduce sexually.

To their frustration and dismay, men discovered that fighting among themselves to see who would win mating rights was no longer enough. Size and superior strength were no longer enough. Unlike other, nonhuman males, they couldn't just warble a great song or grow the most impressive antlers. Men frantically searched for a way to melt women's new sexual reticence.

Concerned perhaps that She had overshot the mark, Mother Nature sprang to his rescue. In an unkind move, She plagued *Gyna sapiens* with a persistent drain of iron. Men seized upon this opportunity. A man discovered that presenting a woman with the gift of meat substantially decreased her resistance to his sexual overtures. Soon, however, even the meat-for-sex exchange was not enough. As women became more self-aware, they raised the bar. Women who had grasped the enormity of pregnancy and the lifelong commitment of child-rearing also noted with keen interest how pliant a man became when his member was stiff.

---

*Some male animals, from tortoises to dogs, seem to be perpetually randy, but the difference between their behavior in the presence of a female that is ovulating and their behavior in the presence of one that is not distinguishes them from the human male.

Women began to demand not only meat and/or resources for this particular tryst, but also the *promise* of meat and/or resources far into the future. A woman often held out the possibility of sexual exclusivity if a suitor complied. Considering how difficult it was for many men to keep finding a willing sexual partner, this idea had a strong appeal. She also wanted him to help her raise the children that often resulted from their sexual union. She desired aid, love, protection, and companionship for her and her brood alone. In short, she wanted a commitment from him to offset her own.

Increasingly, deciding whom she would choose for a long-term mate became ever more complicated. No longer could she judge a man on mere physical traits. The females of prior species were content to mate with the biggest, strongest, fiercest, cleverest, most dominant males. In general, women remained attracted to these traits. But in the human line, a woman added many new ones to the list—wit, kindness, loyalty, honesty, steadfastness, courage, intelligence, poise, resourcefulness, generosity, leadership, ambition, and compatibility. In other species, hard-wired courtship and sexual behaviors do not allow females the ability to assess these qualities before being mounted.* *Gyna sapiens* had to conduct this vetting process while trying to delay having sex with a panting, overeager male. Through many generations of trial and error, a woman learned that her needs would best be met if she accurately gauged a man's character *before* she committed herself. Human females desperately *needed an alternative method, never used before by the females of any other species,* to judge the character of a courting male.

Meanwhile, *Homo sapiens'* frustration level mounted. His stratospherically high levels of testosterone whispered to him urgently, "Close the deal already!" But, unlike the males of other species, he found that this remained a daunting task. He was confronted with increasingly choosy females who were deflecting his sexual overtures, artfully delaying the consummate moment the better to get to know him. The female kept him at arm's length until she had taken the measure of his character. During this time, his ardor would increase to the point where he could barely stand it any longer. It was clear to him that he required a vastly more sophisticated persuasion technique than chest thumping. A man desperately *needed an alternative method, never used before by the males of any other species,* to convince a female to have sex with him.

And so these two very different agendas converged. Men and women both sought an improved process by which they could *relate* to each other. If

---

*A few monogamous species' females may also have this chance. The courtship phase of wolves, for example, is long enough for the female to make a fair assessment of the male's worth before she consents to intercourse.

males could not figure out a way to convince females to consent to intercourse, the species might become extinct. If a female could not figure out a way to judge a male's character while postponing intercourse, she and her future children might die, and the species might become extinct. Perhaps the notion of extinction is hyperbole, but one can say with conviction that no large animal that reproduces sexually muddles the process of procreation as much as humans do. Each sex metaphorically turned to the Red Queen, beseeching her to improvise a new adaptation that would help them out of their seemingly insurmountable mutual quandary. The Red Queen delivered a lulu. Her brilliant solution was speech.

Males of other mammalian species commonly signal their initial sexual intention by either grabbing hold of a female's behind or sniffing her vulva. Once *Gyna sapiens* gained the veto power over sex, a *Homo sapiens* trying this approach would likely find himself summarily kneed in the groin. Gifted with language, a man could initiate a conversation with a strange woman from a safe distance, well outside the perimeter of her personal space.

Some nervous young man tentatively approaching a young woman who had caught his eye most likely stammered the first grammatically correct sentence ever uttered by our species. It was probably an ancestral version of, "Saaay, you look familiar. Haven't we met somewhere before?" Thus was born, on a Pleistocene summer evening, the "opening line."*

A man learned that speech was the well-oiled vehicle he needed to convince a woman, over time, that she should be his. Suave, macho, witty, genteel, charming, intellectual, and rebel-without-a-cause are conversational styles rehearsed by teenage boys in the privacy of their bathrooms, before a mirror, preparatory to trying out these routines in front of a genuine prospect. Perfecting just the right slouch while arching one eyebrow is often deemed as important as inflecting certain words in these fledgling Romeo routines. I suspect that young males practiced similar stratagems eons ago.

It would not have been lost on such a clever creature as *Homo sapiens* that a man who was short and homely but could make a woman laugh was more likely to be a successful wooer than a strong, silent type. We are talking about success at sex here, the subject closest to a man's heart and testicles. Acutely aware of the stakes involved, other men surreptitiously eavesdropped and began to mimic the styles that appeared to work in accomplishing what seemed to many shy men so intimidating a task. Convinced that they must market their virtues, most men believe they have to talk rapidly

---

*Some might object by pointing out that there were more important things to talk about, such as predators, prey, and natural phenomena. Yet other animals interact with each other and deal with the major issues in their lives without possessing the complexity of human language.

and volubly to interest a new woman, not unlike a salesman trying to get his foot in the door.

A male's spoken style and substance proved repeatedly to be a most felicitous key in the art of romance. Men, for the most part, fall in love with their eyes; women, with their ears. Men often do not pay much attention to what a woman says; they are more impressed—some would say blinded—by her appearance. Conversely, the words a man utters to a woman win her heart more often than the way he looks. The ability of speech to facilitate mating began to supersede in importance the language of hunting, the motherese of nurturing, the strategies of social chess, or the give and take of economics.

Speech became the premier sexual lubricant. Humans communicate with each other more often, and over a longer period of time, *before* they first have consensual sex than any other animal. Many couples vocalize *during* sex, and, most certainly, they must converse with each other *afterward* if either of them wishes to repeat the performance. Speech is an integral aspect of courting and a key component to successful mating. Ultimately, a man must be skilled in the art of wooing if he hopes to pass along his genes.

What was true then is still true today. Men, on average, speak about two thousand words in the course of a day; women, on average, speak around seven thousand.[25] During courtship, however, this ratio reverses. On a first "date" (or whatever each culture calls the beginning of a new relationship), usually a man does most of the talking and a woman does most of the listening. Given the almost universal male disposition to compete for a female's attention, it would be reasonable to assume that this male penchant for verbosity upon meeting a new woman has been around since the Pleistocene's Big Bang of human speech exploded.

In short, language evolved primarily because men and women had to *negotiate* sex. Women needed a means to assess whether they *would* say *Yes!* Men, having acquired from the female the gift concerning foresight, began to think in the long term. A man needed to assess whether a woman was more or less likely to bear only *his* children. Along the way, speech became the "language" of love.

Wordplay greatly enhanced the richness of interaction between the sexes. It transmuted raw desire into artistry, creating poetry, literature, haunting love songs, and purple love letters. The sweet murmurings passing back and forth between lovers oblivious to the rest of the world represent a special language, quite distinct from speech's more quotidian functions.

Speech became a critical factor ensuring a man's genetic success. Though triumphal hunting remained important in mate selection, men who were the most convincing talkers increased their chances of leaving offspring. It mattered little that a man could yak endlessly with the other men about hunting.

Nor did it matter that he could out-pious the shaman when praying to the gods, or if he was the only one who could commune with the beasts. If he couldn't get it right when talking with a woman so that she agreed to let him enter her, he was finished genetically. (This assumes that men had not yet structured society to strip women of their power of Original Choice.)

In the game of "Genomic Musical Chairs," smooth-talking DNA made it to the next generation and inept-talking DNA did not. Any male who could not successfully position his sperm to be next to an ovum when the music stopped was eliminated. No chair for him. Agility with words rapidly surpassed looks, strength, display dances, courting calls, and fighting ability to become the most surefire means by which a man could ensure that half his chromosomes would be present and accounted for at the reveille roll call of the next generation. (The relative newness of our species, however, means there will be women who continue to prefer good providers over smooth talkers.)

The evolution of speech also had a major impact on a woman's sexual strategy—both positively and negatively.

On the positive side, speech offered a divining rod with which she could more accurately locate the deepest wellspring from which a man's character courses. From all the male persiflage headed her way, she had to weed out the fast talkers and the sweet talkers from the truth talkers. She needed to evaluate: Was he interested in her just for the sex? Did the character of this man blah-blahing before her possess the staying power to see her through five children, two miscarriages, illness, and wrinkles? Or was he the sort who would abandon her and their brood for younger, greener pastures? Often, a woman had to endure a seemingly endless procession of male encounters. Through many of them, she tried valiantly to appear interested in what the new man was saying, all the while inwardly struggling not to succumb to the MEGO* phenomenon, while she waited for the One.

Then there was that tricky moment when she informed him, usually with a steely resolve in her voice that he had hitherto failed to notice, that from this moment on he must forgo his freewheeling ways and pledge himself to her. Often, a man perceived this as an ultimatum. He balked at what was a major wrench in his genetic agenda of spreading his sperm among as many different women as possible in the shortest period of time. She had to overcome his resistance to a foreign idea that most men believed to be quite onerous and unreasonable. Withholding speech, curiously, became a woman's most effective stratagem in this tug-of-war. (It goes without comment that she would have already withheld sex.) Once a man committed himself by utter-

---

*Mine Eyes Glaze Over.

ing those two simple words, "I do," he learned—often the hard way, as other men over many thousands of years had—that all hell would break loose if he backslid. As the eighteenth-century English playwright William Congreve warned, "Heaven has no rage, like love to hatred turned, nor Hell a fury, like a woman scorned." (The issue of jealous husbands will be given equal treatment in a later chapter.)

Another advantage speech afforded a woman was the ability to compare her judgment of a man with those of other women who knew him. By minutely rehashing a man's every mannerism and statement from the previous night's encounter, a woman could count on the advice of her friends to reach a consensus opinion as to his true worth. She could even discuss her male relationships with women who had *never met* the suitor in question. Women friends can give a woman advice based solely on the descriptions provided to them by the woman seeking their counsel. Using the analytic features of human language to evaluate which male is likely to be the best mate is a skill beyond the ken of other female creatures. Men, in contrast, infrequently avail themselves of this potent function of language.

On the negative side, speech greatly increased the chances that a clever prevaricator might deceive a woman as to his true intentions. Shakespeare warned women in *Much Ado About Nothing:*

> Sigh no more, ladies, sigh no more,
> Men were deceivers ever,
> One foot in sea, and one on shore,
> To one thing constant never.

Though female mendacity poses a similar risk to a male, the stakes of male deceit are much higher. Many a woman has rued the day she met a particularly artful liar. Women needed to become expert at reading between the lines. If a woman garbled the transmission of an unworthy beau, and granted permission for his emission to swim northward in her, she opened herself to a raft of future problems that had *no* simple solutions. Unlike the offspring of the several million other sexually reproducing species, *Gyna sapiens'* children are *forever*. Anthropologists studying hunter-gatherer societies have documented that single mothers fare far less well than mothers who have mates.[26] The same holds true for their children. Any astute observer perusing the daily newspapers can draw a similar conclusion.

More common than outright deceit, a man will often greatly exaggerate his virtues and accomplishments in an effort to impress. This is not unique to humans. Virtually all male animals engage in such puffery. Males want the females upon whom they are temporarily fixated to see them as larger than they really are when they go acourting.*

Gyna sapiens had to learn how to see through a *Homo sapiens'* inevitable braggadocio in order to assess accurately his fitness as a possible partner. (Surely there are a few genuinely shy and modest men, but it is a rare man who could keep a straight face and claim that he had never stretched the truth when trying to inflate a woman's opinion of him.)

Despite its drawbacks, speech was a godsend to the female gender. A woman could evaluate both the content and the tenor of a man's message from a safe distance and decide over time whether she wanted to have sex with him. Since birth-control measures most likely would have been ineffectual or nonexistent at the dawn of our species, she knew her decision was unfairly freighted—grave consequences for her and minimal price for him.

Lest the reader gasp at my audacity for proposing that we began to talk primarily to woo, consider the following. A main reason that males of a wide variety of species initiate communication of any kind is to attract females and entice them to mate. The chirps, howls, bellows, crickets, and whistle clicks of innumerable birds, wolves, lions, insects, and dolphins are principally to convince a discriminating female to choose them. The frog thrumming out his pond's most resonant basso profundo "ribbet," and the bird with the most extensive song repertoire, have the greatest opportunity to mate and subsequently deposit the highest number of offspring into the next generation. Among many diverse species, a good rap is the key to DNA longevity.

Another common reason that a male decides to make a sound of any kind is to alert any other males in the vicinity that he is in charge—or, if he isn't, that he wants to put the resident boss on notice that a challenger awaits him. Maintaining or achieving dominance or protecting or acquiring territory is another common reason males communicate. And what, may we ask, accounts for a male's eagerness to engage in pitched battle? Simply put: A male seeks to maintain or attain a high standing because exalted positions in the pecking order or holding territory is the decisive means by which he can secure access to females.

A corroborating bit of evidence bolstering my thesis is the recent finding of linguist Tecumseh Fitch. He discovered that the male red deer, an animal

---

*This is not universal behavior. Some males try to appear as feminine as possible when they court. A male mountain goat lowers his back, takes tiny steps, and keeps his horns back. Male brown bears slouch and flatten their ears, trying to look smaller, and act playful, presumably so as not to frighten their intended.

previously thought to be mute, is capable of producing a prodigious low-pitched roar when his mating season arrives. A fascinating video capturing the stag's vocalization reveals that, just before beginning his love serenade, the red deer's larynx descends deep into his throat, allowing him to emit a bass mating call that could compare favorably to the sound emanating from an audiophile's high-tech sub-woofer.

Adding an exclamation point to the red deer's swain song, his penis comes thrusting out of its sheath in synchrony with the descent of his larynx. In a coordinated pistoning movement, he vocalizes and thrusts simultaneously, one structure moving down and up as the other moves out and in.[27] I would predict that, could such measurements be recorded in an aroused human male engaged in murmuring sweet nothings into the ear of a female, a similar, nearly undetectable thrusting occurs synchronously, keeping time with the bobbing of his Adam's apple.

Although the impetus to begin talking was similar for each sex, men and women evolved speech for different reasons. Because the two sexes' hormonal systems, brain organizations, and sexual agendas are so different, the gulf between them has proved difficult to bridge. Despite the incredible suppleness of human speech, men and women continue to be confounded by the words and behavior of the opposite sex.

Novelist Margaret Atwood asked women what was their greatest fear concerning men. The most common reply: A man might kill them. When she asked men to confide to her their greatest fear concerning women, the most common response: A woman might laugh at them. The vast disparity between these two anxieties forms the Great Divide inhibiting understanding between the sexes. The potential of dying at the hand of a man is, for women, unfortunately very real. Speech affords a woman the chance to determine in advance (assuming she has time to make this judgment) whether her suitor has the predisposition or intention to harm her. There appears to be no complementary advantage of speech that might prevent a man from making an utter fool of himself in the presence of a woman.

Human language can be used to transfer a veritable host of different messages, and there can be no doubt that each one of its variegated applications was instrumental in driving it to its present state of unfathomable complexity. I propose that the combination of high maternal mortality and low female iron stores contributed substantially in propelling human language into the stratosphere of animal communication.

Parallel with speech's sexual-selection function, language was undoubtedly proving its merit in social relations, toolmaking, child-rearing, and hunting. Once human speech had jump-started, any tiny mutation that came along afterward and incrementally improved speech survived to make it to the next generation. Thus, fluency became the signature feature of the human species.

Steven Pinker, extolling the wonders of human language, observes that information is the sole commodity that a person can give away and keep at the same time.[28] I would add that sexual pleasure is also something that a person can confer on another and personally enjoy simultaneously. The linkage between sex and language can be further divined by noting that the English language tacitly acknowledges that sex was the primary force behind the evolution of speech. I doubt that it is mere coincidence that the word "intercourse" has two common meanings, only one of which refers to speech.

Each of us is a combination of masculine and feminine traits.

Chapter 15

Anima/Animus

He who knows the male, yet cleaves to what is female,
Becomes like a ravine, receiving all the things under heaven.
[Thence] the eternal virtue never leaks away.

—Lao Tzu[1]

The meeting of two personalities is like the contact of two chemical substances.
If there is any reaction, both are transformed.

—Carl Jung[2]

*P*sychiatrist Carl Jung introduced a valuable metaphor when he made the distinction between the masculine *animus* and the feminine *anima*. He further divided each category into a light and dark side. For example, loyalty, heroism, and protection are masculinity's nobler aspects; cruelty, violence, and bullying represent its shadow side. He made similar dichotomies between traits we commonly associate with the feminine. Love, generosity, and compassion illuminate its light side; cunning, masochism, and passive aggression represent the *anima*'s dark aspect. Using Jung's analogy, every individual's psyche, regardless of sex, possesses both a feminine side (*anima*) and a masculine side (*animus*).

As we would expect, women have, in general, a better-developed *anima* than *animus*; the converse holds for men. In any given society, the proportions of these two gender characteristics vary from individual to individual. A similar categorization holds when comparing one culture with another.

Suppose that it were possible for a team of statisticians to tabulate the exact percentages of the *anima* and *animus* coexisting within each individual of

Female

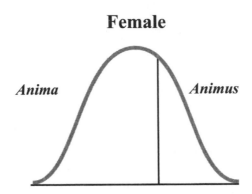

Anima   *Animus*

A hypothetical distribution curve of *anima-animus* ratios within a population of women.

a community and then plot these results on a graph. The dots would trace out two overlapping bell-shaped distribution curves, one curve for each sex. Out at one extreme on the male curve are the few men who seem to be devoid of any feminine aspect. At the other extreme is the minority of men whose feminine side is greater than their masculine. Similarly, there would be a small percentage of women out on one edge of the female curve whose personali-

Male

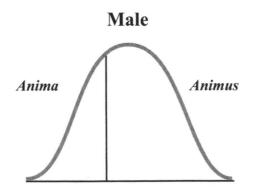

Anima   *Animus*

A hypothetical distribution curve of *anima-animus* ratios within a population of men.

ties expressed their *animus* more prominently than their *anima*. At the opposite extremity would be those women who seem to be entirely feminine without a trace of the masculine. The majority of people would cluster toward the center of the curve, manifesting both an *anima* and an *animus* in generous proportions. Most people would acknowledge that under certain circumstances they have, or know others who have, exhibited the best and worst aspects of the feminine and masculine.

By identifying the *anima-animus* duality, Jung created a useful way to envision human nature. He pointed out that the stunning advancement of the West for the last two thousand years was inspired principally by the *animus*. Working within the framework of cultural psychology, Jung called attention to the glaring lack of a familiar Western symbol denoting balance comparable to the East's yin-yang circle. Evolutionary psychologists rarely comment on Jung's contributions, because there is at present an academic war sputtering between these two disciplines. At the risk of being attacked in the no-man's-land between the two sides' forward positions, I would like to raise a few questions concerning Jung's dichotomy.

Is the *anima-animus* duality common across the evolutionary spectrum? Or is the human species the most prominently psychically split? Did the two-personas-in-one-person arrangement evolve long ago because, in some way, it furthered the survival of some individuals of the human species? Is there a physiological basis for Jung's assertions? Could it be simply a function of testosterone and estrogen?

For the sake of argument, let us assume the masculine-feminine dichotomy has its roots in a biological adaptation mediated by gene mutations. What selective pressure might have encouraged the emergence of such a split personality in this one species?

I take for granted the understanding of how sexual intercourse shuffles the chromosomal deck and that the fifty-fifty random linking of an XX and an XY creates the conditions for the nearly equal debut in the labor-and-delivery room of boys and girls nine months later. I also assume familiarity with the numerous studies showing that administering testosterone to a male or a female animal of any species will make the recipient more aggressive and dominant. The correlation of femininity and estrogen is not as clear-cut, but the preponderance of evidence points to a similar feminine-behavior-enhancing effect.

Among various species, there are few males that exhibit strong feminine qualities, yet all females possess generous proportions of both masculinity and femininity. A mother's ferocity when her babies are threatened can be compared to the most aggressive male display. Few alpha males would dare tangle with an angry mother protecting her brood. Predatory females such as lionesses, tigresses, and leopardesses kill their prey as mercilessly as any male. Domesticated felines appear to enjoy toying with their captured prey before bringing the *Totentanz** to its inevitable conclusion. Among wild pigs, some deer species, lemurs, and elephants, the female, not the male, is dominant. The others in her pack, troop, or herd, both males and females, look to the alpha female for leadership. The presence of both an *anima* and an *animus* coexisting in one animal has a long tradition among the females of all higher animals.

The most extreme example of a female with balls, literally, is the spotted hyena. The female is fiercer, larger, and more dominant than the male. Typically, a female leads the pack and is the more aggressive hunter. Her circulating testosterone is so high that she has external genitalia resembling a scrotum and penis, making it nearly impossible for all but a few specialists to determine the sex of a hyena from a cursory inspection. Globular fat deposits mimicking the appearance of testes round out the deception.†

Should anyone doubt the propensity of testosterone to incite aggression, consider this. Hyenas often conceive twins that are each born with a full set of teeth. The pups, having been marinating in their mother's testosterone throughout their gestation, immediately engage in mortal combat after birth until one defeats and kills the other. There are only a few other species in which immature siblingcide occurs; there is no other known mother that has such high testosterone levels.‡

Is there a male vertebrate's *anima* that would correspond to the female's hyena's *animus*? Although infrequent, examples exist. A male emperor penguin sits for months on his mate's egg, protecting it from the subzero temperatures of the Antarctic. During this time, he forgoes eating. He engages in this trying and selfless act because the female, having initially sat on her egg for a long stretch, has dangerously depleted her body's fat and nutrient stores. She must abandon her perch to replenish her reserves or she will die. At the moment of her forced departure, the male promptly takes over. Should he leave his station for even an instant, the chick within its shell will freeze. Typically, the male loses half his body weight in performing this traditionally female

---

* *Totentanz* is the German word for "dance of death."
†Birth is excruciatingly difficult for the female hyena, not because her pelvic outlet is narrow, as it is in humans, but because she must extrude her fetus through a vagina that has become an elongated narrow tube inside her fake penis. Despite the difficulties associated with hyena birth, however, hyenas do not experience nearly so high a rate of maternal mortality as human mothers do.[3]
‡In some species of sharks, fetuses engage in mortal combat in utero. In a few species of eagles, siblingcide occurs immediately after birth. The stronger pecks the weaker one to death.

task.* After the fledgling has hatched and grown, however, it is nigh unto impossible for an observer to tell whether the male emperor penguin cares about the fate of his offspring.

The largest majority of instances in which the male shares the burdens of child-rearing occur in birds. Only 10 percent of males of mammalian species exhibit this kind of nurturance, and only 3 percent of male mammals enter into monogamous relationships. When males of either phylum help the females feed and protect their young, the males' concern for their helpless offspring rarely includes other fatherly behaviors. The most one could say about these sporadically helpful dads is that their *anima* intermittently appears and disappears.

For the majority of male animals, interest in females rarely extends beyond sex. From the male's observable behavior, he does not seem to express concern or curiosity about a female's state of health, happiness, or economic security. When she is sick, he does not appear solicitous. When she is injured, he does not appear to be helpful. If she finds food, he will frequently take it away from her. And when she has little ones, he expresses little or no interest in them even if they are his.

Prior to mounting her, the typical mammalian male does not take into consideration her physical, emotional, and mental fitness to raise the offspring that he is about to engender. Either he hasn't a clue as to the purpose of sex, or he flat-out doesn't care. He zeroes in on the part of her anatomy that interests him and rarely lets his attention stray from it.

In the majority of species, foreplay, to be distinguished from mating dances and displays, is nonexistent. For those in which it appears that a male is behaving somewhat tenderly toward the female he is about to mount, his solicitude would be judged woefully inadequate by human standards. One of the few exceptions is the occasional male chimpanzee who will spend extensive periods grooming a female, even though she is not in estrus, in the expectation that she will favor him the next time around—a strategy that is usually successful.

Among the 270 species of primates, only a handful can be called monogamous. The most notable example is the gibbon.† A couple's connubial comity, however, stands in contrast to their unprimatelike lack of sociability. Gibbons do not form social groups larger than couples, and each pair expresses extreme hostility toward any other strange male or female. Although

---

*In some species of the tiny seahorse, the male performs the yeoman's duty of nurturing the young while the female, after releasing her eggs, is free to go her own way.
†Marmosets are another.

monogamous, gibbons are relatively asexual: Females come into a short period of estrus only once every two or three years. Primatologists conjecture
that the only way gibbons survive is that two members of the opposite sex
declare a lifelong truce. This cease-fire does not extend to their young for
very long. As soon as a young gibbon can take care of itself, the parents drive
it away. Other than a temporary nuclear family, there is little in a gibbon couple's life-style that can be compared to any other primate's.

In other monogamous mammalian species, such as wolves, coyotes, and
prairie voles, each member of a pair bond seems to take a genuine interest in
the welfare of the other. At the death of one, the other grieves. But these
examples are few and far between. Generally, among social mammals, males
are fissiparous—that is, they form and dissolve temporary coalitions and
alliances periodically, as they fight among themselves to establish dominance. Females with their offspring associate in groups with the other
females. The nuclear-family unit for the majority of mammals is a mother
and her young. Evolutionarily speaking, fatherhood is a very recent—and
wobbly—development.

Most male mammals exhibit precious little *anima*. Take, for example, the
lion. With his circular ruff of mane and imposing stature, a male lion exudes
the pith of masculinity. A mature male lives for the moment that he can challenge the alpha lion and defeat him in a thunderous fight. Sometimes he enlists allies to accomplish the dethronement, but these cabals are usually
temporary. Not uncommonly, one usurper takes over the pride and gains
sexual access to all of its females.

His first act upon achieving this much-sought-for victory is to find and
kill all of the pride's cubs. Instinct has built in this cruel (to us) behavior, because it increases the chances that the victor will pass his genes along to the
next generation. Despite the attempts of lionesses to protect them, the cubs
are defenseless. As he dispatches one cub after another, the new king remains indifferent to the cries of their mothers. Their little ones' deaths cause
all the lionesses to come promptly into heat.

Driven by her powerful sex drive, a lioness approaches the recent mur-
derer of her young and proceeds to "sleep with the enemy." Observers have
reported up to fifty mountings with a single lioness within a pride of twenty
females, each impatiently waiting her turn. Lionesses return repeatedly, urgently demanding another encounter. The alpha male, with or without allies,
is visibly harassed and loses weight, but he never flags. Becoming the alpha

male of any species always boosts a male's testosterone levels, and he appears eager to rise to the occasion.

The leonine sexual act is not tender. He mounts her roughly and thrusts mightily. The entire affair is over in less than a minute. His penis contains a small barb at its tip, which ensures that it will not become dislodged during his bumptious pistoning. When the lion withdraws, the lioness cries out in evident discomfort. His response to her distress is to bite her on the back of her neck, presumably to hold her still.

Though lionesses are chiefly responsible for bringing down a group kill, the alpha male feeds first and has the choice of the most delectable parts. So selfish is he that if one of his young sons encroaches on his prerogatives the alpha may kill him. A survey of other males of other mammalian species reveals that charity, mercy, tenderness, and nurturance, the characteristics traditionally associated with the feminine, are in meager supply. One does not pair the words "fuzzy," "affectionate," and "friendly" with bulls, stags, or boars.

Humans share many attributes with other social mammals. Males ally with males, and females bond with other females. A mother and her children remain the stripped-down nucleus of the nuclear family. However, men behave very differently from other social animals in several fundamental ways.

A professional wrestler, bulging with visible testosterone, can change a baby's diapers with the utmost tenderness. A Marine Corps drill sergeant may weep watching a touching film. A samurai can compose exquisitely sensitive haiku poetry. Hitler, the grotesque epitome of the hypertrophied dark side of the *animus*, adored dogs and little children.

Men (with very few exceptions) possess the capacity for love, empathy, kindness, and compassion. A feeling of awe in the presence of magnificent natural beauty swells the heart of a fierce warrior as much as it lifts the soul of a caring mother. The question arises: Why do males of the human species have such a well-developed feminine side?

A cursory review of any history text confirms that the human male would win any contest when pitted against any other animal for his ability to perform—nay, pleasure in performing—the most vicious and sadistic acts against other members of his species. No other animal tortures its own for the pleasure of it, and no other animal runs torture experiments on other animals. And the human male can sustain such acts of cruelty for incredibly long periods.

How is it, then, that this same male can be found engaging in the most

No male of any other species can match a human male's
penchant for sadism.

charitable acts of kindness to strangers? He can tend to an ailing wife, an elderly mother, or a disabled child with the utmost solicitude. And he can sustain his pathos over a substantial span of his life. Mirroring his high standing
in the torture category, he exceeds every other species' males in the "empathy" department by a similarly wide margin.

Many would argue that these behaviors are entirely a function of culture.

And no male of any other species can match a human male's capacity for tenderness.

Boys can be taught kindness and charity. Warrior cultures, like the Spartans of classical Greece, inculcated their boys with the exact opposite values. Though there can be no doubt that culture plays a significant role in personality development, there is, I believe, a more fundamental neuro-rewiring scheme underlying these two extremes.

Assume for the sake of argument that the cruelty-kindness ensemble is the result of changes that occurred to *Homo sapiens'* nervous systems at the dawn of our species. Assume also that the modules necessary for each behavior pattern evolved because they increased *Homo sapiens'* survival and reproductive potential. What would have compelled Natural Selection to design such an anomalous male? Why so cruel? Why so kind? The simple one-word answer is: sex. The savagery-charity antipodes were parts of a two-step human-male mating dance, the first step of which required steely-hearted dispassion and cruelty.

Homo sapiens inherited a male primate's nervous system, wired for over fifty million years to react to danger as would a typical prey animal. In a rather abrupt makeover, *Homo sapiens* had to switch sides, so to speak. The stalked became the stalker as meat protein increasingly became part of his diet. Millions of years after his direct ancestors developed a taste for meat, *Homo sapiens* discerned meat's power as an aphrodisiac. Goaded by his sex drive, he had to be crueler in order to stalk, chase, and kill more elusive and dangerous prey. The benefit to him was not more meat to eat, but more meat to trade.

Lacking the usual predatory pro-forma weaponry with which to kill, he needed to supplement his puniness with a combination of cleverness and raw savagery. Cold-bloodedness became an indispensable feature of his armamentarium.

Picture the following scene. A hunter has just seriously wounded a deer that lies on the ground immobilized and writhing in pain. The hunter must finish the deer off before butchering it. Suppose, however, he is overcome by a spasm of compassion. Suppose he decides to bring the deer back to the base camp and nurse it back to health. This St. Francis of Assisi approach would ensure that he would later come acourting empty-handed.

The other men would look upon him as either a fool or a slacker. He would have to compete with other hunters who were not so tenderhearted. If he regularly repeated these acts of kindness toward prey, he would fail in his role as a provider and would, most likely, not get the girl, either.

But consider his dilemma if he was a rough fellow without a shred of empathy. He dispatches the deer with a skull-crushing blow. Returning to the base, he thrusts a hunk of his bloody prize unceremoniously in the face of the woman he desires to impress. He arrogantly expects that his offering will be all that is necessary. To his consternation, she quickly disabuses him of these expectations. After she spurns his graceless overture, he is left standing flatfooted and slack-jawed as he watches her stalk off. He has discovered what all men eventually learn: Presents alone will not achieve his aim.

Besides a gift in his hand, he must also have gentleness in his heart. Here, then, is the crux of the male dilemma. *Homo sapiens* has two entirely paradoxical aspects to his mating display: First *kill*, then *court*. One demands a prodigious *animus;* the other requires an ample *anima*. Keeping his natural-born killer aspect separate from that part of him that must be a gentle suitor will become one of *Homo sapiens'* greatest challenges.

Generally, a man's masculinity, his animal magnetism, is the quality that initially attracts a woman. But machismo will mesmerize a woman for only so long (or not at all). As *Gyna sapiens* overrode her libidinous urges and understood the long-term implications of sexual intercourse, she began to look for qualities beyond rugged he-manism. A mutual child-rearing relationship requires that the two parents live in fairly constant proximity. A woman's *anima* must connect with the feminine side of a man in order for the relationship to take root. In other species, intimate long-term relationships are not necessary for genetic continuance. Among monogamous species, only humans must mesh each sex's identity so thoroughly with the other.

Because of the recentness of our species' evolution, sex remains highly influenced by instinctual drives and hormonal concentrations. Pheromones, detectable below conscious awareness, still attract the opposite sex. Gestural clues, the archaic sexual body language, still intrigue. A tongue running lightly and slowly along a glistening lower lip of a slightly open mouth can wordlessly jump-start a sexual adventure between two strangers.

But even though pheromones and body language may bring a man and woman together in the short run, they are not enough to fuel the long haul. For a relationship to progress beyond the initial mutual sexual-attraction phase and continue over decades, the pair must be able to talk to each other.

And so we return to that necessary adjunct of mating, speech. Conversation is the primary means of negotiating a short-term sexual relationship or cementing a long-term parenting relationship. And as in all other aspects of sexuality, a considerable difference exists between the sexes.

For a man, killing to get meat was the first step. Talking his way through a woman's natural resistance was the second. By developing a charming conversational style, a man could entice a woman to let down her guard. This latter skill was as tricky to learn as the first. *Homo sapiens* had to discover, through trial and error, what he could say to please a woman. The best way to accomplish this was for him to *think* like a woman. The only creature on earth that thinks like a woman *is* a woman. Therefore, there had to be a part of him that *was* a woman.

Not only did *Homo sapiens* have to hide the cruelty he found indispensable for hunting, but also he had to assume the persona of a woman when courting. He had to *imagine* what she would like him to say. He had to *imagine* how she would like him to smell. He had to *imagine* how she would like him to dress. He had to *imagine* a pleasing manner he could adopt when presenting her with a gift. If he had failed to study how women respond to male overtures in general, and if he had not paid attention to how this particular woman reacted to his various past actions, he would have imagined incorrectly. A seriously flawed womanly imagination doomed his suit and negated his efforts at killing game for her. There was a high likelihood that she would reject his gift and seek another suitor. All men understood that *Gyna sapiens* possessed an enormous power. He proposed; she disposed.

Homo sapiens dreaded hearing her utter the one word that would completely deflate his ego. That word is not *No!* More often than not, *No!* encouraged a man to redouble his efforts to overcome her resistance. The word he most feared meant that the game was over and he had lost, she wanted to move on. The dreaded word was *Next!*

Courting among humans became an elaborate chess game. *Homo sapiens* aimed to disarm *Gyna sapiens'* queen and render her emotionally defenseless; *Gyna sapiens* sought to capture *Homo sapiens'* king and claim his heart. The difference in the strategies employed by men and women is the reason the relationship game so often ends in stalemate.

Homo sapiens learned that there came a time in his courtship when, the more feminine he became, the more likely he would gain a *Gyna sapiens'* assent. He learned that bringing a woman flowers and composing poetry pleased her immensely. He discovered that his crying, a behavior that would be deeply humiliating in the company of men, paradoxically reassured her.

Those males who had been programmed to learn the lesson well left more descendants than those who did not. Selective pressures favored men who possessed a generous *anima* over those who did not. Through their exercise of choice, women brought about this change. Slowly, the gene pool filled with men who had both strong *animus* and *anima* genes. Among male animals, this was an extraordinary evolutionary development.

*   *   *

Within the psyche of most men, a war rages. A man's *animus* strongly opposes engaging in the kind of shilly-shallying that is the modus operandi of his *anima*. It lusts to treat a woman as a sexual object, something to satisfy his urges so that he can move on. The impetus for sex arises deep within his reptilian brain and is activated by pelvic reflexes carried by fine filamentous nerves buried beneath the coils of his bowels. Despite the inaccessible location of his sexual circuitry, a man's sex drive is closer to the surface than a woman's.

Sexual tension resembles an appetite, not unlike hunger. Before fulfillment can be achieved, both require a man to be courageous and aggressive. As a hunter, he learned to objectify prey. Many men transfer techniques they use for the hunt to the chase known as *Cherchez la femme.** Young men further objectify women by subconsciously borrowing the language of the hunt to refer to the women they are "pursuing," calling them "foxes" or "birds."

Unfortunately, a small percentage of males confuse the different behaviors needed to kill and court. Hunting routines become wooing techniques, transforming these men into "stalkers" who believe delusionally that the use of fear and obsessive persistence will advance their suit. A woman must call upon all her intuitive skills to identify whether a man has these tendencies. Harassment and possible death await the woman who fails to make this assessment accurately.

Sex researchers Nancy and Randy Thornhill posed the following question to a large sampling of men: "If you could rape a woman, knowing with certainty that there would be no chance you would get caught and no one would ever find out, would you commit the act?" Unfortunately, in one of their surveys, 35 percent of the respondents answered that there was some likelihood they would commit rape.[4] This hypothetical question becomes frighteningly real during the conditions of war. Susan Brownmiller documented the high incidence of rape during wartime.[5] Otherwise "civilized" soldiers commit rape under the conditions that the Thornhills' survey hypothesizes.

Although rape is far more common in humans than it is among other animals, it remains a relatively infrequent male sexual strategy. The reasons are many. Among the foremost, rape is extremely dangerous for a man. Some women are as physically strong as some men, and a few women are more powerful than many men. Ranked just beneath fear of losing one's life, repulsion of a rapist often activates in a woman a superhuman strength, enabling her to mount a fearsome resistance. Given these facts, rape is not a particularly effective sexual strategy for a male to pursue.

The human male's genital assemblage is extremely vulnerable to injury. The penis is not encased in a retractable protective sheath, as it is in most

---

*"Search for the woman" in French.

other male animals.* Nor do strong ligaments anchor it to the muscles of the abdominal wall, as they do in every other male mammal. Pendulously swinging when he walks or runs, the human penis is attached primarily by mere skin alone. And it differs from other mammalian members in that it is not strengthened and supported by a rigid bone (bacculum) running down its length. Instead, *Homo sapiens* must depend entirely on a cockamamie system of trap doors opening and closing synchronously in the blood vessels coming and going to his inflatable appendage to sequester the requisite amount of blood under sufficient pressure for him to maintain an erection.† Performance anxiety is never a concern for a mammal equipped with a stiff bone in his penis.

The human male's testes hang loose in an unprotected bag of exceedingly thin scrotal skin. In contrast, the genitalia of male birds and other male mammals are generally safely tucked either inside the abdomen or under and between sturdy haunches, out of harm's way. When the hominid line adopted a bipedal stance, what had been hidden now adorned the male's most forward-advancing part. When a man is aroused, his penis edges out his nose to become his unprotected leading edge. Although a man can inflict serious injury and even kill a rape victim, he risks serious trauma to his manhood in the process. In earlier eras, a rapist risked alerting predators to his raw power play because they would be attracted to the victim's distress calls. Other men (the kin of the rape victim, or the majority of men who abhor rape for moral reasons) would likely rush to the aid of the victim, and the rapist risked a severe beating or even death from this quarter.

In addition, as mentioned earlier, the resting pH level of a woman's vaginal lining creates an environment deadly to sperm. Responding to a lover she desires, a woman secretes a lubricating fluid that has as one of its primary properties the ability to neutralize her vaginal pH level's spermicidal effects. The more foreplay, the more lubricated a woman becomes, the stronger the likelihood that a man's sperm will survive. A rapist's sperm would enter a

---

*The penis was an innovation that first made its appearance two hundred million years ago. Earlier, the female released her eggs into the environment, and the male came along and sprayed his sperm in the general vicinity of her eggs. Use of this hit-or-miss method meant many eggs and sperm never made contact, and predators in the area had a particularly filling dinner during mating times. Birds have a cloaca, an external opening through which a male can eject his sperm and into which a female can receive them, and where her eggs reside. Male birds lack a penis and rely on rubbing the cloaca up against the females'. A penis ensured that the male's sperm would be concentrated in the location most favorable for insemination and that the fertilized conceptus would enjoy the safety of the female's interior during its critical early stages. The penis was an ingenious adaptation. Just ask any man.

†The large amount of sequestered blood necessary for a man to maintain an erection has led many women to conclude that, because a man has but one brain and one penis, he has just enough blood to operate one or the other organ but not both at the same time.

killing field. The effort a considerate lover expends arousing a woman pays off in a longer life span for his sperm. In short, the chances of a man with a well-developed *anima* passing along his genes exceed those of a rapist possessed only with an overbearing *animus*.

We lived as hunter-gatherers in intimate groupings for 99.9 percent of our hominid history. Unlike in contemporary urban environments, where meeting potentially dangerous strangers daily is the norm, rape within extant hunter-gatherer bands is exceedingly rare.

Electrical appliances such as radios and television sets contain a vital component called a capacitor, a flat piece of copper that stores electrical charges. Electrical energy flows down a wire leading to the capacitor as water courses down a narrow stream. The capacitor resembles a placid lake into which various streams empty. Over the course of time, the capacitor, like a lake, fills to the limit. The stored energy must be discharged; otherwise a fuse will blow. Or, as in a full lake, water will burst a dam or flood over the banks. The metaphor of a capacitor can aid in explaining why men and women stay together for so many years even after the intense phase of their joint child-rearing duties is finished. Many sociologists and theologians have weighed in on this issue, but I would like to present what I believe to be the primary reason.

Each morning, after a good night's sleep, men and women arise fresh and ready to face the day's tasks and interpersonal contacts. Each small event and interaction generates a charge that incrementally accumulates on their internal capacitors. By the end of a day, their capacitors are brimming and need to be discharged. The best way to drain a human capacitor is through talking with another. In conversation, the emotional tension (analogous to an electrical charge) that has built up gradually leaks away.

Women often have one or more women friends who serve as good listeners with whom they share the events of their daily lives. Men do not usually discuss the minor details of their day among themselves. When mates reunite after a busy day, they *need* each other in order to mutually discharge their respective capacitors. The ensuing small talk serves to maintain the mental health of both partners.

Even though a woman may not find the man's discussion of his workday inherently interesting and vice versa, they both pay attention, because each understands that this gossip and small talk about the minor maintenance details of life becomes an increasingly important element that eventually supersedes sex as an attractive force binding the couple together throughout the years.

If sexual attraction is the major fugue of romance, small talk is the minor. Couples who are unwilling or unable to engage in seemingly inconsequential conversations will find the music slowly fading away. A healthy relationship begins by building on sexual attraction, common interests, mutual delight, warmth, and support. Chitchat plays an indispensable role in sustaining it.

Men discover that running their daily happenings past their mates provides them with access to a feminine point of view. Similarly, a woman who recounts the events of her day will often receive advice steeped in a strong male perspective. Each member of a couple gains access to the opposite side of his or her own personality, enhancing the minor side.

A common observation is that couples involved in long relationships begin to look alike. It is no wonder, since a similar process is occurring inside their minds and psyches. Couples attuned to each other begin to speak the same sentences at the same moment. Slowly, their opinions begin to converge on both central and peripheral matters. The two partners, by blending their respective *anima* and *animus* with the antipode of the other partner, are, in essence, becoming one person. Relationships ignite in youth fueled by the passion of sexual attraction; the aura of intimacy and familiarity emanating from an older couple engaged in small talk resembles the glowing embers of an old fire that provides a steady warmth even though the flame is barely visible.

*   *   *

A man is highly stimulated by his visual sense. A woman need say little to arouse him. By slowly and artfully removing her clothes, a woman can thoroughly flummox a man. If asked, a man will say that he is "attracted" to a particular woman without precisely knowing why. For many men, this attraction fixates on external appearance.* Generally, after his visual assessment has piqued his interest, a man becomes drawn in by the expression of a woman's *anima.* Belatedly, and in many cases to his regret, he is made aware of the potency of her *animus.*

In contrast, a woman expresses keen interest in both sides of a man's psyche at the very outset. His *animus,* like a woman's *anima,* is easily apparent. She can judge it by his size, strength, health, and the way he walks, dresses, or cocks his head. A man's *anima,* however, requires more discernment. She, more than he, needs to become a sleuth.

---

*Anthropologist Bobbi Low posits that a man unconsciously calculates a woman's hip-to-waist ratio, then uses this number to determine her fitness to bear children, decide if she is currently pregnant, and calculate whether or not she has borne children to another man.[6]

A man reveals his *anima* through actions and words. Gestures of generosity and attentiveness, facial expressions of concern, and demonstrations of interest in her as a person are signs that a woman can use to gauge his "soft" side. If he speaks in an engaging manner, if he asks questions about her life and listens to her answers, and if he says considerate things, she can mine his speech, both the content and inflection, for clues.

Empathy is the ability of one animal to imagine and to experience vicariously how another feels emotionally. Intuition is the skill that allows one animal to reconstruct in his or her mind what another is thinking. Females are considerably more adept at both these arts than are men. The most effective way for a female to understand what a male is feeling or thinking is to try to imagine herself as a man. To accomplish this best, a part of her would have to *be* manly. She would need to know how her mate interpreted the world over the course of his life, not just when he was courting. *Gyna sapiens*, too, needed a neural makeover. She required not only an *animus* with which to protect herself and her babies but also one that would provide a window into a man's castle keep.

*   *   *

The setting for Plato's *Symposium* is a ribald drinking party of Athenian elite who are debating the meaning of love. The playwright Aristophanes tells the gathering the following parable. At the dawn of creation, there were three sexes of a single, very strange creature. These precursors of mortals each sported four arms, four legs, and two heads. Aristophanes identified three different sexes: males, females, and hermaphrodites. The males were men on either side. The females were women on either side. And the hermaphrodites were male on one side and female on the other. All three sexes moved by cartwheeling on all eight of their extremities.

Zeus gathered the gods in council to express his concern that these unusual creatures could one day challenge their hegemony. He was loath to exterminate them with his thunderbolts, though, because then there would be no one to bring the gods offerings. He solved the problem by putting each creature into a trance and then splitting it down the middle. Upon awakening, each half only dimly remembered what it had been prior to being cleft in two. Zeus explained to the assembled gods and goddesses the cleverness of his scheme. These creatures would no longer pose a threat to the gods, because they would dissipate their considerable energy by spending the rest of their days searching for their missing halves.

Aristophanes used his parable to explain gay, lesbian, and heterosexual love. And yet there is something about the story that Jung would have undoubtedly found intriguing. In poetic terms, it explains the dilemma at the heart of each person's *anima* and *animus*. Men and women seek each other out to complement that part of their dual natures that needs shoring. A man with a strong masculine side and a weak feminine one often marries a woman who has a strong feminine side and a weak masculine one. There are numerous other combinations that can be annealed to make a whole Aristophanean being.

A well-matched couple represents an organism that is greater than the sum of its parts. Like Aristophanes' mythical chimeras, they reunite to become a single organism with four legs, four arms, and two heads. When conditions are right, such a union can indeed challenge the hegemony of the gods.

Gay and lesbian couples are ubiquitous throughout human culture and history.

Chapter 16

Gay/Lesbian

We are the only species where a small proportion of individuals are exclusively and consistently oriented toward the same sex. Biologically this is a profound puzzle.   —Malcolm Potts and Roger Short[1]

Homosexual relations offer a clear window on the desires of each sex. Every heterosexual relationship is a compromise between the wants of a man and the wants of a woman, so differences between the sexes are minimized. But homosexuals do not have to compromise, and their sex lives showcase human sexuality in pure form.   —Donald Symons[2]

*T*he generous portions of *anima* and *animus* allotted to both men and women resulted in each sex's displaying considerable traits of the other. Psychic hermaphroditism, the two sexes in one individual, appeared at the very outset of the human species, resulting from selection pressures that were radically altering human mating patterns. These in turn were necessary because of the sharp division of labor imposed upon the sexes resulting from *sapients'* energy-sapping prolonged childhoods. For procreation to proceed and children to survive, men had to be killers and lovers simultaneously. For procreation to proceed and children to survive, women had to be cunning and caring simultaneously. As psychic hermaphroditism became an entrenched feature among humans, it contributed to the emergence of another trait unique in the degree of its ubiquity—homosexuality.

A species is a group of animals or plants sharing certain characteristics that set them apart from all others. All species, therefore, are unique. Two of the many human traits that fall into this category—in the fulsome degree to which they have manifested—are psychic hermaphroditism and homosexuality. Though ethologists have identified other species in which members exhibit some homosexual behavior, there is no other species that expresses same-sex exclusive preference as often and as unwaveringly as do members of the human one.

To understand the roots of homosexuality, we must first examine how sex itself is determined. Mammalian embryos pass through a stage during which their sex is indeterminate. Subsequent male and female distinguishing features arise from anatomic precursors that have the potential to develop into either sex. For example, the tissue destined to become the clitoris in the female can become a penis in a male. The labia majora in a female can turn into the scrotal sac in the male. Programmed in humans by X and Y chromosomes, the levels of circulating hormones secreted in the very early weeks of a pregnancy determine which anatomical sexual feature will appear.

All fetuses will develop into females, regardless of whether they contain XX or XY chromosomes, unless they are exposed to high doses of testosterone in utero in the early weeks. *The female is the default mode for all mammalian fetuses.** In normal male embryogenesis,† a male's budding testes will begin secreting prodigious amounts of testosterone a mere six weeks after conception. (At this time, the future person is thirteen millimeters in length and resembles a plump worm.) Genes on his Y chromosome orchestrate the deluge. The thirty-fold rise in testosterone flooding the interior of the tiny amniotic sac will command the fetal XY's body to choose a penis instead of a clitoris and a scrotum instead of labia.[3] This androgenic hormone ("androgen" literally means "man-maker") is also responsible for wiring his developing brain so that later he will see, think, and experience life as a man. Contrary to standard male triumphalist teachings that have been around since antiquity, the mammalian male is derived from the female, and not the other way around. The Genesis rib story has it backward. A clitoris is not simply a stunted penis; rather, a penis is a transformed clitoris. Although each is derived from precursor antecedents, the female body plan rules unless the male essence intervenes.

True hermaphrodites exist in the plant and animal world. Some creatures can literally change sex depending on environmental conditions. The common whip-tailed lizard of the American Southwestern deserts is able to

---

*This is the reverse of the situation among birds, reptiles, and fish. In these phyla, fetuses begin as males and have to be acted upon by estrogen in order to develop into females.
†The development of an embryo.

change from male to female and back again. This ability to sex-switch stops at reptiles; birds and mammals are incapable of the feat. The whiptail is an example of *sequential* hermaphroditism—that is, the creature is a male in one season and a female in the next. The human, in contrast, manifests a *simultaneous* hermaphroditism, in that each person combines elements of male and female all the time.

The imperative that each man and woman be fully emotionally and psychically hermaphroditic was an unprecedented evolutionary development. And it led, I propose, to an unusual variation among humans—a persistent tendency to manifest homosexuality in both sexes.

Historically, homosexuality has been an enduring part of the human condition. Ancient Mesopotamian texts refer to *sag-ur-sags,* a phrase that linguists have deciphered to mean a person of indeterminate sex or one who is gay. Murals from ancient Egypt depict scenes of gay behavior. Inca ceramic statuettes from Peru, dating from around A.D. 500, show men engaging in anal intercourse. Many Greek and Roman vase paintings depict graphic same-sex acts.

Despite the many treatises and books that have been written about its root causes, few satisfying theories have emerged to explain *why* homosexuality effloresced in the human species in so robust a form. Recent research has belatedly begun to identify examples of homosexual behavior in a variety of other species. To date, biologists have identified over two hundred species, from insects to mammals, that engage in homosexual behavior. But these activities are qualitatively different from those that occur between same-sex-preferenced humans.

Many male dogs, sheep, chimps, and mallards have been sighted mounting one another, but no observer has ever witnessed the sexual act actually being consummated. Animal behaviorists hypothesize that these mounting behaviors denote dominance display rather than pure sexual desire. For example, baboon males often approach each other and gently grasp each other's penises. Primatologists refer to this behavior as "diddling." A tension-reducing gesture, akin to the human handshake, it reciprocally assures that neither party has aggressive intentions.[4] Similarly, by extending an open palm upon greeting each other, men signal from a respectable distance that their dominant hand does not conceal a weapon.*

In some of the more complex animal species, both males and females will occasionally use sex to defuse tense social situations, gain advantage, make

---

*Ancient Romans would conclude a transaction by placing their testicles in each other's palm while mutually reciting an oath—the ultimate expression of trust. The words "testimony" and "testament" both derive from this ancient practice.[5]

allies, and barter for food. Bonobo chimps of both sexes engage in behavior that could be construed as homosexual, but the animals involved do not limit their sexual advances to members of the same sex. Bonobos are thoroughly bisexual. Humans, too, may engage in sex for any number of different reasons, but no other single species has yet been identified in which a significant number of its population prefers sexual congress *exclusively* with the same sex.

From an evolutionary point of view, homosexuality is a supreme paradox. Assume it is driven by genetics and one would quickly conclude that the gene controlling it guarantees the trait's extinction. How could a homosexual gene survive if the person possessing it does not desire to reproduce? A nonreplicating "unselfish" gene is an oxymoron. In theory, such a gene should be gleaned from the genome within a few generations.

Yet researcher Dean Hamer and his co-workers reported in 1993 that they had identified a gene that plays a critical role in determining homosexual behavior.* Research in this field is in its infancy, but increasingly scientists view the predisposition for homosexuality to be in large part genetically determined.

"Homosexual" literally means "same sex." The "homo" in the word is Greek for "same," not, as one might assume, "man." Because the word "homosexuality" is both so confusing and so emotionally freighted for many gays and straights alike, in the following discussion I will replace it whenever possible with the phrase "exclusive same-sex preference," hereafter designated by the acronym ESSP. I will also use the term "gay" in referring to males, "lesbian" for females, and "straight" in place of "heterosexual." I should mention that there has been a paucity of scientific studies on lesbianism, compared with what has been learned about gays. In the following discussion, more space will be devoted to gays than lesbians because of this evidential bias.

Intriguing evidence has emerged that the birth rank among siblings may influence ESSP. Researchers conjecture that the presence of a male child in the uterus somehow alters the uterine environment so that subsequent male fetuses have a higher statistical probability of being gay than firstborn sons.[7]

Another unexpected finding is that parents conceive a higher number of gay males in times of great stress—for example, the generation of children born during World War II. During their pregnancies, the mothers of these children would have had higher levels of circulating cortisol, the anxiety hormone.[8] Cortisol competes to some extent with the sex hormone testosterone. Researchers hypothesize that somehow this influences the sexual orientation of boys born during these periods, even though there is no difference in the

---

*Tracing families with the gene, Hamer and his co-workers found an incidence of male homosexuality in 13.5 percent of male relatives, compared with only 2 percent in the control group.[6]

levels of testosterone between male heterosexuals and male homosexuals, or in the levels of estrogen in female homosexuals and female heterosexuals, and both sexes have the correct amounts of pituitary gonadotropic (ovary-and-testes enhancing) hormones regardless of their sexual orientation."[9]

Richard Swaab, a Dutch researcher, believes he has pinpointed a clump of cells (called the BSTc) within the amygdala of the brain that he believes may determine sexual orientation.† Straight men have a larger BSTc than straight women, and this area is half again as large in women as it is in transsexual women (who believe they are women trapped in men's bodies). Smaller still is this minute portion of the nervous system in gay men.[10] Simon LeVay, a gay neuroscientific researcher, has also claimed progress in identifying where in the hypothalamus gayness resides.[11] Hamer may provide an answer to the question of how, and Swaab's and LeVay's research may give us insight as to the where of gayness, but neither provides a compelling evolutionary theory of why.

For many years, it was commonly held that psychological factors played the leading role. Many theorized that a disconnected father and an overbearing mother caused boys to become gay. Recently, this Freudian view has been stood on its head. Instead of being the *cause* for boys' becoming gay, some have speculated that it may be the *result* of boys' expressing their gayness. Many fathers, observing that their sons are less manly than they expect them to be, distance themselves emotionally. The boys' mothers, to compensate, become more protective. Blaming mothers for deficient nurturing has been a convenient excuse that fathers have often used to deflect from themselves any possible responsibility for their sons' sexual orientation.

Many in fundamentalist religions believe homosexuality is a curse visited on gays by a vengeful god who is displeased by what these self-righteous people consider to be a sin against nature. Since fundamentalists claim that their God is omniscient and omnipotent, it is not clear from their arguments why a God who possessed such power and foresight would create mortals who were born to sin against Him. Renaissance humanist Erasmus considered this line of reasoning to represent the direst blasphemy. He believed that such a God would be a monster, unworthy of worship.[12] The tortured arguments used by many fundamentalists to justify their intolerant public de-

---

*Freudians would interpret the same data differently, positing that it was the absence of fathers during wartime that was the primary causative factor.

†"Amygdala" is the Greek word for "almond." A part of the hypothalamus, this centrally located, small, but very important almond-shaped segment of the limbic system plays a vital role in regulating the emotional behavior of mammals.

nunciations of gays and lesbians proves only that the Dark Ages have not entirely dispelled.

Until very recently, the gay-and-lesbian life-style was considered to be a mental disease in Western societies. Not until 1973 did members of the American Psychiatric Association, in an exceedingly contentious session, vote to remove homosexuality from its list of pathological mental conditions. The nature-nurture debate seesaws back and forth without a clear-cut resolution. The reasons why ESSP evolved so exuberantly in only the human species remain as puzzling as ever.

Perhaps the installation into each sex's nervous system of both a robust *anima* and *animus* provides the missing clue. Let us revisit the two overlapping bell-shaped curves on page 210 that graphically plotted each sex's masculine and feminine aspects. Combining both the male and female curves into one and exaggerating the extremities of both creates a graph as represented below. Both the masculine and feminine sides contain outliers on either end that manifest a skewing of one trait over the other. I believe that it is in these rarefied sections that a partial answer to the puzzle of gays and lesbians resides. Let us examine first the extremes of the male *anima-animus* distribution curve. At the left-side extreme are those males who possess an *anima* so well developed that it overshadows their *animus*.

The ability of ESSP individuals to form bonds is a function of a further du-

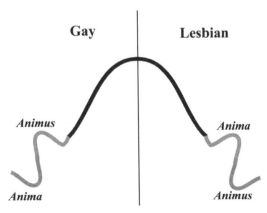

A speculative graph of the edges of the *anima-animus* curve within a society.

ality existing within this left-sided group. Half of these males possess an assertive *animus,* while the other half has a diminished sense of masculinity. This additional dichotomy creates a subset that contains complementary antipodes. Sexual attraction in all species depends on the tension generated by polar opposites. Having one half of the male outliers on this edge of the curve combine an exuberant *anima* with an anemic *animus,* and the other half combine a still-strong *anima* that must coexist with a slightly stronger *animus,* creates the necessary conditions conducive for male-male relationships to come into existence.

A gay male possessed of an assertive *animus* generally bonds with a male who has an excess of *anima*–and vice versa. This mini-bell-shaped curve within a larger bell-shaped curve provides the fractal layering necessary for gay men to form pair bonds in which lust, sex, love, and caring between two men can thrive. Because we are dealing with the subtle shadings of distribution curves, a continuum forms in which a preference for bisexuality also fits into this scheme.

The same dichotomy exists on the right-hand edge of the female curve and can be used to explain the attraction between lesbians. Half of the women at this extremity possess an *animus* stronger than their *anima,* while the other half's feminine manages to exceed in expression the woman's masculine. Within this female dichotomous subset, desire between two women can blossom. In both gay and lesbian relationships (as in straight relationships), it is the rule rather than the exception for one member of the dyad to play the dominant role while the other plays the role of the submissive. The tension necessary to attract opposites would not be possible unless each half of an ESSP population contained disproportionate elements of both *anima* and *animus.*

Located at the other extreme of the male distribution curve in the figure on page 234 are those few males whose character is all *animus,* bereft of an *anima.* These men lack the fundamental social skills sufficient to allow them to interact gracefully with members of the same sex. They are especially ill-at-ease when trying to relate to women. Their lack of empathy and intuition, combined with their unrestrained *animus*-driven penchant for aggression and violence, results in their disconnecting from society. The community, in turn, ostracizes them for their failure to blend in, and they become outcasts. The ranks of rapists, criminals, and murderers are filled with hyperanimus, antisocial, hypoanima males. Psychiatrists diagnose these right-sided male outliers, positioned at the opposite pole from gay men, as psychopaths and sociopaths. (The woof and warp of human behavior is so textured that psychopathic and sociopathic gays also exist.)

Schizophrenia, autism, and the latter's milder form, Asperger syndrome, are psychiatric conditions in which the individual, most commonly a male, seems to be missing the part of his nervous system that promotes healthy human relationships.* Someone afflicted with autism prefers to be alone, lacks empathy, and displays a single-minded pursuit of solitary interests. Men with these conditions can be characterized as having a very weak *anima*, and they would generally cluster at the opposite end of the curve from gay men. In general, habitual criminals have too much *animus*, and autistic men and those diagnosed with Asperger syndrome have too little *anima*.

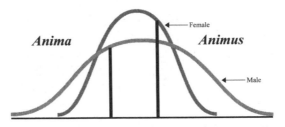

The male *anima-animus* distribution curve is flatter and longer than the female one. There are more extremes of male behavior than there are of female behavior.

Mother Nature's delicate task of creating a male who could be both a lover and killer was exceedingly difficult. The male bell-shaped *anima-animus* distribution curve is therefore flatter and wider than the female one, resulting in greater numbers of individuals who exist out at either extreme. There are more gays than lesbians. There are more male psychopaths than female psychopaths. There are more autistic and schizophrenic men than there are autistic and schizophrenic women. Asperger syndrome afflicts approximately one male in three hundred. It is exceedingly rare in females. When the two male extremes are compared with all other species, certain aspects are thrown into sharp relief. The behavior humans attribute to rapists and murderers is exceptionally rare among wild animals. So, too, are behaviors consistent with the symptoms of autism and schizophrenia. Exclusively homosexual males are nearly nonexistent in the wild.

It seems that the consequence for human culture of Mother Nature's attempts to create a balanced male is a peculiar pairing of opposites. Whenever the Red Queen introduces a talented Oscar Wilde, She unfortunately also

---

*An Asperger child has difficulty trying to understand what others are thinking and feeling, causing him to have problematic human interactions. The syndrome describes the behavior of many male "loners."

seems to plague culture with a Jack the Ripper.* For every creative, aesthetically inclined, relationship-attuned gay man, there appears, it seems, a nongay inchoate male schizophrenic.

Corroborative evidence that Mother Nature had difficulty titrating the male *animus* and *anima* can be observed in the behavior of postpubertal males. At a time when male testosterone surges, boys struggle to establish their gender. Young males are far more confused about their sexual identity than are young females. They exhibit more gay behavior than older men, and their exhibitionism exceeds, by a wide margin, lesbian exhibitionist behavior among young women. Paradoxically, because of this gender confusion, young straight men are far more homophobic than are older men.

Homophobia is rampant in the majority of societies because every man is a blend of masculine and feminine. In cultures that venerate the former values over the latter, many men cannot come to terms with their *anima*. Not knowing how to deal with a part of themselves they consider foreign, they project their unease onto gay men, who have obviously accepted what they cannot. A man who acknowledges both his *anima* and his *animus* is not threatened by the existence of gay men.

Sociologists have conducted numerous studies attempting to determine the proportions within a society of gays and lesbians. Virtually all of these studies depend on voluntary surveys conducted by questionnaires. A hidden political agenda motivates many of them. The scientific surveys most often cited are the Kinsey Report and the work of Masters and Johnson, both of which relied on the forthrightness of the respondents.† A recent study by Laumann and his associates reported the numbers of males who acknowledge that they are ESSP to be 2.8 percent. In his study, only 1 percent of women responded that they were lesbian.[13]

Gays and lesbians, however, smile when told of these statistics, because they know better. The evolutionary biologist Edward O. Wilson estimated in his 1978 book, *On Human Nature,* that in any given male population approximately 8–13 percent of males and 5 percent of females have strong ESSP inclinations.‡[14] Wilson drew on the research of Alfred Kinsey, who took great pains to point out that there was a sliding scale between exclusive heterosexuality and pure ESSP.[15] There remained many shades of gray. Some men experienced ESSP only for a brief period of time in their lives. Others experimented

---

*The serial killer Jack the Ripper terrorized London in 1888. He acquired his moniker because after he murdered a woman he mutilated her by removing her uterus, which he kept.

†Remember, during the fifties, when the Kinsey surveys took place, psychiatrists still believed that homosexuality was a pathological condition, and many states had statutes making consensual sex between same-sex partners a crime. A few still do.

‡Faced with quite divergent sets, I choose to use Wilson's statistics, since I believe they are more accurate.

on both sides of the scale, and Kinsey also defined the parameters of bisexuality. Kinsey concluded, after spending twenty years studying the sexual habits of Americans at midcentury, that "one in ten men has had strong homosexual tendencies for at least three years in his life between the ages of 16 and 55."[16]

Wilson estimates that the lower limit of his 8–13 percent figure has been surprisingly stable throughout history. The number of men willing to openly acknowledge their gayness depends, however, on prevailing attitudes within the larger culture. In classical Greece, for instance, it was socially acceptable for a man to express his same-sex sexual preferences. Homer's *Iliad* repeatedly extolled male friendships, such as the one between Achilles and Patroclus, as more noble and pure than heterosexual relations. From historical accounts, classical Greece contained many gay men, and bisexuality was far more common then than it is today. At the battle of the Plain of Chaeronea, in which the Thebans defended their city against the invading Macedonians in 338 B.C., Theban warriors fought alongside each other as pairs. The Sacred Legion consisted of gay soldiers paired with their lovers. (To many Greeks, a soldier could hardly be expected to risk his life for another unless he loved him.) Today, there are many nodal points in the world that follow the example of ancient Greece, from San Francisco's Castro district to the Greek island of Mykonos.

On the other hand, the Protestant Reformation and the Catholic Church's Counter-Reformation forced gay alliances underground at a time when even an offhand accusation could lead to burning at the stake, a not uncommon fate for gays in Calvin's Geneva, Pope Paul IV's Rome, Isabella and Ferdinand's Spain, and elsewhere. Many a so-called witch was incinerated because others had accused her of being a lesbian. Three centuries later, in Victorian England, the brilliant, witty playwright and married father of two, Oscar Wilde, suffered a debilitating prison term for publicly acknowledging "the love that dare not speak its name."

The attitudes and practices regarding sex in the gay and lesbian communities provide a window into the intricate, accommodating dance straights must perform. In general, gays and lesbians are freer from the various traditional male-female constraints.

Prior to the AIDS epidemic (which dramatically altered the sexual behavior of gays), multiple partners and/or multiple acts in an evening were not uncommon in many of the bathhouses that flourished in metropolitan centers.

The randiest straight male could rarely match the number of encounters available to the sexually adventuresome gay.

Gay sex has a much higher incidence of anonymity or near anonymity than does straight sex. One particularly hypersexual gay man recorded forty-eight anonymous sexual encounters in a single evening.[17] These may occur in assignations so brief and casual that any kind of emotional commitment is nonexistent. Of course, not all gay men choose to behave in this manner, but enough of them do to provide a stark contrast to the sexual habits of straight men.

The actions of these members of the gay male community suggest that if straight men could have their druthers many would also prefer to indulge in sex in this manner. However, straight men are far less likely to find partners eager to engage in what the novelist Erica Jong in her 1973 book *Fear of Flying* called a "zipless f—k." A straight man would find it exceedingly difficult to locate a group of enthusiastic women to join him for an orgy of anonymous sex. Explicit sex clubs, and peep shows with private booths for straights, have cropped up all around the world. But, unlike his gay counterpart, a straight male must pay money to receive these benefits. (Gay prostitution also exists primarily in the context of age differences between consenting gays. Almost without exception, the older man pays the younger one.)

In contrast, lesbian women are rarely interested in anonymous sex, nor do they exhibit the hyperkinetic frenzy that is evident in some of their gay male counterparts. More often, they form long-lasting committed relationships. Among themselves, anthropologist Meredith Small reports, lesbians wryly refer to the phenomenon they call "lesbian bed death."[18] After an initial frenzy of passionate lovemaking, it is not uncommon for a lesbian couple to settle into a routine in which sex plays a remarkably minor role.

The principal reason for the marked differences between the sexual encounters of both gays and lesbians and those of straights is the former's lack of the need to compromise when negotiating sex. Further, each partner brings to an encounter the same sexual equipment, ensuring that each knows how the other's operates. Each partner in a male-male or female-female sex act has more-or-less similar expectations, because they are of the same sex. Contrasting the differences between gay sex and lesbian sex throws into stark relief the compromises each straight man and woman must make to accommodate his or her partner.

Nowhere are these negotiations more clearly evident than in the "world's oldest profession." The extraordinary scope and persistence of prostitution attests to many males' desire to seek sexual relief without any emotional strings. In discussions among themselves, many men express the belief that exchanging money for sex may actually be the least expensive way—emotionally, psychologically, and financially—to alleviate the male's near-constant sexual burr.

But it is not only sex that men purchase. One young sex-worker, in a conversation with her older and more experienced madam, marveled at how men were willing to part with their hard-earned money for what to her seemed so brief a sexual release. The older woman explained, "You are mistaken if you believe that your john is paying you only to have sex with him. He is also paying you to *go away after the sex is over.*"

Hormones, sexual attraction, and love continue to play prominent roles in sexual union, as they have since the outset of our species. Factoring these out of the equation, men often attempt to achieve intercourse with a woman without a commitment, whereas women rarely acquiesce prior to obtaining it. There are many gradations of commitment, but a woman most commonly seeks a man who will offer her resources to initiate sex. She also must determine if he would be willing to "provision" her if she gets pregnant. And, most important, he has to demonstrate that he has the character that would motivate him to remain to help raise their child, and to continue to grubstake his family with a steady supply of resources. Add to this laundry list that he would enter into such an arrangement willingly, displaying eagerness and joy. The archetypal two-parent family remains the best social instrument for ensuring that a child will mature in an emotionally healthy and economically stable environment. And the creation of offspring that mature to beget offspring is the primary force propelling the process of evolution. That being the case, could there be other, subtler reasons why homosexuality persists in our species?

The human species manifests features of each sex within each individual.

Chapter 17

Same Sex/Hermaphrodite

A capon will sit upon eggs, as well as, and often better than a female. This is full of interest; for [there are] latent instincts even in the brain of the male. Every animal is surely hermaphrodite.   —Charles Darwin[1]

In each of us two powers preside, one male, one female; and in the man's brain, the man predominates over the woman, and in the woman's brain, the woman predominates over the man. . . . If one is a man, still the woman part of the brain must have effect; and a woman also must have intercourse with the man in her.
—Virginia Woolf[2]

*A*lthough poorly understood, selective pressures in the human species have ensured gay behavior in every generation and in every culture. ESSP's ubiquity suggests that it is a very old trait and was likely present at the dawn of our species. Perhaps this unique development arose from the need to enhance the survival of a primate species that had precipitously switched its diet from roots and shoots to haunches and hamhocks in order to provision burdened mothers and their children with intelligence-enhancing brain food. The insertion of ESSP into the human genome was one of four adaptations that Natural Selection cleverly slipped in among the chromosomes to assist a newly minted *Homo sapiens* hunter in his deadly competition with other species for survival. A dollop of the same mysterious formula was also ladled into *Gyna sapiens* chromosomal potpourri, so that lesbians would also be mixed into each generation. Before discussing the wherefores of lesbianism, let me propose an additional hypothesis to explain the evolutionary whys of gays.

I shall call my thesis the Theory of Eights.* Four unique human traits appear in any given *Homo sapiens* population, and each one uncannily hovers around the stable level of 8 percent of the males. The four are ESSP, color-blindness, left-handedness, and baldness. Eight percent roughly equals about one out of twelve men. I believe that these four traits taken together represent a constellation of genetic adaptations that enhanced the success of the original human male hunting band. Let us examine each one of these, beginning with ESSP.

Evolutionary biologist William Hamilton proposed that a brother (or a sister) is more invested in helping his sister's children than he would be in assisting unrelated children. In a series of experiments, Hamilton reaffirmed the adage that "blood is thicker than water." A man's niece and nephew carry one-quarter of the same genes as he does. If he aids them and they live long enough to reproduce, then some of his genetic material wiggles into the next generation, even if he himself does not father any offspring.[3] Anthropologist Robert Trivers greatly expanded Hamilton's theories and used them to explain the prevalence of human altruism.[4] Building on Hamilton's and Trivers's work, E. O. Wilson proposed that the emergence of human homosexuality increased the likelihood that ancestral children would survive.

Raising a child to maturity is a perilous, difficult, and energetically expensive exercise. Having uncles or aunts who do not have children of their own is a plus for any child. Along with a mother, father, and grandparents, an additional set of undistracted avuncular relatives who have the child's interests at heart increases the chances that the child will pass on the many different genes of disparate members of the earlier generations. Gays incrementally increase the fitness of a species that carries heavy child-care responsibilities.[5] Before I discuss the crux of Wilson's argument concerning the advantage ESSP conferred on the hunting band, another area of anthropological research must be examined.

Using brain measurements and making comparisons to extant hunter-gatherer tribes, anthropologists Leslie Aiello and Robin Dunbar have correlated the size of the hominid cortex with the ideal size for an archaic hunter-gatherer society. They estimate that the approximate number of the optimal *sapient* group—including babies, children, adolescents, elders, and

---

*Lesbianism is not so prominent a feature of our genome as is male gayness, and I do not have a female version of the Theory of Eights.

infirm—would be somewhere between 100 and 150.* Within each Pleistocene band, the core group of hunters contained from eight to twelve men in their prime.

The template of an eight-to-twelve-man hunting band is still discernible in contemporary and historical societies. Ten soldiers constitute a squad, corporations consider twelve adults the ideal size for a board of directors, there are nine players on a baseball team, eleven members on a football team, nine Supreme Court justices, twelve members of a jury, twelve apostles, twelve deities formed the Golden Circle in ancient Greece, and ten Jewish men comprise a prayer minyan. It takes approximately nine to twelve vigorous adults to undertake a communal project, lend to it their combined wisdom, energy, and cooperation, and see it through to a successful conclusion.

Wilson argued that if there was one man out of the twelve who did not have a wife and small children to feed, then a hefty portion of his share of the kill could be distributed among the remaining hunters' families. Assuming that the ratio within ancestral populations was similar to contemporary ratios, the presence of a gay hunter among eight to eleven straights would have increased the meat supply available to women and children by roughly 8 percent.

Even though the hunting group no longer forms the essential core of society, gays still contribute to the general welfare by playing a similar life-enhancing role within the culture at large. Meat segued into resources, which, in turn, have transformed into capital. Excess capital is one of the reasons gay contributions to the larger community are disproportionate to their numbers.

Additionally, gays enrich the *frisson* of their culture's life because they seem to possess, on average, aesthetic sensibilities and creative abilities in greater proportions than are present in the average population. The reasons for this skewing relate to the *anima-animus* discussion from the previous chapter. The connection between beauty, creativity, and openly gay males is plainly evident in the extraordinary blossoming of the historical periods we deem "Golden Ages"—classical Greece and the Italian Renaissance, to cite two, both of which had many high-profile ESSP contributors.

Gay representation in the arts and literature has ennobled human culture throughout the ages. But these are not the only fields in which gays have made their mark. Architecture, music, religion, the military, science, mathematics, academia, and many other fields can claim gay geniuses. Human culture would be grayer and less sumptuous had the following men been interested in heterosexual pursuits and channeled their energies into a

---

*Aiello and Dunbar point out that 150 seems to be the limit contemporary people would call friends they know well enough to invite to a wedding. Most personal address books do not contain more than 150 names whom the owner would categorize as friends.[6]

workaday life: Socrates, Plato, Aristotle, Tchaikovsky, Leonardo, Newton, Milton, Michelangelo, and Nietzsche.*

This increased sensibility to aesthetics, harmony, and creativity could be the result of differences in the organization of the gay brain. The corpus callosum is the broad band of connecting fibers joining the two cortical hemispheres. In a gay male, it is approximately 15 percent larger than that of a straight male, and there is less specialization between the two lobes in gays than in straights.† Language centers are more evenly distributed. Ambidexterity, an indication of less left-hemisphere dominance, is also more common among gays and lesbians. The degree of differences discovered so far between the size of the corpus callosum and specialization of the hemispheres in gay men lies midway between the variance present between a right-handed heterosexual male's brain and a right-handed heterosexual female's brain.[8] We are just beginning to understand the subtleties of how the brain functions differently in these subgroups.

Whether these differences in brain organization are responsible for what appears to be the enhanced aesthetic and creative spirit in gay men is still unknown. Not every gay is more creative and sensitive than every straight male. In general, however, few would dispute that gays more often choose to work in the arts and other fields that require appreciation of beauty and harmony than do straight males. Gayness, a trait that originally evolved to put more meat in the mouths of mothers and babes, has advanced to making contributions that enhance the overall quality of the culture.

Another 8 percent condition found in males is commonly referred to as "color blindness." Technically, true color-blindness (achromatopsia) is an exceedingly rare and debilitating condition. The more familiar inherited color-deficiency is the inability to distinguish between two of the three primary colors, red-green defects being more common than red–other-color defects. Color-deficient men are so little inconvenienced by their disability that many are not even aware they have a problem. Most commonly, a woman calls it to the man's attention by pointing out to him that he is wearing two different-colored socks. The gene responsible for this flaw resides on the X chromosome and is recessive. Therefore, the condition is exceedingly rare in women.[9]

Few species of animals see colors. Some fish, amphibians, and reptiles

---

*Controversy exists concerning the orientation of some of the members of this list. The weight of circumstantial evidence suggests that they were all gay.
†A considerable debate over these findings is presently raging in the neuroscientific community. The preponderance of evidence suggests that these findings are real.[7]

have color vision, as do the majority of birds, but polychromatic vision is poorly represented among mammals. Most of the planet's creatures see the world in gray scale, but their lack of color vision does not pose a handicap because smell (or some other sense) is so superior. Primates possess extraordinary color vision. The vital need to judge distances accurately when swinging through trees made the discernment of slightly different shades of color critical to an arboreal, wingless creature's survival.

A significant number of primates are frugivorous, meaning that their diet's primary staple is fruit. Being able to distinguish a brightly colored orange mango or yellow plantain among distant foliage and gauge its degree of ripeness by its tint is an indispensable visual skill.

Since color vision was so crucial to our ancestors' survival, why would 8 percent of males, but not females, fail to develop it to its fullest potential? Is there a male-centric activity for which a visual color defect could be turned to an advantage?

One of the best defensive shields an animal possesses is an appearance that blends in with its natural environment. Some creatures have perfected this camouflage so well that even when they are pointed out, they remain difficult to isolate from their surroundings when they stand still. But spotting an animal with a coat that is shaded to blend in with its familiar habitat is considerably less difficult for a hunter who does not see the same colors as the others in his band. One color-deficient male within a hunting party would be a valuable asset indeed.

His worth in the Pleistocene can be extrapolated from the way color-deficient soldiers were deployed in recent history. Intelligence corps on both sides in the two world wars actively recruited them to serve as front-line spotters because they, better than their color-visioned comrades-in-arms, could see right through the enemy's efforts to camouflage guns and tanks. Having one fellow out of twelve who could easily spot predators lurking, prey hiding, or a camouflaged enemy would incrementally increase the success of that group.

Over time, selective pressures would have encouraged the spread of a male color-deficient gene in a small segment of the male population. A point of diminishing returns would soon be exceeded if more than one out of eleven or twelve men were color-deficient—then the hunting effectiveness of the entire band diminished—hence the persistence of the 8 percent ratio. Because women rarely engaged in full-scale cooperative hunting or warfare, no selective advantage existed for them to evolve a color deficiency.

The third anomalous skewed masculine quirk is that 8 percent of men are left-handed. One would expect that handedness would be evenly distributed, with 50 percent preferring the right hand and 50 percent preferring the left. This is decidedly not the case. Surveys from widely differing cultures confirm the odd 92-to-8 ratio of righties to lefties the world over.*

Many theories attempt to explain why we, as a species, have such a strong predilection to use the right hand. The one most commonly accepted has to do with the connection between language and handedness. It served Natural Selection to position speech (which is a highly sequential fine motor activity) in the hemisphere that also controlled the preferred hand, which would execute highly sequential fine motor movements—overwhelmingly, the left hemisphere. The need to manipulate the environment concretely through the use of handmade tools and abstract words was economically packaged in the hemisphere that best sequenced information in a linear fashion. The right hemisphere, nearly devoid of language skills, could then devote its entire attention and storage space to processing information in a mode opposite to the left. The right hemisphere typically excels at those tasks that require gestalt, all-at-once awareness, such as face recognition or map reading.

If right-handedness conferred a significant advantage for humans, one might expect that all humans would be right-handed. But the pesky problem of having to explain the 8 percent who are left-handed persists. Left-handedness, like ESSP, color-deficient vision, and baldness, is more common in males than females. But here the disparity between the sexes is not as great.[11] Most surveys find that 8 percent of men are left-handed compared with only 5 percent of women. Nevertheless, this gender skewing remains consistent enough that an explanation for the inequality should be sought.

As I have hypothesized about the previous two male traits, the answer lies in hunting and survival. The presence of one left-handed male hunter among eleven or twelve other, right-handed hunters would have markedly increased their overall success.

Archaic hunting required three essential skills—location of prey, accurate aiming, and effective throwing—all of which depend on excellent vision. A feature of the human visual system would have made the presence of one left-hander out of twelve a distinct asset. The right and left eyes, separated by the bridge of the nose, see the same scene from ever-so-slightly different angles. Because of the partial crossing over of fibers from each optic nerve

---

*The ratio of left-handedness to right-handedness seems fairly constant across history and geographical areas. Hovering around 8 percent, it varies between 15.5 percent in England to 1.5 percent in Hong Kong.[10]

within the brain to the opposite cortex, however, the image reconstructed in the right and left visual cortex located in the back of the skull is markedly different from what the right and left eyes see.

The right visual cortex reconstructs what is before both eyes using information presented *only* to the *left-hand side* of their full-on visual field. Conversely, the left visual cortex processes *only* visual information presented to the *right-hand side* of their full-on visual field. Right-handed people detect movement, judge distances, and estimate measurements best in their right visual cortex, *which is far superior in visio-spatial skills to their left hemisphere.*

Although the brain of a left-handed person is not simply the mirror image of a right-handed person's, it comes close enough so we can make the following observation. Left-handers judge distances, orient spatially, and distinguish visual clues best in the half of the world appearing on their right. *Their left hemisphere is far superior in visio-spatial skills to their right hemisphere.*

An advancing group of right-handed hunters would judge distances, orient spatially, and distinguish visual clues best in the half of the scene that appeared on their left. They would be able to target prey more accurately by throwing right-handed toward their left than by throwing right-handed toward their right. (Observing a right-handed quarterback attempting to hit a receiver who is running to his right demonstrates how awkward and inaccurate throwing in that direction can be.)

A small but significant advantage would accrue to a band consisting of twelve hunters working in concert if one member of the band had a left visual

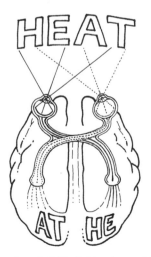

Imagine a vertical line dividing each eye's field of vision into a right side and a left side. The left hemisphere's visual cortex in the back of the brain processes visual information from only what both eyes see to the right of the two vertical lines; the right hemisphere processes visual information from only what both eyes see to the left of the two vertical lines.

cortex that saw best what was approaching (or fleeing) to the group's right. Adding to his value would be this hunter's ability to throw left-handed toward the right with great accuracy. (Although it may have been more advantageous to have six right-handers and six left-handers in the hunting group, Natural Selection had to balance the demand that language localize in the left hemisphere against the success of the hunting band. The solution was to have only 8 percent of males prefer their left hand.)

In swordfights (or any other hand-to-hand combat), a left-handed swordsman has a slight advantage over a right-handed one, because the latter has trained to fight against an opponent who typically uses his right arm. Right-handed swordsmen called those using their left hand "sinister," from a Latin word that is also the source of the English word meaning left-sided, "sinistral." Its alternative meaning is "evil" and "deceitful." If the hunting band were a war party about to engage in battle with other mostly right-handed adversaries, a left-handed warrior would be their secret weapon.

In baseball, right-handed batters fear the slight advantage a southpaw pitcher enjoys over them. Thirty percent of Hall of Fame major-league pitchers are southpaws. The same edge holds true for a batter hitting lefty against a pitcher throwing righty. Half of the Hall of Fame batters are either left-handed or switch-hitters.[12]

Our species is very young as species go, being only 150,000 years old. Not enough time has passed to alter significantly traits that served to increase our survival when we lived in a hunter-gatherer mode. The addition of one left-handed hunter-warrior in hunting bands could explain the persistence of this ratio among males all across the spectrum of cultures, as well as explain why left-handedness is more prevalent in men than it is in women.

Baldness is the last peculiar 8 percent trait that almost exclusively affects human males. Over their lifetime, 40 percent of men will experience a significant loss of hair from the crown of their head. Eight percent of men, however, experience extensive loss of hair by the time they are in the prime of their lives.* Typically, they lose the entire crown, leaving a fringe around the sides and rear to form what has been called a "Hippocrates wreath." Hair loss, with a few very rare exceptions, does not involve the eyebrows, beard, or mustache, nor is hair lost from a man's legs, chest, underarms, or pubic area. No

---

*Unable to locate reliable statistics on this matter, I conducted an informal survey of my own in crowds sampled from my travels. Approximately every twelfth man I encountered in a crowd was both in his prime and bald.

one that I am aware of has come forth with a credible theory for why this trait appears so often in men, and why baldness is almost (but not quite) unknown among women.

The prime of a man's life is the period between twenty and forty-five, when he applies earlier-acquired critical skills to his primary work in life. The primary work of men for a vast span of our history as a species was hunting. It seems a strange coincidence that the 8 percent of men destined to lose their hair did so within the phase of their life when they would have become the most skilled and experienced hunters. And their baldness was confined to the very top of the head.

Male patterned baldness is virtually absent in other mammals.* Why don't healthy dogs, cats, seals, horses, and cows experience baldness? A trait as prominent and persistent as male patterned hair loss, I propose, must be present in the human genome for a reason related to either survival or reproduction.

Sexual selection is the other powerful force molding species besides the drive to survive. Hair frames the face, a human's most distinguishing visual feature. Men who lose their hair are obsessively concerned that they will be less attractive to women. Women, for their part, prefer men with a full head of hair, but they do not discriminate against bald men as much as men believe they do. Men either wear hats or often resort to rugs, plugs, or drugs to conceal from women that they have lost their hair. Yet bald men do not have less chance to mate and have children than men with a full head of hair. (If they did, the gene controlling baldness would be leached out of the genome.) Sexual selection does not seem to be the reason that 8 percent of men in their prime have a dramatically altered appearance.

Testosterone levels have been linked to baldness. If this was the culprit, then one would expect baby boys to emerge bald, because they have had such high levels in utero. Instead, the shock of newborn hair is the same for boys as it is for girls. If testosterone was a factor, why wouldn't teenage boys, experiencing the greatest surge in testosterone after puberty, be at the greatest risk for developing baldness? Instead, a male's hair is never more luxurious and full than when he is adolescent. Testosterone levels are essentially the same between bald and nonbald men. The peculiar distribution of male patterned baldness, the similiar reproductive rates between bald and nonbald men, its unisexual distribution, and its peak in a man's prime of life are features that suggest it was built into the male genome because it somehow advanced the survival chances of the species.

Yet baldness would have been extremely disadvantageous to a man try-

---

*An occasional chimp will exhibit partial baldness.

ing to compete on the Pleistocene's Serengeti. The brain is the most metabolically energetic part of a human's anatomy. It generates more heat than any
other organ, thus requiring the evolution of a sophisticated cooling system to
prevent it from overheating. The scalp has an extensive network of small arteries and veins, as anyone trying to stanch the bleeding from even the smallest laceration of the scalp can attest. The blood supply to this patch of skin is
particularly robust. Dean Falk and her co-workers hypothesize that the
overabundance of veins and arteries in the scalp present in humans serves
the same function as circulating radiator water does in cooling an automo-
bile engine. By constantly running cooler blood over the scalp, the blood vessels carrying "coolant" reduce the temperature of this vital organ.[13]

Besides the miles of both large and small scalp veins, the peculiar human
distribution of body hair also plays a critical role in moderating the brain's
temperature. The densest patch of hair growth on the human body is on the
crown of the skull. When our ancestors stood up, they markedly cut down
the amount of body surface exposed to the glare of a noontime sun. Unlike
animals that can distribute the sun's heat over the length of their body, humans have the very top of the head receiving the full brunt of sunlight during
the hottest part of the day. Selective pressure provided ample protection to
this vulnerable area by covering it with a thick tuft of crown hair. The abundant amount of air trapped under and between individual hairs provides a
natural form of insulation against the heat. The density of scalp hair also
serves to slow the evaporation of scalp perspiration and adds to the efficiency of the brain's natural air conditioner.

Peoples living near the equator have frizzier hair than people who live in
cooler climes. Its microscopic structure ensures that more air will be sequestered closer to the scalp, thus better insulating their brain from excessive ambient heat. When the temperature drops, scalp hair serves the same
function by protecting the brain against excessive cold.

The loss of scalp hair would expose a bald person to the very real risk of
heatstroke, not to mention the less serious but troublesome problem of protecting the exposed skin from ultraviolet light. Bald men do not venture out
in the hot sun without some form of protection. One would predict that the
gene for baldness would have been culled out of the gene pool, because men
who had it would be more likely to die of heatstroke than men who retained
their hair. This would have been a much more important factor when our ancestors left the shade of the forest and began making a living on the shim-
mering yellow immensity of the African savanna.

Yet baldness occurs in nearly every population of males around the
globe, indicating that it is a very old adaptation that must have been performing some beneficial function for the human species. What significant advan-

tage did male baldness confer on the human species that offset the removal of nature's key protection for the brain against overheating?

*   *   *

Whatever the reason our hominid ancestors decided, five million years ago, to rear back on their haunches, stand up, and start heel-to-toe walking, achieving greater speed was not among them. There is virtually no fish, bird, amphibian, reptile, or mammal that a bipedal human can chase and easily catch.* If eating the flesh of other animals became the key to increasing human intelligence, then *Homo sapiens* would first have to get close enough to kill something before his prey sensed his presence and bolted.

In the cat-and-mouse game played for 99.9 percent of hominid existence, male hunters had to sneak up on unsuspecting prey. The bow and arrow did not come into use until fifteen thousand years ago. Stealth was the critical component of early *Homo sapiens'* hunting strategy, since only very crude weaponry was available to inflict a fatal wound.

Prey, over time, must learn to recognize their enemies, for the simple reason that those that cannot distinguish between friend and foe will not live long enough to reproduce. Therefore, one of the most effective defenses a prey animal can evolve is the instinct called "flight distance." Attuned to the smell, sound, or sight of a familiar predator, prey animals flee at the approach of one. If part of a larger group, an alert individual prey can sound the alarm signaling the others to escape en masse. Indisputably, the most dangerous predator is a human. Animals that did not learn to recognize and avoid us paid the price of being hunted to extinction.†

Among the many explanations proffered for why humans adopted bipedalism is the Theory of Sentinel Behavior. African meercats, squirrels, and rabbits, along with many other wary four-footed prey animals, exhibit what is called sentinel behavior to increase alertness. Habitually locomoting on all fours, they periodically pause to stand on their hind legs. The vertical position allows them to see over obstacles and gain a larger field of vision. Many primates—for example, gorillas and baboons—have descended from their habitat in the trees and spend most of their time on the ground.

---

*Tortoises and sea mammals sunning themselves on land are among the few exceptions.
†Because of its isolation from the other continents, Australia was home to an exotic and diverse megafauna 50,000 years ago. And then the first humans arrived. Rapidly spreading over the Australian continent, the precursors of the indigenous Aborigines precipitated a mass extinction of large birds and mammals. The primary reason for the ease and rapidity with which this wholesale slaughter took place is that the animals in Australia had never seen a human and therefore lacked the necessary flight-distance instinct that African birds and mammals possess, given their uneasy coexistence with hominids over a much longer period of time.

Ground-dwelling primates typically engage in frequent sentinel behavior. Temporarily standing on two legs serves as an adjunct to survival.

Any animal desiring to peer over an obstacle must first show the top of its head. The peerless expert at sentinel behavior is the only truly bipedal mammal, *Homo sapiens*. Animals that have survived the predations of humans have been programmed by instinct or have learned by experience to identify the typical silhouette of a human head peeking from behind a shrub. Minus a periscope, there is no way that a hunter can sneak a peek without first having to first expose his pate.

Anthropologists observing the techniques of present-day hunters such as the !Kung San of the Kalahari note the elaborate lengths tribesmen go to disguise themselves as they try to get within striking distance of prey. Hunters know to approach prey downwind, crouch low, and maintain silence. If the prey identifies them too soon, escape will surely follow, and the hunters will return empty-handed. They must do everything they can to prevent any stimulus that will activate their quarry's flight-distance response. Experienced African hand Michael Crawford comments, "The idea of an upright primate scoring by being able to peer over the tops of the grasses is an appealing one—to anyone who has no experience in hunting. In reality, the main difficulty facing any hunter is not spotting his prey, but preventing his prey from spotting him. The art lies in the stalking."[14]

Imagine that you are a prey animal lucky enough to have survived to the present. You made it this far because you and your ancestors learned to recognize the typical features of your dreaded human nemesis. The most striking characteristic of *Homo sapiens,* making him hard not to miss, was his outline—watermelon head perched on a pole resembling in silhouette an outsized lollipop. Other distinguishing features facilitating your ability to identify the savanna's Public Enemy Number One: distinctive tangy aroma of his perspiration, strangest damn form of moving about you had ever seen, odd squeaking noises, bald-bodied, yet sporting a hairy crown.

Suppose you found yourself in the predicament of being among humans' favored quarry. You would need to evolve a critical instinct that would identify the part of this dangerous adversary that you saw first when one of these critters was attempting to ambush you. Over time, you evolved a mutation that allowed you to associate the sight of what looked like a mop rising slowly over yonder rock with the imminent attack by the bizarre-looking, squeaking lollipops heaving sharp sticks. Whenever you espied that peculiar configuration or caught a whiff of the odor that emanated from the mop, you knew to sound the alarm and head for the hills. Inevitably, a pair of round eyes soon appeared after you sensed the mop, and you instinctively knew that next all hell would break loose. You and the fellow members of your species came to rely on this early warning to stay alive.

Suppose that, among a group of twelve hunters, one had a crown that was anomalous in that it was shiny, and lacked the typical hairy outline of a human head. Suppose this hunter was the one chosen by the rest to peer over the top of a rock to check on your position. The first thing you would see rising slowly over the crest of a nearby boulder would be strange and unfamiliar. A band of hunters, employing the bald-man subterfuge, might gain a small but incremental advantage by delaying the activation of your flight-distance instinct just long enough for the others to move in for the kill.

A bald hunter would have a slight advantage over the others in that he would most likely be able to get the closest to a skittish prey before it would flee. Proximity increases accuracy when trying to hit a target. Success in hunting would increase a man's reputation with the women. Perhaps this is the basis of the ineradicable myth that bald men are more virile. Slightly better hunting success would offset any loss of physical attractiveness to the opposite sex a bald man might experience.

All of the four features catalogued—ESSP, color-deficient vision, left-handedness, and baldness—might very well be spandrels. They might have no bearing whatsoever on the survival or reproductive success of the human species. And yet it does seem odd that all of them boil down to the 8 percent ratio, all appear in every human culture, all affect males far more than females, and all would serve to make a hypothetical Pleistocene band of twelve hunters more deadly and efficient.

One way to test my Theory of Eights indirectly would be to search for correlations between these four features. In a general population, a higher percentage of gay men are left-handed than are straight men.[15] Left-handed men are more likely to have color-deficient vision than are right-handed men. Color-deficient men are more often gay than are non-color-deficient men. There is a suggestion in some studies of a link between left-handedness and baldness, and baldness and color deficiency, but I can find no study linking baldness and ESSP.*

---

*An unexpected side effect of using the hair-enhancing drug minoxidil is the reported increased incidence of color-vision problems. There are many side effects of taking drugs of any kind. Color-vision problems are exceedingly rare. It seems a peculiar coincidence that a drug that affects hair growth can also diminish one's color vision, suggesting some link between the biochemistry of hair growth and color vision, which in turn would point to a gene controlling the enzymes involved in both processes.

Though my Theory of Eights may provide a partial explanation for same-sex preference in males, how do evolutionary biologists account for a lesbian gene? Throughout history, lesbians have maintained a lower profile than gays for a variety of reasons primarily related to the masculinist thrust of the majority of the world's cultures, with its repression of the feminine principle in general and women in particular. From a mixture of scientific studies (few) and guesstimates (many), the proportion of lesbians seems to have remained fairly constant, hovering around 5 percent.

Wilson's theory of the advantage to the hunting group of having one hunter in twelve who was gay can be applied to the one woman in twenty who was a lesbian. In a group of women burdened with child-care duties, there is a benefit to having one woman who prefers women to men. A lesbian's inclination would make her less likely to have offspring (although many do). She could then choose to use her surplus time and energy either directly or indirectly assisting the other mothers. Nieces and nephews benefit from having an aunt without children of her own. Wilson's theory accounting for lesbians dovetails with the earlier Grandmother Theory, discussed in chapter 8, because it provides an accessory adult female to assist mothers. Many lesbians are in the helping professions, and their efforts in both direct and indirect ways often alleviate the workload of straight mothers. Lesbians are also in a better position to make notable contributions to culture. The world would be a poorer place without the considerable talents of Sappho, Gertrude Stein, Josephine Baker, Anaïs Nin, Ethel Waters, Mary Cassatt, Florence Nightingale, Emily Dickinson, Willa Cather, and Virginia Woolf.*

*   *   *

Mythology is a projection of a community's collective beliefs onto the scrim of the cosmos. Therefore, it should come as no surprise that mythical characters who fully manifested both their *anima* and *animus* were considered the wisest seers and most powerful deities in ancient cultures. Originating in the myth of the union of Isis and Osiris in Egypt, hermaphroditic sagacity has remained a constant thread connecting many civilizations throughout the ancient world. The symbol of this wisdom in ancient Egypt was two entwined snakes. In the Mesopotamian myth of the Garden of Delights, two serpents coil around the trunk of the Tree of Life. In the Greek myth of Hercules, en-

---

*Determining the sexual orientation of a historical figure is highly speculative. The weight of evidence—from public knowledge, personal biographical details, and information gleaned from these renowned women's writings and correspondence—suggests that these women were lesbians.

twined snakes guard the tree bearing Zeus's golden apples on the isle of Hesperides.

Hermes, the Greek god of magic, transformation, and wisdom, was an androgynous god who combined both the essence of the female and the virility of the male. Appropriately, Hermes was the god of the hermaphrodites (Hermes and Aphrodite were the parents of Hermaphroditus, who combined an equal mixture of masculine and feminine in one body). Hermes' aegis was the caduceus, two entwined snakes, one female and one male. It has come down to us through the ages as the symbol of life and vitality and now signifies the medical profession.*

In Sophocles' plays, Tiresias, the blind hermaphroditic prophet, warns first Oedipus in *Oedipus Rex,* then Cleon in *Antigone.* Tiresias' gift of foresight allowed him/her to see the character flaws of mortals and predict events that had not yet transpired.

The myth associated with how Tiresias became a hermaphrodite reveals much about relations between the sexes. One day, Zeus and Hera began to bicker over who derived the most pleasure from sexual intercourse. Soon, all the other gods and goddesses on Mount Olympus joined the contentious debate. Cleverly, the goddesses protested that men did, not wanting the men to know the truth, that women actually enjoyed sex more than they. The gods, seeing through the ruse, insisted women did. As the fervor of the controversy intensified, it became clear that neither side would ever convince the other, since sexual pleasure is a purely subjective experience.

To settle the question, Zeus selected a poor shepherd, Tiresias, to change into a hermaphrodite. The unsuspecting young lad came upon two snakes copulating and, striking them with a stick, suddenly found himself transformed by Zeus into a woman. After Tiresias had lived ten years in this mode, Zeus summoned the hermaphrodite before an assemblage of all the deities and asked: "Now that you have experienced sex both as a man and a woman, who has more pleasure?" Tiresias replied, "Of the ten parts of pleasure there are in sex, the female partakes of nine and the male is left with the remaining one."

Hera was so furious at the truthteller for not supporting her position that she struck Tiresias blind.† Zeus, immensely pleased that Tiresias had revealed the goddesses' ploy, could not reverse Hera's terrible punishment, but he endowed Tiresias with the gift of a seer's foresight. Thus, Tiresias the her-

---

*Originally, the icon was the staff of Aesculapius, a single snake entwined around a staff. Somewhere in the last few centuries, the medical profession adopted the caduceus.
†Since in all cultures men must exchange something of value to obtain a woman's consent to sex, Tiresias' revelation that women enjoyed sex far more than men did threatened the balance of power between men and women; hence Hera's fury.

maphrodite would play the important role of warning heroes of what was to come in Greek tragedy.

Among Hindus, Lord Shiva is an androgynous god who is both the creator and destroyer of worlds. His feminine aspect is Durga. Many of the paintings depicting this hermaphroditic deity portray Shiva on the right side and Durga on the left. The Hindu caste system has a special class, called *hidjra,* for homosexuals, transvestites, and intersexes. In Mesopotamia, cultic dancers dressed in costumes to emphasize the hermaphroditic nature of their deities. The right side of their costume signified the male, and the left side represented the female aspect.[16] The ancient culture of Oman recognized a third sex they called the "zaniths."[17]

Farther east, originating in the same distant era, people began to worship a hermaphroditic deity in China. Quan Yin, born a man but transformed into a woman, is the god/goddess of wisdom and compassion.

In North America, many Pueblo Indians—for example, the Zuni and the Hopi—had a special reverence for a man who manifested a strong *anima.* White settlers disparagingly called a man who lived and dressed like a woman a "berdache." The other members of his native tribe often believed a transvestite or gay male possessed magical and intercessionary powers. Often he was the tribe's shaman.[18]

But hermaphroditic wisdom did not prevail. Three thousand years ago in the West, the powerful idea of monotheism transformed the world. Patriarchy became a defining characteristic of Western culture and religion. Soon after, an unsettling question began to trouble the faithful. If there was but one God, then of what sex was this singular deity? Over the years rabbis, clerics, and mullahs attempted to evade this question by claiming that God has no sex. But this stance contradicted the sacred scripture of their singular God that states, "And God created man in his own image, in the image of god created he him; male and female created he them" (Genesis 1:27). To have created man and woman "in his own image" means that the god of Moses, Jesus, and the Prophet would have to have been a hermaphrodite.

Further confusing the issue is a second, older Genesis story of Adam, Eve, and the serpent. In this parable, Yahweh is singular and obviously very male. Although the Bible does not actually state the deity's gender, all of the adjectives and names that subsequent Biblical writers used for Yahweh (Lord, Ruler, King of the Universe, Host, Adonai, and Elohim) are masculine. Even in this story, however, there are intimations of Yahweh's hermaphroditic persona. In Genesis 3:22, He states, "And the Lord God said, Behold, the man is become as one of us, to know good and evil. . . ." The obvious question is, who are the entities constituting an "us"? Why does the monotheistic deity refer to Himself as an "us"?

The banishment of goddesses from Western religions was a wrenching change from all that had gone before. The deity of the three Western religions—Yahweh, God Almighty, and Allah—came to represent a lopsided *animus*. It is notable that the serpent in the Genesis story, once twins of opposite genders in prior legends, was reduced to one.*

Beginning with the ancient Hebrew Old Testament and continuing through Christianity's New Testament and Islam's Koran, Western culture declared war on the feminine. Men emphasized their deity's vengeful and wrathful character. Eve, Mary, and Fatima, by comparison, were all mortal women whose stature was nowhere near as numinous as that of a masculine godhead.

Worshipping a God perceived to be a Father but not a Mother presents a series of confounding problems. Within the three Western religions, there have been periodic attempts to reconcile an *animus* monotheism with the notion that the deity embodied masculine *and* feminine traits.

In the first century A.D., a time when Hellenistic values were pervading all Mediterranean cultures, rabbis, uncomfortable with the Torah's masculine bias, introduced the concept of a Shekina. She represented the essence of the divine *anima*, which they believed was the missing aspect of God. Around the same time, the Gnostic Christians began venerating a deity that possessed both feminine and masculine traits. They elevated the divine Sophia to a place coequal with God. Christian mystics held that the Trinity consisted of a Holy Father, a Holy Mother, and a Holy Son. The Shiite branch of Islam honored the feminine essence of Fatima, the Prophet's daughter. Even today, the veneration of Mary in many Latin American cultures borders (some might say has become) goddess worship. All of these movements have attempted to soften the male monotheistic deity by converting Him into a hermaphrodite. In every case, these attempts have been defeated by those who want to believe that a patriarchal singular male deity reigns supreme. Religious movements that honored the divine feminine have tasseled the fringes of the West's dominant religions from time to time, but they have never been central to any of them.

At present, Judeo-Christian denominations (with the exception of a few hard-core fundamentalists) are rushing to expunge their liturgies of passages that harshly denigrate the feminine. Women are becoming priestesses, masculine pronouns are being replaced by more egalitarian ones, and the feminine aspects of the deity are increasingly emphasized as mercy and compassion replace vengeance and punishment as His/Her key features. Many leaders of the West's religions seem to acknowledge, at long last, that

---

*Biblical scholars have traced the origin of this myth to Babylon.

the failure to reform their doctrines will consign their religions to irrelevancy in the near future. The essential truth for both men and women is that those of each gender must acknowledge and encourage their psychic hermaphrodite within in order to achieve their intellectual, creative, emotional, and spiritual potential. The poet Coleridge opined that all great minds are truly androgynous. Only when fertilized by elements of both the masculine and the feminine can someone fully become wise and creative. Cultural institutions such as religion, governments, and educational systems undergoing the same realignment will also prosper.

In 1953, James Watson and Francis Crick won the Nobel Prize for discovering that the molecular configuration of DNA was a double helix. Life, they demonstrated, derives from the intertwining of one strand of DNA from the father and one from the mother. The double helix instantly became the modern symbol of the combination of the *anima* and *animus*. But the ancients had already intuited this truth. Watson and Crick confirmed that, at its most fundamental level, life's most compelling icon is Hermes' caduceus, the quintessential symbol of the hermaphrodite.

Death and Paternity

Part IV

Women were the first to achieve the knowledge of time. They did so at considerable risk to their life and health. Now it was the men's turn to experience extreme anxiety.

Chapter 18

Mortality/Angst

Everyone *once, once* only. Just *once* and no more.
And we also *once*. Never again. But this having been
Once, although only *once*, to have been of the earth,
Seems irrevocable.
                                          —Rainer Maria Rilke

It takes sixty years of incredible suffering and effort to make a unique self-conscious individual, and then he is good only for dying.
                                          —André Malraux

*B*y using her new split brain to make the connection between the moon's periodicity and her monthly menses, *Gyna sapiens* gained entry to an entirely new realm. Her insight profoundly altered the course of human evolution. Like Alice falling through the looking glass, she tumbled headlong into the tunnel of time. And, *Homo sapiens* fell in right behind her.

As with all vast gifts, learning time's secret carried a terrible price that first women and then men had to pay. Not long after women crossed time's threshold by endangering their health and life, it would be the men's turn to be discomfited. Before exploring the feature of time that so unnerved them, let us tally the considerable benefits *Homo sapiens* incorporated into his survival kit by gaining the ability to think in deeptime.

Homo sapiens began to put this new skill to use by first discerning certain recurring patterns in nature. He began to merge his gradually increasing life span with his prodigiously expanding memory capacity in order to study and then *remember* the habits of prey and predator. Soon, he felt confident enough to organize larger and more ambitious hunting forays. Success in these endeavors increased the available quantity of high-quality brain food. A diet rich in supplemental iron, animal fats, and amino acids increased the band's survival odds and produced slightly smarter and healthier babies. The *Homo sapiens* hunter reveled in his maneuverability both in the canyons of the past and on the misty plains of the future, exultant in the knowledge that he was smarter than any other creature.

When game was not plentiful, both men and women stored the knowledge necessary to differentiate poisonous flora from edible flora in the deep pouches of their new long-term memories. Slowly, over time, they identified and *remembered* plants that had medicinal properties. To their delight, they also discovered and *remembered* plants that altered their state of mind. There seemed to be no end to the largesse spilling forth from time's cornucopia.

Communication between humans expanded rapidly once it was reinforced by the concept of linear sequence. The addition of words such as "after," "before," "long ago," "next," and "fortnight" made speech increasingly supple and sophisticated. Syntax and semantics were the twin factors enriching language. Adding tenses to verbs and time-related adverbs, adjectives, and prepositions greatly expanded speech's power.

This expanded sense of time greatly facilitated a new mode of thought called "reason," which in turn was dependent on a simple two-step mental equation: The *if* = *then* algorithm. Calling upon his considerable memory reserves, *Homo sapiens* began to place to the left of the equals sign a hypothetical situation. He conjured scenes based on similar situations that he had previously encountered. Then he constructed an imaginary scene of what he would *predict* might happen in an epiphenomenal scenario that existed only in his mind.

Eventually, he tested this novel approach to problem solving in the field. Many times it failed, giving the process the appropriate name of "trial and error." But sometimes it worked, and through the sustained application of reason, early *Homo sapiens* sought to achieve a special kind of knowing called wisdom. Mother Nature repeatedly humbled him, forcing him to discover the trying lesson that all subsequent humans have been pained to learn: Good judgment is based on experience, and experience is often based on poor judgment.

Over time, he became increasingly adept at controlling extraneous factors in his *if* = *then* predicting machine, and Mother Nature rewarded his accuracy with kernels of knowledge about how the world worked. Thus, the storehouse of communal wisdom steadily enlarged.

Communication between individuals further enhanced the stockpiling of valuable information in an ectoplasmic silo called culture. Any member of a tribe, having learned a useful fact, could place his or her contribution in the invisible tower by exhaling a few controlled puffs of air in the direction of another's ears. The expenditure of calories was minuscule: Human language was an incredibly energy efficient transfer and storage system.

The *if = then* equation was linear and sequential. The dimension of time forms the very core of cause-and-effect logic, and it issues forth primarily from the left hemisphere of right-handed men and women. Because problem solving using this technique is mental rather than "real," it falls under the rubric of "abstract" thinking—another primarily left-brain function.

A core principle of abstract thinking—*reductionism*—greatly facilitated the *if = then* equation. An opposable thumb and forefinger enabled *Homo sapiens* to indulge his insatiable curiosity. Like many primate species before him, he was able to pick apart things to learn what was inside. Abstract thinking allowed him to take this skill to another level—a virtual one.

Now, without actually dissecting something, he could use his imagination to break down a whole (system, object, or pattern) into its component parts. This mental manipulation gave him a huge advantage over every other animal. Reductionism is synonymous with analysis, and it advances in a measured step-by-step process that depends heavily on the concept of linear sequence. *If this, then that; if that, then this* . . . Extending the line requires the knowledge of deeptime. It is the underlying mechanism that drives reason, logic, rationality, and ultimately art and science.

The artist looks at nature and breaks it down into its component parts, reassembling them in a novel and compelling way. The scientist does the same, but is interested in understanding how the parts relate to the whole. When the scientist reassembles the parts in a novel way, it is called technology. The artist uses reductionism and synthesis in the service of aesthetics; the scientist uses reductionism and synthesis in the service of advancing knowledge. The artist employs the images and metaphors of nature to interpret the relationships of reality; the scientist imposes number and equation on nature to express the relationships of reality. The writer Vladimir Nabokov observed, "There can be no science without fancy and there can be no art without facts."[1] The revolutionary artist and the visionary scientist are both fundamentally engaged in investigating the essence of reality.

The *if = then* equation played a crucial role in the invention of new tools and weapons. It had an immense bearing on *Homo sapiens'* rapid ascent in the standings of nature's survival-of-the-fittest contest. Rationality boosted his ability to imagine, construct, and deploy ever-superior weapons with which to kill prey and, later, perceived enemies. For all but the final 0.008 percent of the five-million-year bipedal-primate experiment, his weapons of choice were rocks, bones, pits, and sticks. Then, beginning somewhere between forty thousand and thirty thousand years ago, he invented the spear thrower, or *ataltl*, a small, hand-held, leveraged device that a hunter could attach to the base of his spear. It greatly increased the speed, thrust, and accuracy of a hurled projectile. *Homo sapiens* invented the bow and arrow fifteen thousand years ago, the iron sword five thousand years ago, the chariot four thousand years ago, flaming naphtha three thousand years ago, gunpowder one thousand years ago, the airplane one hundred years ago, and the atom bomb fifty years ago. The Doubly Wise Man had transmogrified into the formidable Weapons Maker Man.

Imagine a ladder. On each rung sits a predator; the higher the rung, the more dangerous the predator. Emerging from a long primate tradition of chewing nuts, roots, and leaves, *Homo sapiens* began to ascend the ladder rapidly, successively displacing more fearsome predators at each step. In the astonishingly brief span of 150,000 years, *Homo sapiens* attained the top rung, and from there he claimed the title of undisputed king of the jungle ... and sea, plain, desert, tundra, forest, and mountain. Just as Genesis had foretold, humans had gained "dominion over the fish in the sea, and over the fowl of the air, and over every living thing that moveth upon the earth" (Genesis: 1:28).

Gripping the top rung of the predators' ladder, *Homo sapiens* surveyed all that he had conquered. No single creature could withstand the juggernaut of his onslaught. Just about every plant and animal simmered as an ingredient in one or another flavorful *sapient* recipe. He had succeeded in stalking, harvesting, capturing, foraging, taming, plucking, domesticating, and killing the entire outpouring of billions of years of evolution. His sovereignty was secure. Nowhere on the horizon was there an animal that posed a threat to his dominion.

And yet, despite his stunning and rapid success, he increasingly experienced a gnawing sense of anxiety. His unease resulted from his having discovered one of time's terrible certainties. Let us attempt to reconstruct the moment of his insight that forever changed the human species.

Somewhere, sometime, a *Homo sapiens* sat alone on a rock lost in contemplation. Let us name this particular individual Adam. Although physically in an attitude of repose, he was actually intensely exercising the expanded new circuits in his brain's left hemisphere, challenging the limits of how far he could peer into the future.

Each time Adam tried to extend his vision, he was blocked by an obstruction: an impenetrable black wall. Placing his fingertips against his temples, he increased his concentration. Despite his most focused efforts, he could not go through, around, under, or over the obsidian barrier that loomed before him. And then, in a searing insight, it dawned on him: He was going to die. The black monolith stood for his own personal death. He could imagine nothing beyond it. His discovery filled him with terror. Here, then, was the terrible price he paid for following *Gyna sapiens* into the mysterious vortex of time's tunnel.

Death is an all-too-common occurrence in nature. It exists side by side with the exuberance of life. Animals see, hear, and smell it all around them. Zebras nervously whinny as they witness one of their own being torn to pieces by hyenas. A vixen mews as she listens to the weakening cry of her young kit dying of disease or malnutrition. A distressed baby monkey cringes in horror as an eagle snatches his mother, who, only a moment before, was at his side grooming him. "Nature," as the poet Alfred Tennyson remarked, "red in tooth and claw." Psychoanalytically inclined philosopher Ernest Becker summarized the situation thus: "Creation is a nightmare spectacular taking place on a planet that has been soaked for hundreds of millions of years in the blood of all its creatures. The soberest conclusion we can make is that the planet is being turned into a vast pit of fertilizer."[2]

For our closest relatives, the chimpanzees, death appears to be an unfathomable enigma. When a chimp suddenly dies by falling from a great height, the others clamber down and gather round the corpse. Among much panting, hooting, and obvious agitation, they peer and prod. But finally, after what we would consider an astonishingly short period of concern for a fallen troopmate, they nonchalantly resume their daily activities. In all but a few extraordinary cases, it would appear that the matter has been quickly forgotten.*

Paleontologist Richard Leakey and Roger Lewin, in their 1992 book, *Origins Reconsidered,* observed:

> In all human societies, the awareness of death has played a large part in
> the construction of mythology and religion. There seems, however, to be
> no awareness of death among chimpanzees. Females have been known to

---

*This is not always the case. Jane Goodall has observed the relationships of chimpanzees in the wild over long periods of time. She kept track of them by assigning personal names to each chimpanzee. Flint, the son of Flo, was inordinately attached to his mother. When Flo died, seven-year-old Flint would not leave her side for days. He disappeared shortly thereafter, and Goodall believes, "Flint died of grief."[3]

Primatologist Frans de Waal has a remarkable photo in his book *Good Natured* of a young elephant that returns regularly to the site where her mother was shot and killed by poachers. The youth gently rolls the skull that is all that remains of her mother, a behavior consistent with a daughter still mourning for her mother. Still, this behavior does not indicate that the young elephant knows that she, too, must someday die.

carry around the corpse of an infant for a few days after its death, but they seem to be experiencing bewilderment rather than grief. More important, other mature individuals appear to offer no condolence or sympathy to the bereaved mother. The emotional experience seems to go unappreciated by others, and unshared. So far no observer has seen reliable indications that chimpanzees have an awareness of their own death, the extinction of self.[4]

Dogs obviously mourn the loss of their owners. But it is not evident from their sorrowful keenings and whimperings that they are aware that the same fate awaits them. As death approaches for some higher animals, they appear to behave as if they were anticipating this final event. Elephants seem to know when one of their herd is going to die. They appear to become more protective and sympathetic, as if trying to alleviate the suffering of the old one. Observers report that elephants sometimes cover a dead companion with leaves and branches before they move on.[5] There is, however, nothing in their observable behavior to suggest that earlier in their lives they confronted the inevitability of their own demise.

We may never know with certainty what other animals know about their own death, but it appears that an absolute precondition for being able to anticipate personal extinction is the capacity to think in an extended time frame. Humans seem to be the only animals to have mastered this complex feat—and perform it at an early age.

The jolt of adrenaline that activates the fight-or-flight response when any animal confronts mortal danger is among the most basic survival in-

The human confrontation with death's finality altered the course of every subsequent culture.

stincts. However, there does not appear to be any animal other than a human that seems fixated on the prospect of its death in the far-distant future. Only a human clearly understands the harsh truth that death is inevitable, inescapable, and nonnegotiable.

Because humans are animals, we experience instant fear whenever danger suddenly thrusts itself smack into our awareness. In addition to these sudden episodic spasms of acute fear, we became the first animal to be plagued by an ever-present, free-floating anxiety resulting from an awareness of death's omnipresence. Becker, aware of his own impending demise, wrote poignantly:

> Anxiety is the result of the perception of the truth of one's condition. What does it mean to be a self-conscious animal? The idea is ludicrous, if it is not monstrous. It means to know that one is food for worms. This is the terror: to have emerged from nothing, to have a name, consciousness of self, deep inner feelings, an excruciating inner yearning for life and self-expression—and with all this, yet to die.[6]

The psalmist expressed the same idea poetically:

> As for man, his days are as grass: as a flower of the field, so he flourisheth.
>
> For the wind passeth over it, and it is gone; and the place thereof shall know it no more. [Psalm 103:15.]

Herodotus recounts an incident from the fifth century B.C. Xerxes, the Persian despot, was leading a vast army to invade Greece. As he neared the storied ruins of Troy, he felt a strong desire to visit the site. Surveying the scene of the ancient battle deeply moved him. The next day, he told his staff that he wanted to review his troops. Xerxes had a throne hastily set upon a high knoll so that he could view on the plain below the panorama of a million men on the march accompanied by their engines of war.

At first, he exulted vaingloriously in the pageantry of so many virile men parading before him. Then he began to weep. His aide-de-camp, witnessing the king's distress, asked why he cried. "I was thinking," Xerxes replied, "and it came into my mind how pitifully short is human life—for of all these thousands of men not one will be alive in a hundred years."[7]

One could safely assume that Xerxes was weeping for himself as well. Death would be his fate, too. As he stood at the height of his power, Xerxes understood that life's impermanence made all his strivings seem a preposterous vanity. It is extremely doubtful that any other animal might have been similarly overcome with emotion. Coming to grips with the finality of death was one of the major milestones in the evolution of our species, and it remains a key insight in the maturation of each individual.

*   *   *

The child psychologist Jean Piaget reported that infants as young as six months can make a distinction between animate and inanimate objects. Late in the second year, a toddler ambles past a mirror, stops, touches it, and then begins to make funny faces, an activity we share with chimps that confirms that the child has a concept of self. The awareness of chronological time becomes refined in children between the ages of three and three and a half, when they begin to use the tenses of verbs correctly. Properly conjugating past, present, and future indicates he or she has acquired an understanding of time. At the beginning of the fourth year, a child understands that other individuals can hold beliefs different from his or her own. This constitutes one of life's major discoveries. For the first time, children relish experimenting with a new behavior mode they will put to extensive use for the rest of their lives—lying. Leo Tolstoy wrote, "From the child of five to myself is but a step. But from the newborn baby to the child of five is an appalling distance."[8]

Yet, in the early years of every child's life, death is an incomprehensible concept. At around the age of six or seven, children finally become aware that dead birds, goldfish, and hamsters will never return. Then, often after the death of a grandparent, they come to the realization that someday their parents will die. Soon afterward, they formulate the key insight that the same fate awaits them, too.

The Catholic Church considers seven to be the age of moral understanding and uses it as the milestone for when a child can receive first communion. Confucian followers believe seven is the age of the beginning of wisdom. Many other cultures have used this age as a dividing line between innocence and the beginnings of a mature mind. The reason, I suspect, that the great traditions acknowledge seven as a turning point in one's life is that this is the moment one grasps the inevitability and finality of death. Advance knowledge of one's own personal final exit is uniquely human. Eve took a bite from an apple and gained godlike knowledge. She then offered it to Adam. Neither noticed that there was a worm in the core.

Homo sapiens emerged as a distinct species 150,000 years ago. Paleontologists have not identified a single major physical attribute that has changed since that time. An ancestral *Homo sapiens* dressed in modern clothing and walking a crowded street in New York today would not attract the slightest attention.

For the first 110,000 years of our species, there is no consistent record of art, burials, or grave goods.* Then what author John Pfeiffer calls the *Creative Explosion* began. A magnificent awakening mysteriously emerged nearly all at once approximately forty thousand years ago, during which the above-mentioned cultural markers unique to humans habitually made their appearance in areas as disparate as Australia, Morocco, Siberia, and, especially, Europe. Thereafter, they have become staples of every human culture. Since paleontologists have not identified any increase in brain capacity, what discovery could *Homo sapiens* have suddenly made that engendered the practice of the ritual disposition of dead relatives? What encouraged them to relinquish perfectly usable goods and place them in a grave alongside the deceased? Why adorn the dead with painstakingly crafted adornment? And most intriguing: What compelled *sapients* to begin creating art in abundance after the passage of tens of thousands of years without experiencing a similar urge?†

The "Great Leap Forward," as Jared Diamond calls the Creative Explosion, is also referred to by others as the "Upper Paleolithic Revolution." Many different theories besides the Big Bang of language have been advanced to explain this forty-thousand-year-old unsolved evolutionary whodunit. These include a "restructuring of social relations," the "appearance of economic specialization," and an as yet to be identified "technological invention."[9]

Stephen Mithin, in his 1996 book, *The Prehistory of the Mind*, proposes that early *Homo sapiens'* cognition was divided among separated "domains of knowledge." Although each domain had been expanding, the technical intelligence concerning how to make tools was isolated from natural-history intelligence, which in turn had little interaction with the domains of social intelligence, general intelligence, and language. Mithin believes that the Creative Explosion occurred when some event caused the walls to come tumbling down. Like Gabriel's trumpet blast at Jericho, an unknown factor smashed through the barriers erected in the brain that kept the various domains from knowing what another knew.

---

*Archeologists have excavated what some among them would characterize as reverential burials that date back to 60,000 years ago among both *Homo neanderthalensis* and *sapiens*. Others disagree. Not until 40,000 years ago, among only *sapiens*, does it appear that burials and funerary rites became a defining characteristic of human culture.

†The oldest item of high-quality representational art thus far recovered is an ivory carving of a half-lion, half-man unearthed in southern Germany and dated to 33,000 years ago. Despite the near universal urge to create art, anthropologist Margaret Conkey has identified a few cultures that do not produce art.

In 1866, the biologist Ernst Haeckel proposed his Biogenetic Law: "Ontogeny is the short and rapid recapitulation of phylogeny." "Ontogeny" in humans is the process by which a fertilized ovum develops into a full-term baby; "phylogeny" is the process by which simple life-forms evolve into more complex ones. Phylogeny took billions of years to occur; ontogeny mimics (recapitulates) the journey in nine months. Each process parallels the other. Stated simply: The successive developmental stages that a human conceptus-embryo-fetus-neonate runs through in only nine months of gestation repeats the entire 3.8-billion-year evolution of all animals.*

The instant a sperm and an ovum unite, the future human has all the characteristics of a single-celled organism, similar to an amoeba. Within hours, the fertilized ovum begins to rapidly divide. The spherical multicelled blastocyst it becomes in four days is nearly indistinguishable from other primitive multicellular organisms, such as molds and sponges. The journey from single-celled organisms to multicelled ones took evolution several billion years to make. Within three weeks, the human embryo takes on the appearance of a worm, and then, in another week, it develops gills to mimic those of a fish. Several weeks later, lungs replace gills, a process that took millions of years in evolution. Shortly afterward, it morphs into a reptile, complete with a tail. Finally, the fetus begins to manifest all the characteristics of a mammal, and then those of a primate. With the disappearance of the fetus's tail, a fetal ape appears. Finally, when the changeling sheds its outer layer of dense intrauterine fur (called "lanugo"), he or she differentiates into a human in the remaining weeks of pregnancy. The fifth-century-B.C. Greek philosopher Empedocles was aware of the process when he wrote in *Purifications*, "Already have I once been a boy and a girl, and a bush and a bird, and a silent fish in the sea."

Jean Piaget suggested that human evolution follows the schedule of a single child.† Think of the life cycle of the entire *Homo sapiens* species as analogous to the life span of an individual. The average human life expectancy is ap-

---

*Biologist Ernst von Mayr disagreed with Haeckel's formulation: Humans do not go through each of the adult stages of evolution, as Haeckel originally proposed, but mimic the embryonic stage of each life-form. Though there has been much tinkering with Haeckel's original formulation of his flawed law, it is beyond dispute that, at various stages of human gestation, the evolving future individual sports gills, a tail, and fur.
†Stephen Jay Gould defended Piaget's ideas after they met a cool reception among anthropologists and archeologists.[10]

proximately eighty years. Our species is 150,000 years old, and we do not know how long, in the natural course of events, it will last before going extinct or become the precursor to an entirely new hominid successor.

Using a similar analogy, let us superimpose a child's schedule of "insight" milestones onto what we have proposed was the timetable for the human species' major quantum leaps of awareness. Was the point our species attained forty thousand years ago analogous to the moment in a seven-year-old-child's life when the child understands for the first time the full implications of death?* If it was, that implies that, even with their new big brains, the earliest *Homo sapiens* did not collectively comprehend personal death for the first 110,000 years of our history.† I propose that the detonator igniting the Creative Explosion was a critical mass of *sapients'* minds individually realizing, "Omigod! I'm gonna die!" Terror of death was an insight so powerful that it set off a chain reaction, and this in turn spurred the rapid advancement of human culture—as it still does.

A woman most likely was the first to have this revelation, since it was she who first embarked into the dimension of the future. But a woman's attitude toward death is markedly different from a man's. In general, women tend to accept its inevitability with greater equanimity. When the poet Dylan Thomas urged his failing father, "Do not go gentle into that dark night. / Rage, rage against the dying of the light," he inadvertently called attention to this gender gap.

I base the conclusion on my personal observations. As a surgeon for thirty-five years, I have had to inform many patients after an operation that they had a terminal illness. On average, men, in all age groups, take this news much harder than women do. They more often insist on heroic and drastic measures to stave off the inevitable. Men fear death more—though, of course, there are many exceptions to this generalization. I have witnessed many women fight tenaciously to forestall death, and I have attended many men

---

*If seven years is to 110,000 years, as 80 (the average life span of an individual) is to x (the expected life span of our species), then, according to a whimsical exercise, our species would last to the ripe old age of 1.2 million years before going extinct (barring some self-imposed catastrophe before then!). In this exercise, the current state of civilization suggests it has attained the equivalent maturation of a ten-year-old. Observing the daily events in the newspaper or on television, one would might well concur with this assessment. Our species, unfortunately, does behave much like a ten-year-old. Not mature, but beginning to show signs of it. Gaining control over some human instincts, but still relatively unsocialized. And he has become just strong enough to become dangerous to himself and others.[11]

†The finding of occasional burials among earlier *Homo sapiens* and perhaps *Homo neanderthalensis* suggests that the insight concerning death was made sporadically among disparate individuals and bands. Some as-yet-unidentified factor, however, inhibited the spread of mortuary practices. I suggest that not enough *sapients* had reached the species age of sophistication in these matters. Just as there are precocious children who learn lessons faster than others of the same age, so, too, the species also had its precocious members. Another coincidence: 40,000 years ago is also the breakpoint at which Marshack identified what appears to be the first lunar calendar. This suggests that the discovery by women of the nature of deeptime set the stage for men to learn the awful truth about their own mortality.

during their gracious and courageous last days. Nevertheless, art, custom, culture, religion, mythology, and literature provide ample supporting evidence of this gender difference.

Asked to create a metaphor for time, a woman will usually draw a circle. Ask a man the same question and he will, as often as not, draw an arrow. Perhaps women, being closer to the cycles of life, view death as an ineluctable phase in the turning wheel of fate. Men, on the other hand, tend to conceptualize life as moving along on a linear trajectory with a beginning and an end.*

Women face death on a more intimate basis than men. Every woman is aware of the danger of child-bearing—a truth more trenchant in ancestral times than in the present. Each menses to a woman trying to conceive represents a *petite mort*. She knows the pain of the death of her children, both before and after birth. She lives with the knowledge of death from the moment she knows she can give new life. Men must seek out death experiences actively, as they do not confront them on a routine basis. Mini-deaths do not greet them monthly whether they want them or not.

Women attempt suicide more often than men. Men, for their part, kill others more often than do women. The first statement implies that fear of death does not deter women as much as it does men; the second statement implies that men much prefer that someone else do the dying. The World War II commander General George Patton told his soldiers on the eve of a great battle that the point of war was not to sacrifice your life for your country but, rather, to force the other dumb bastards to sacrifice *their* lives for *their* country.

Men risk greater harm to life and limb than women do in many endeavors. They use their flirtation with death to test their courage, to tempt *finis* and return to tell about it. The point of these exercises, however, is not to satisfy a death wish but, rather, to get so close to death that one can thumb one's nose at it, thus proving one's manliness to both oneself and others.

Because men do not bear children, they are disconnected by nine months from their transitory act of insemination and the birth of their child. Women feel more connected to the rhythms of their cycles and the turnings of the seasons. Through their act of birthing life, they know that they can live on through their children, an option only lately understood by men. This subject will be more thoroughly discussed in the next chapter.

Recently, geneticists identified the gene that causes Huntington's chorea,

---

*Nearly all men, at least at some point in their lives, aspire to be great hunters (or the modern equivalent). The universality of cops-and-robbers or cowboys-and-Indians play emphasizes how early in their lives boys aspire to become the most accomplished killer. For an ancestral hunter, the death of his prey was among the most transcendent triumphs in his life—especially if he was stalking an animal that had the capability of killing him. Should the hunter die while in pursuit of a dangerous quarry, all his companions would consider this turn of fortune to be a noble death, but a part of them would also likely consider it an ignominious defeat for the deceased.[12]

a fatal hereditary illness that typically strikes adults in their early thirties without any warning. Members of families that carry the gene do not know as they age whether or not they have the full-blown form of the disease until the first symptoms appear. There is no cure or treatment, and death inevitably follows after the illness's appearance. By locating the gene, scientists are able to offer people who are at risk the opportunity to learn if they are doomed to die an early death.

The response of those contacted has been revealing. Men are three times more likely than women to decline the offer. The inescapable conclusion to be drawn from this gender skewing is that men feared discovering that they were destined to die more than the women did. The men preferred ignorance over knowledge in so crucial an area as their life span. Women expressed more concern than men about passing on the gene to their offspring, and by agreeing to take the test indicated that they were less fearful than men about learning the dreaded truth.[13]

Similar differences between the sexes regarding fear of death became evident in the early years of the AIDS epidemic. At a time when there was no effective treatment for the disease, middle-class women, who were at minimal risk to have contracted it, were eager to learn whether they were carriers. In contrast, many men, who were far more likely to test positive because of their risky past behavior, behaved as if in denial. Men, much more than women, simply did not want to know they were fatally infected.

*   *   *

The correlation between her menses and the moon led *Gyna sapiens* to appreciate the significance of a month. After she had vaulted over a bar set at 29.5 days, she raised it until she cleared nine months, and by so doing finally connected sex with birth. Men learned the art from the women, and, to their eternal dismay, they also learned that their life was limited.

Sex (marriage), birth (children), and death (funerals) are the three most central rituals in the human life cycle. Burned into our memory banks are the details of our first sexual adventure, the pomp and ceremony enveloping our marriage, the drama attendant upon the birth of our first child, and the circumstances surrounding our parents' deaths. Sex and death percolate through all the interstices of human existence.

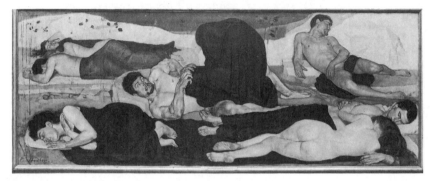

Many religious convictions arose from a yearning to believe in an afterlife prompted by the terror of death.

Chapter 19

Superstition/Laughter

With me the horrid doubt always arises as to whether the convictions of man's mind, which has been developed from the lower animals, are of any value or are at all trustworthy.
        —Charles Darwin[1]

We should take care not to make the intellect our god; it has, of course, powerful muscles, but no personality. It cannot lead; it can only serve.
        —Albert Einstein

Gyna sapiens' awareness of the cause and implications of pregnancy constitutes the first major breakthrough *sapiens* experienced as a result of acquiring a knowledge of deeptime. The second momentous insight—the certainty of death—would be the prime mover behind a spray of unique mental states and behavior patterns. Some have appeared in rudimentary form in a few scattered species; others have never been observed within any animal's cultural repertoire.* Chief among these states and behaviors are self-delusion, superstition, irony, humor, greed, anxiety, ambition, suicide, heroism, and that most enigmatic of human endeavors—art. Since their advent, each has become an integral part of the human condition.

  The first on this list is the peculiar mind-set called "self-delusion." The vast majority of creatures do not have an awareness of self, and therefore are incapable of deluding themselves. Those few that do possess self-awareness

---

*Other social animals do have cultures, in that members learn certain ways to behave that are not instinctual behaviors. Geese, lions, and chimpanzees are examples of animals that have cultures.

(e.g., apes, dolphins, elephants) do not possess sufficient mental acuity to engage in serious self-deception. Observers have documented numerous instances of chimpanzees employing sophisticated subterfuge in their social interactions, but they do not appear to turn this behavior inward. Only humans, endowed with a hyperactive (and, in some cases, riotous) imagination, can fool themselves.

We are the only creature that relies heavily on the *if* = *then* algorithm for our success. Philosopher William Irwin Thompson warned, "Even a positive thing casts a shadow. Its unique excellence is at the same time its tragic flaw."[2] The ability to reason was among the major factors that propelled our species so rapidly up creation's steeple. Unfortunately, behind shining reason lurks a malignant defect that could ultimately knock us off our high perch.

All syllogisms begin with a set of premises. Premises, unfortunately, are often based on unreliable beliefs, and therefore are often not put through the same rigors as the reasoning process that follows. If the premises of a particular *if* = *then* algorithm are flawed, then the answer will be false. A human usually sets the premises in place *before* activating the logic process.

Faulty logic, leading inexorably to self-delusion, gained an easy foothold in the evolving human mind because of the overarching need to mitigate one fear in particular. Lucretius, the Roman philosopher, opined, "Fear was the first mother of the gods. Fear, above all, of death." Once having discovered the end of their own personal time, individuals were no longer able to accept the deaths of others as extraneous events of no particular interest to them personally. "And therefore never send to know for whom the bell tolls," wrote poet John Donne. "It tolls for thee."

Unlike other animals, the new deeptime primate was under considerable psychic pressure to insulate himself somehow from this newly gleaned and very spooky epiphany. To alleviate his angst, reason was put to an inventive use for which it was not originally intended. Ancestral humans, continuing right up to the present, concocted fantastic stories about what happens after death. Through the transmission from one generation to another, these phantasmagorias eventually took on a reality all their own. Thus was born superstition, the shadow side of logic. Ironically, reason became the primary vehicle for this complex self-deception. All superstitions "make sense": If this, then that. If that, then this. Given fallacious premises, an algorithm may begin to sound plausible: *If* you step on a crack, *then* you'll break your mother's back. The psychoanalyst Sandor Ferenczi once said, "Pure intelligence is in principle madness."[3]

Each superstition hastily patched a newly discovered hole in the fabric of *Homo sapiens'* ever-expanding knowledge of nature. For the first animal that began to understand how the world worked, ignorance of the reasons behind events generated anxiety. If the event happened to concern one's personal well-being, then *not knowing* became unbearable. Thus, superstitions were increasingly deployed as protection, as insurance, and for peace of mind. If one could perform a ritual, intone a chant, clutch a talisman, participate in a rite, or make a wish—devoutly believing that these activities could influence the outcome of future events—then one gained a modicum of control over a mysterious and unpredictable world.

Philosopher Nathaniel Shaler observed, "The ancestral men who experienced the most terror were those that were the most realistic about their situation in nature, and they passed on to their offspring a realism that had a high survival value."[4] The *Homo sapiens* that emerged is the modern man as we know him: a hyperanxious creature who constantly invents reasons for anxiety even when none are present. But here was the quandary: Not to invent superstitions would bring about another form of anxiety. Blaise Pascal commented, "Men are necessarily mad because *not* to be mad would constitute another form of madness."

Excessive anxiety interferes with the normal mentation necessary for survival and reproduction. In numerous studies, psychologists have demonstrated that stressed animals do not live as long or reproduce as prolifically as contented ones.[5] Superstition can be viewed as a novel adaptation that humans evolved to mask unpleasant truths with which no other animal has ever had to come to terms.

The puzzle of death was the most vexing of these unpleasant truths. Stone Age people observed that when a warm, animate, rosy-cheeked person expired, he stopped breathing and became very still. Color drained from his face, and his body turned cold. He donned what the poet Homer described as a "garment of clay." Within hours, the body began to decay, and the vividness that had characterized life slowly rotted away. After the passage of years, all that remained were disconnected bones and a hollow-socketed skull, the sight of which continues to give the living the heebie-jeebies.

Terrified by the prospect of personal annihilation, the human species (especially men) used its collective wits to embark on a witless errand. Approximately forty thousand years ago, ancestral humans, whistling bravely in the dark, arrived at the astonishing conclusion that no one really died. Despite clear, incontrovertible evidence to the contrary, they declared death entirely illusory. A belief in the afterlife seems to have became embedded in the human psyche around then, in the period when the consistent use of mortuary rituals first appeared in the archeological record. Early peoples conjured an invisible component that resides internally and that did not expire with the body, an incorporeal essence they called a soul. This ectoplasmic entity departed the body at the moment of death and began a journey to another realm. So entrenched is this belief that in the majority of cultures, when a man dies, others consistently refer to him as "the recently departed."

Because the soul was invisible, no one seemed troubled that no one had actually witnessed it departing a body. That is one of the convenient aspects of self-delusion. Hell, Hades, Purgatory, Avalon, Valhalla, Paradise, Limbo, Reincarnation, the Netherworld, the Happy Hunting Ground, the Land of the Dead, and the Kingdom of Heaven were all places invented by humans, particularly men, to assuage their fear of death. These insubstantial, ephemeral worlds arose to obscure the unacceptable evidence of their senses—to wit, that death has the distinct look, unmistakable smell, and cold and waxen feel of permanence.

Homo sapiens embroidered this core self-delusion by enshrouding death in elaborate ritual. Undecided as to whether the body might come in handy as an adjunct to the soul when it traveled to the next realm, ancestral people hedged their bets. Corpses, they decreed, should remain intact, no longer abandoned where the unfortunate had died, to be torn apart and scattered by wild dogs and vultures. Before being buried deep in the ground or in tombs, the dead were laid out in finery and often oriented toward the rising sun. Limbs were carefully arranged prior to rigor mortis. The favored position was the fetal one, presumably because people wished to believe that rebirth followed death. "Tomb" and "womb" became synonymous in people's minds.*

The body was washed and dressed, and in many cases bedizened with ivory beads, perforated seashells, and finery unlike any that had ever adorned the deceased during his or her life.† No other animal engages in similar behavior. Notes the Spanish novelist Miguel de Unamuno, "The gorilla, the chimpanzee, and the orangutan and their kind, must look upon man as a feeble and infirm animal whose strange custom is to store up his dead."[6]

A long journey requires elaborate preparation. Practical items were placed alongside the corpse to accompany the soul. Food, valuables, and weapons

---

*Some cultures have come full-circle. Mongols were known to leave their dead on the steppes, and some sects in Tibet and India allow vultures to feed on the dead.
†In one group burial of a male and two children from Siberia dating 30,000 years ago, archeologists counted an astonishing number of polished sea shells strung as beads, a belt made from fifty fox canine teeth, and ivory beads numbering over 12,000. Researchers estimate that 13,000 hours of labor went into these incredibly elaborate funereal decorations.

were consigned to the grave. In many early-historical cultures, whole armies of slaves, attendants, relatives, wives, concubines, and/or guards were buried alive so that they might accompany a deceased ruler on his mysterious journey.* Imagined spirits were conjured to convey the dead. And the dead themselves often became supernatural beings whom the living imbued with supernatural powers.

Most important was the need to invest the passing of a life with communal ritual. Shamans recited incantations to help ease the travails of the dead through what would surely be their frightening gauntlet. Communal mourning alleviated the survivors' sense of loss.

Early peoples fashioned an elaborate Land of the Dead out of the whole cloth of their imagination. It was natural that they would use human society as the template for this complex spirit world. Some spirits controlled destiny, just as parents dictate the fates of their young children. The living endowed supernatural beings with omniscient and omnipotent powers, just as absolute rulers in the world of the living exercise powers over their subjects. People readily accepted these potent beings as the cause of natural events that, to them, were inexplicable. And they could plead with and try to influence these "higher powers" to help them negotiate the hardscrabble of daily life. It would be natural for the living to summon those spirits with whom they were most familiar, their dead relations. Ancestor worship is an enduring feature of all early and many contemporary societies. An alternative spirit world has been an abiding feature of the human condition; it came into existence when *Homo sapiens* learned the awful truth that he was doomed to die. As the writer José Ortega y Gasset observed:

> For life is at the start a chaos in which one is lost. The individual suspects this, but he is frightened at finding himself in this terrible reality, and he tries to cover it over with a curtain of fantasy, where everything is clear. It does not worry him that his 'ideas' are not true. He uses them as trenches for the defense of his existence, as scarecrows to frighten away reality.[7]

To comprehend fully the true nature of the human condition means, unfortunately, to be full of fear and trembling.

One cannot say with absolute conviction that a belief in an afterlife is a form of self-delusion. One can say with assurance, however, that no one knows

---

*Until the British banned the practice in the late nineteenth century, the Hindus practiced *suttee,* the ritual suicide of a wife on her husband's funeral pyre. Her death was required, the priests demanded, so that the man might have her at his side in the next world.

for sure, because in all of recorded history no living mortal has returned to tell those who remained among the living exactly what happens on the other side. Given the current state of our knowledge, it would be appropriate to maintain an attitude of hopeful expectancy; to be dogmatic concerning one's belief in the existence of an afterlife would be self-delusional. As Bertrand Russell observed, "The fact that an opinion has been widely held is no evidence whatever that it is not utterly absurd."

Once humans learned to deceive themselves about death, they began to apply this novel talent to other areas of their lives. Unfortunately, denying the evidence of the senses blunts the sharp edge of reason—*Homo sapiens'* most effective weapon. Self-delusion eventually threatened to become a factor that could undo the species. An attribute that diminishes an organism's ability to compete in the race for resources reduces its overall survival chances. Hunters blithely charging a dangerous rhinoceros because they believed that an amulet flopping from a thong around their necks would magically protect them from harm substantially increased their likelihood of being injured or killed. As self-deception became embedded in human behavior, its more florid examples invited the destruction of its possessor. An example from history illustrates this point.

Early in the sixteenth century, the Aztec ruler Montezuma commanded a vast army and presided over an empire that extended from Mexico to Honduras. By an extraordinary coincidence, the Spanish explorer Hernán Cortés strode into his court on the date that Montezuma's astrologers had predicted the return of the Aztec god Quetzalcoatl, who had made his last appearance 476 years earlier. The god, the priests had foretold, would appear in the form of a bearded white-skinned man. Montezuma believed that Cortés was Quetzalcoatl.

Despite a modest contingent of troops, Cortés quickly took advantage of Montezuma's delusional belief. He conquered the Aztecs and nearly exterminated them. The king and his subjects could not be disabused of their faith in Cortés' divinity until it was too late.

A cursory reading of history confirms countless examples of how self-delusion wreaked havoc on the social fabric. The witch craze of the Renaissance, the Nazi Holocaust, or the religious and "ethnic-cleansing" wars that continue to plague the planet expose its dark aspect. Fear of individual death leads some individuals to revel in someone else's death. They seek power in order to destroy. To feel alive and make sure that they are not doing the dying, these death-drenched individuals order the slaughter of others for the purpose of mocking death. Fear of death became a factor shaping the evolution of the human species. Self-delusion evolved primarily as an antidote to ameliorate the anxiety created by the wrenching suspicion that death is final. The now eyes-wide-open-with-fear hominid was in serious need of an anti-antidote to counter the antidote.

Fortuitously, irony and humor evolved in parallel to balance the bane of self-delusion and irrationality. Irony is the juxtaposition of two circumstances that lead to the opposite of what might have been predicted. For example, an ambulance rushing to the scene of an accident is itself involved in an accident. Not until an animal developed the capacity to anticipate what might occur could a sense of irony emerge. That other animals read irony into events they see happening around them is highly doubtful.

The perception of irony bootstrapped the evolution of humor. Behind the door leading to irony, absurdity lies in wait, ready to leap out and startle us into laughing aloud. When we suddenly apprehend a cause-and-effect conjunction between two events that are wildly incongruous, we burst out laughing. The punch line of a joke is hilarious because it is so surprisingly removed from what we might have predicted. If we can correctly guess where a joke is headed, we will not laugh as hard, or will not even laugh at all. The more absurd and unexpected the joke or situation, the funnier it is.

Seeing a man who walking down the street suddenly slips and falls on a banana peel may strike many as funny because it is not what we have come to expect when we watch someone walking. Our expectation is that walking in the future will follow the pattern of walking in the present. The disruption of our prediction so surprises us that we laugh. As Mel Brooks deadpanned, "Tragedy is when I cut my finger. Comedy is when you walk into an open sewer and die."

Joy and mirth are emotions experienced by other higher animals, but humans are the only ones who can be incapacitated by the sudden intrusion of absurdity. Rolling on the floor, holding our sides, and trying to catch our breath while we are temporarily blinded by tears is a uniquely human behavior.* There is neuroanatomical evidence to suggest that laughter is the antidote to self-delusion. Reason and logic are left-brain functions; laughter and humor are decidedly right-brain ones. Laughter complements reason and is an evolutionary mechanism to reduce anxiety. When we are in stressful situations, we tend to laugh harder and longer than when we are relaxed.

---

*Chimpanzees are the only other animals that have been observed to laugh. They do not, however, achieve the level of incapacitation that humans can when convulsed in a fit of laughing.

Ridicule remains the most potent weapon to shock or shame someone out of a false belief or delusion. Attempts to point out logical inconsistencies to a true believer are notoriously unsuccessful. Laughing at another's aggrandizing self-inflation is a persuasive social tool used by the many to control the few who might believe they are entitled to more power. But laughing at oneself also plays a critical role in mental health. A dollop of self-deprecating humor, and the awareness of how often we humans can be gulled, are necessary adjuncts that maintain mental balance. Margot Fonteyn, the great ballerina, remarked, "One should always take one's work seriously, but one should never take oneself seriously."

Laughter is a key defensive measure a *sapient* can deploy against its most implacable foe, death. During the bubonic plague that harrowed the fourteenth century, comedians were in great demand, and "gallows" humor continues to remain the rampart behind which we can stick out our tongues and waggle our fingers at our ears, mocking death.

Despite the benefits conferred on one with a well-developed sense of humor, there remains a disquieting irony between the feeling of well-being associated with the glowing health of the prime of life, and the dread associated with the somber realization that everyone is destined to age decrepitly and then perish. This dissonance can induce the mental state of depression, to which humans, more than any other animal, are susceptible. Sometimes, the human awareness that we are destined to die no matter how we act or what we do leads a few to be unable to justify going on living. Apathy and loss of libido, two prominent manifestations of depression, interfere with an organism's ability to survive and reproduce. To counter this susceptibility, selective pressures honed a human's sense of humor. Laughing easily at the exigencies of life and indulging in a modicum of self-delusion remain the two best defenses against existential despair.

An exceptionally severe depression can lead to another unique human behavior—suicide. We are the only animal that can be so overwhelmed by a deep sense of life's futility that under certain circumstances afflicted individuals will voluntarily end their life. The rise of despair impelled the evolution of another uniquely human adaptation to mitigate these potentially devastating feelings. Along with humor and self-delusion, the search for meaning and purpose evolved.

Albert Camus, the French existentialist novelist and philosopher, began his elegiac paean to futility, *The Myth of Sisyphus*, with these lines: "There is but one serious philosophical problem, and that is suicide."[8] Sisyphus was a legendary king of Corinth whom the gods punished for his overweening hubris. His sentence was to roll a huge boulder up a hill in Hades, only to have the boulder roll down again as soon as he reached the top. He was condemned to repeat this meaningless labor over and over throughout eternity.

Camus used the myth as a metaphor for human existence. Why strive? At some point in almost every person's life, he or she becomes convinced that no matter what he or she may accomplish, death will intervene to make it all pointless. After the car Camus was in swerved and hit a tree while he was driving on a dry road in good visibility, those who knew him well speculated that he had somehow contrived to live—and die—his philosophy.

Upon learning of the death of his wife, and knowing that his turn must come soon, Shakespeare's Macbeth utters this most famous of mournful passages:

> Tomorrow, and tomorrow, and tomorrow,
> Creeps in this petty pace from day to day,
> To the last syllable of recorded time;
> And all our yesterdays have lighted fools
> The way to dusty death. Out, out, brief candle!
> Life's but a walking shadow, a poor player
> That struts and frets his hour upon the stage,
> And then is heard no more: it is a tale
> Told by an idiot, full of sound and fury,
> Signifying nothing.

Does a member of a lion pride, resting in the noontime shade, ponder what exceptional things it must accomplish to justify its existence? Do cows chewing their cud ruminate on the meaning of it all? Other animals do not, I suspect, expend time or energy lost in contemplation about why they are here.

Only humans demand an answer, because we are the one animal that has glimpsed the land of the future. Humans created the unique construct that their striving to make a difference gives meaning to their lives. Thus, human ambition has been the force advancing culture, rapidly propelling our species away from its animal roots. Individual ambition has been behind virtually all of the great strides forward in art, law, science, technology, and philosophy. And, unfortunately, the kind of "vaulting ambition" Shakespeare attributed to Macbeth has been behind history's most spectacular follies and tragedies.

The deep need for a purpose to justify one's life fuels the desire to accomplish something that will both be self-satisfying and win the approval of others. These goals often translate into the acquisition of sex, money, and power, the iron triangle that goads the vast majority of human strivings. (The pursuits of beauty, knowledge, service, or spirituality, though qualitatively differ-

ent, are also uniquely human quests.) But there is a more subtle reason behind the human penchant for setting goals and striving to succeed. We
believe that if we accomplish something extraordinary then we will be remembered. *The Guinness Book of World Records* concerns posterity. In his
poem "Carmina," the poet Horace, reflecting on his fame, exulted in his triumph.

> I shall strike the stars with head sublime. . . .
> I have completed a monument
> more lasting than bronze and far higher
> than that royal pile of Pyramids,
> which the gnawing rain and furious
> north wind cannot destroy, nor the chain
> of countless years and the flight of time.
> My death won't be complete.
> I shall not altogether die. . . .

Striving, ambition, and heroism are, in one sense, our attempt to try an end
run past death.

Collectively, men convert their fear of death into history-making. Hegel
developed the startling idea that civilization results from what men do with
death. Among nonhuman animals, there is no history because there is no
sense of individuality, and, without time awareness, no knowledge of death.

Previously, higher organisms could only pass on something of themselves
through DNA. With the development of a sophisticated human culture, a new
way, independent of DNA, became possible. Deeds could outlive a person's
life span, safely preserved in the culture's collective memory. Nobel laureate
Peter Medawar observed, "In human beings, exogenetic heredity—the transfer of information through non-genetic channels—has become more important for our biological success than anything programmed in DNA."[9]

In the *Iliad*, Homer repeatedly alludes to the contract between the community and its warriors. If a man performs heroic deeds, the bards will sing
his praises for untold generations to come. Through the retelling, the hero's
name and his acts of heroism will be beyond death's grasp. Innumerable college freshmen can attest to the thoroughness with which this ancient covenant continues to be fulfilled. The obituaries in the *New York Times* and
entries in the *Encyclopædia Britannica* are testimony to the power of deeds,
posthumous legacies that outlast one's allotment of days.

A considerable gender gap exists concerning this feature of humanhood.
Women who raise their children to maturity often satisfy their need to leave

a mark through motherhood. Men's relationship to their offspring in our species' early history was unknown or poorly understood, and a man's identification with his children (who may or may not be his) has, in general, tended to be more tenuous than that of their mother's.

The male ambition to be a hero occurred at a fortunate juncture. To convert a vegetarian simian into a fully operational predatory anthropoid, the *Homo sapiens* nervous system required a huge infusion of courage. The most common early act of heroism revolved around hunting. Risking one's life to perform daring feats is not a trait for which plant-eating animals are particularly noted.

The cascade of events might have been as follows. When men discovered the existence of the future with its accompanying booby trap of death, they longed to do something for which they would be remembered. Both a strong belief in an afterlife and the burning desire to leave a mark boosted their courage. Bold hunters greatly increased the quantity and quality of meat they brought back to the home base. Women applauded and rewarded male bravery. Courage became a defining male trait that increased the likelihood of women and children surviving. The entire species surged forward, as each interlocking evolutionary event reinforced the others.

*   *   *

Greed is a consequence of humans possessing an opposable thumb. Designed to grasp vines and branches, the primate hand evolved into an appendage like no other in the animal world. A primate can use its hands to hold and carry a variety of inanimate objects considerable distances. A bipedal primate can carry the most the farthest. For most other animals, the preferred method of transport is to use the mouth as a pouch, and teeth or beak to maintain a grip. In the majority of cases, the object carried is either a squirming tyke or a dead meal. For a variety of obvious reasons, this style of conveyance is inherently inefficient. Further, there would be no pressing need for a non-tool-using animal unaware of the concept of deeptime to hoard a cache of inanimate objects.*

Animals that lack hands generally do not covet possessions. They fight fiercely for sex, food, and territory, but neither copulation, sustenance, nor a valley filled with tasty tidbits can be categorized as "objects." Besides a few of the apes, animals do not engage in barter.† With the exception of the flesh

---

*Bowerbirds, pack rats, and ravens are several examples of creatures that hoard inanimate objects.
†Frans de Waal has eloquently detailed the complicated social relationships that exist among higher primates, some of which include primitive forms of barter.[10]

and sinew housing life itself, there is nothing tangible that one animal desires of another. In sharp contrast, among humans there exists no single word in the lexicon of any language that has caused more suffering and death than "mine." Rousseau summed up this strange human obsession: "The first person who, having fenced off a plot of ground, took it into his head to say *this is mine* and found people simple enough to believe him, was the true founder of civil society."[11]

The avaricious surround themselves with accouterments to increase their pleasure and security. Surplus goods and money translate into dominance over others. A component of greed, however, also relates to mortality. People frequently become very miserly with money and possessions as they approach old age, ardently believing that stocks and jewels can be a bulwark with which they can keep death at bay. Money equates with vitality. To many of the aged, spending money is equivalent to wasting life.

Among the variety of reasons that humans create art, one of them relates to fear of death. The impetus to make art arose from the confluence of several different urges, one of which was the desire to create something that would outlive its creator. The over nine hundred stenciled handprints in the cave of Gargas in southwestern France, created over thirty thousand years ago, give the impression that their creators made them for the generations to come. The making of a mark that will live past one's allotted three score and ten is uniquely human, and embodies the longing of a primate equipped with its sense of death and time to touch the hem of immortality. As poet Emily Dickinson wrote, "When this you see, remember me." Nearly everyone experiences a tingle of excitement when holding in his hand an artifact from a bygone age. Knowing that it was fashioned by someone very much like us links us to the past. Art serves as an umbilical cord connecting us to past generations, and can be seen as a novel form of DNA that transmits cultural values.

Hominid species have been in existence for over four million years. During a vast stretch of that time, hominids did not create art. Does it not seem like an extraordinary coincidence that art as a distinct human endeavor effloresced nearly simultaneously throughout the world approximately forty thousand years ago—the same time that the first lunar calendars also appeared? The advent of art also coincides with widespread mortuary practices. Art is deathless. So, too, is the human striving for immortality.

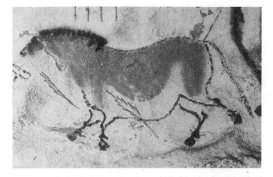

The universal urge to create art exploded in the final third of our species' 150,000-year history. Why did it arise and what took it so long?

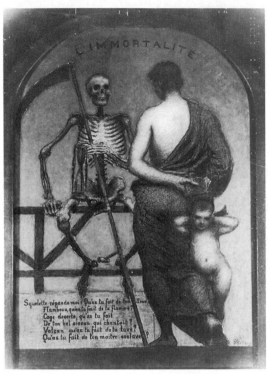

Having an heir that they could name and that they were sure belonged to them gave men their best chance to achieve a kind of immortality.

Chapter 20

Father/Mother

We do not marry for ourselves, whatever we say, we marry just as much or more for our posterity.   —Montaigne[1]

Men desire offspring more than women do, and they generally want more children than women want.   —Mary Batten[2]

Men have to learn to want to provide for others, and this behavior, being learned, is fragile and can disappear rather easily under social conditions that no longer teach it effectively.   —Margaret Mead[3]

*A*s I have speculated, the first key insight to mold human culture profoundly concerned deeptime. Women noted how their monthly cycles coordinated with lunar ones and began to mark off the days of a month. Soon afterward, they made the connection between the sexual act and birth. The next key insight was the recognition of individual mortality. Evidence in the archeological record suggests that the third insight, of equal influence on the relations between the sexes, occurred nearly simultaneously in widely separated regions during the relatively short time span of a few thousand years.

Although the insight could have occurred in places we now call South Africa, the Middle East, Asia Minor, and Australia, let us assume the first place that this thunderbolt struck was in Southwestern Europe, toward the end of the last Ice Age. Let us arbitrarily fix the time at somewhere around forty thousand years ago and grant that a human alive at that time was as intelligent as we are today. Human language, by this late date, was fully operational and as supple as present-day tongues. And let us assume that there had to have been one person who first experienced the third epiphany. For reasons that will soon become obvious, we will also speculate that the discovery made its deepest impact on a male.

The first male who had his mind convulsed by this third insight belonged to a relatively new species called *Homo sapiens,* whose progenitors had established themselves in Eastern Africa 110,000 years earlier. Modern humans, as *Homo sapiens* are called, soon began emigrating from Africa, and pockets of them reached the Middle East approximately ninety thousand years ago. From there, some bands of these intelligent hunter-gatherers pushed north to Siberia, while others moved east to Asia. Fifty thousand years ago, a band of *Homo sapiens* crossed open water to inhabit the island continent of Australia, half a world away.

For reasons that remain unclear, they avoided Europe until the relatively late date of forty thousand years ago. Perhaps one motive delaying their entry was the presence there of very large, big-boned hominid cousins, the Neanderthals, who had been entrenched in Europe for over two hundred thousand years. Within ten thousand remarkably short years after *Homo sapiens* arrived in Europe, the Neanderthals vanished. No one knows with certainty why.

Imagine an enormous outcrop of limestone jutting from the upper slope of one side of a lush valley. The huge overhang protects a wide deep platform from rain and wind. The intervening wall between the overhang and the platform is honeycombed with limestone caves. Their entrances provide further protection from the elements.

At this rock shelter, a band of *sapients* has encamped for the season. Sight lines from the base provide a stunning panorama of the cliffs opposite the entrance and egress from the valley below. Down the center runs a clear river whose banks teem with game. From their strategically located perch high on the valley's side, hunters can monitor migratory herds entering or leaving the narrow valley.

Let us attempt to reconstruct a crucial moment in the history of our species. Some of the thoughts I will attribute to my fictitious character may well have been discovered by many different people over many generations. In a bow to storytelling, let us suspend disbelief and imagine that they all occurred to one particularly insightful person during one long afternoon approximately forty thousand years ago.

On a rock in the shade of the overhang sits a Cro-Magnon man.* He is a handsome specimen, tall and well proportioned. Sleek and well nourished,

---

*"Cro-Magnon" is an alternative name for the modern humans who eventually settled Ice Age Europe.

he looks vaguely familiar. Of course . . . it is Adam, the same fellow we had encountered perched on a rock in an earlier chapter. He is the first person to understand that he is doomed to die, and the revelation has made him morose. Since making his doleful discovery, Adam has suffered periodic bouts of existential angst. On this particular day, once again, he is lost in his own thoughts.

While he is in this pensive mood, a young mother chasing after her toddler breaks his concentration. The little boy has just learned how to run and is testing the limits of his mother's speed and patience. He squeals with delight at his newfound freedom. Annoyed by this intrusion, the Cro-Magnon glances absentmindedly at the boy.

But then his eyes suddenly narrow. He notices, for the first time, that the child resembles him in an uncanny way. Peering closer, he concludes that there can be no mistaking the similarity. The boy has a mop of curly carrot-red hair; Adam's hair is bright red and curly. The boy has a startling shade of green eyes; many women had previously commented to Adam regarding the arresting hue of *his* green eyes. The other boys had always kidded Adam about his freckles, dimples, and cleft chin when he was young; this child has freckles, dimples, and a cleft chin. But the clincher is the little boy's bright-red birthmark that Adam caught a glimpse of on the back of the boy's right shoulder; he, too, has a birthmark in the very same spot! Even the toddler's impish behavior reminds him of his own sense of independence and adventure when he was a child.

An unfamiliar thought begins to form tentatively in Adam's mind: "There can be no mistake. . . . This child looks just like me. We men have often puzzled over how much some children resemble particular men in the band. I wonder how this magic can have happened? Most confusing . . . What does it all mean?"

His thoughts drift to the child's mother. "Her, I know well. . . . The woman called Eve, how she beguiled me several snows ago." Lost in a delicious reverie, he recalls the many instances when he had had pleasure with the boy's mother, bundled up in the warm furs he had given her from his hunts. "How I enjoyed the hours of delight we experienced together. . . . Hmmm . . ." Adam's brow furrowed, and then he started, almost speechless. "Could there be . . . ?"

In an explosive insight destined to change all future relations between the sexes, he grasps that he played a critical role in creating the squirming bundle of energy who had just been scooped up by his mother.

"But what has pleasure to do with creating a child?" he asks himself. He keeps pondering the point. "Somehow my essence gets inside the woman. But how?" Then it strikes him. "The fluid . . . It's the fluid I squirt at the height of my pleasure!" he exclaims. "My essence lives in the milky white liquid." His next conclusion is so obvious he doesn't understand how he could have overlooked it before. "That must be it—my essence gets inside of the woman and it mixes with her essence when we have pleasure together. The combination of the two makes her belly grow large and round with a baby!" Dumbstruck by his insight, but in a moment of prehistoric genius, he makes yet one more cerebral leap. Since he was the only one who had pleasure with her before her belly began to swell, then it was *his* essence that helped create *this particular child* who looked so much like him.*

Adam hurriedly set up his new *If* = *then* tool. "*If* my vital sex fluid somehow contributed to siring this child," he reasons, "*then* it follows that this child is a part of me," he concludes. "Hmmm . . . I knew my brothers and sisters came from our mother, but I did not understand just how exactly a man fit into the web of kin. *If* there is a special connection between a man and a child, *then* I can have a special child who is a part of my blood. That child would have more of my essence than any other man. He would be closer to me by blood than even my brothers."

Energized by his train of thought, he begins to pace. "*If* this child and I were entwined by blood, *then* when I die a portion of my life force will continue to live on. *If* I leave behind an heir, *then* I could snatch a small victory from the clutch of death!" Like two exposed high-voltage wires coming into contact, a blinding flash of sparks explodes in Adam's head. Two insights, one about sex and the other about death, intersect and synthesize into a third—the concept of paternity. To assist him in gaining control over this astounding concept, Adam invents a new word to describe this new kin relationship: "I will name this new role for men 'father.'"

The symbol-making primate's thoughts come in a tumble as he runs through the myriad implications of his discovery. "The story the shaman told me about the soul's journey to an afterlife was comforting, but now I *know* I will live on." Now that it has sunk in that his immortality depended on his child's survival, the intrepid Cro-Magnon concludes, "I must see to it that my son has every advantage so that he will not die. If he dies before he becomes a man, the part of me in him dies with him." Adam exults, "As long as the child lives, I will *not* really die after all."

---

*The people who settled in Europe experienced greater variations in eye, skin, and hair color than any other humans. They also had the greatest variations in the structure of their hair and the pattern of their skin (freckles). I surmise that a man or a woman living in Ice Age Europe would have found it easier to identify that a child was the offspring of a specific man than peoples who did not bear these helpful distinctions.

Knowledge of their fatherhood made men more loving.

Further implications of his insight continue to flood his consciousness. It occurs to him that if he had an heir he could confer his name on the boy. The power to name, he knows, is a magical act that confers on the namer a mysterious claim on the named. Once a name insinuated itself into the fabric of a culture, it took on a life of its own often outliving its namer. "Suppose my child lives long enough to have a child of his own? That child would *also* bear my name. Even when I am no longer physically here, my name will live on!" The very thought makes him shudder with excitement.

Adam continues in his reverie. "And wisdom—what about all the knowledge I have accumulated in my life? All those stupid mistakes I made, especially those that nearly got me killed. If this child were my blood, then I could *teach* him how to avoid these errors.* The secrets and lessons I have learned in my life will not die with me. They will be treasured and guarded by someone whose blood is linked to mine. And when my son gets older, we will be

---

*There is no male of any other species that expends valuable time and resources in teaching its offspring. One must make a distinction between learning through imitation and active teaching, as when our Cro-Magnon dad takes the time to teach his son to throw spears, stalk aurochs, and set snares. An occasional mother of baby predators will sacrifice the opportunity to eat a prey she has disabled in order to bring it back to her young. The soon-to-be-lethal predators practice killing the prey and thus acquire the gist of hunting. No one has observed a male of a nonhuman species expending time in actively teaching skills of any kind to its offspring. There are surely fathers among other animals, but none exhibit behavior similar to what humans call fatherhood.

constant companions. Why, having a son will be like growing my very own friend!" Adam looks out on the valley as the shadows of dusk begin to lengthen, and a broad smile lights up his face. For the first time since our hero discovered that he was doomed to die, the gathering darkness does not seem quite so filled with gloom.

Adam has to sit down as his mind races through the other ramifications of his discovery. "And what about my weapons? All the time and effort I put into fashioning my spear thrower! How lovingly I carved, polished, and decorated it. I will bequeath it to my son. When he uses it, it will be as if a part of me is in on the hunt, as if my spirit permeates the objects I created or owned. Perhaps, each time he notches it with his spear, he will think of me."

Another thought occurs to him. "And what about my other possessions? Instead of having them divided among the others when I die, I will pledge them to him. My son will cherish these objects because they link us together. And when he grows old or dies, he will pass them on to *his* sons with pride."

While relishing the idea of having a son, an unexpected thought suddenly confronts him. Not all babies are boys! "Holy Auroch! . . . Suppose my child was a girl child?" The elation he was just experiencing gives way instantly to a sense of dismay and disappointment. But then, as he rolls this new idea around in his mind for a while, he gradually reconsiders. The women call the young girls that issue from their bellies "daughters." Daughters, Adam thinks, maybe would not be so bad after all. "A young woman who grasps that she owes me her life will be indebted to me.

"My goal has always been to have sex with as many young women as possible." He brightens. "When she begins to bleed, I shall start having sex with her." Then his brow furrows as he tries to think this through. "Can there be harm in that?" he wonders. "What am I to do about having sex with my own daughter? This is hard thinking."

As he ponders the startling prospect of having sex with a woman he helped to create, a dark foreboding begins to intrude. The more he turns the idea over in his mind, the more he concludes that there is something inherently wrong about it even if he can't exactly put his finger on what it is. He recalls how mothers absolutely refuse to seek pleasure with their sons, and sons seem to instinctually abhor the idea of approaching their mothers for sex. After a convoluted internal argument with himself, Adam rejects the idea of having sex with his daughter. But it is a close call.

After coming to terms with this temptation, Adam moves on. "Because she is my daughter, I will teach her, too. Hmmm. There are not nearly so many lessons for me to teach a girl child. It is unlikely she will ever use the secrets I could teach her concerning how to balance my spear thrower nor will my tricks for how to track reindeer be of much use to her. On second thought, having a girl who feels connected to me, and I to her, might have other bene-fits." Soon, another thought brightens the prospect of having a daughter. "Maybe, when she grows up, I can control her pleasure. I shall have a say in choosing the man who gives her his essence. That way I'll make sure that he is a brave, skilled hunter. I will also use her kinship to me to forge advanta-geous alliances with men who want to have pleasure with her."

It is then that the thought strikes him that, if he had something to do with the creation of the little boy who looks like a miniature version of himself, then there must have been one particular hunter who joined with *his* mother to give *him* life. "If I can discover which one of the old ones gave me his essence, I will care for and protect him in his old age." He resolves, "I will begin a tradition to be handed down through the ages. I have always respected my mother, and saw that her final years were as pleasant as I could make them." Adam realizes that there would be an added benefit to his newfound obliga-tion regarding his father. "If I set an example by caring for my father, then my son or daughter will care for me!"

As his thoughts turn to the future, he ruminates, "And illness—who better than my own child to care for me when I am sick, old, or feeble?" Pondering his inevitable infirmity leads him ineluctably to the ramifications of his own de-mise. "And then there's death, that mysterious journey I must embark on into the unknown. I dread even thinking about it. Yet having a child that shares my very essence is a comforting thought. Instead of friends or kin, I can rely on my children to carry out the rituals that are so important, especially that crucial moment when my spirit is still imprisoned in my flesh and my body is unable to obey my soul's instructions. My children will position my limbs, close my eyes, wash me, dress me in my finest, and provision me with the things I need for my journey. They will not send me to the other side unpre-pared. A man's children must attend to his funeral rituals with respect and tenderness. They must solemnly recite incantations in his honor.

"And who better to honor me in the seasons that follow one another after my death than my own children? Each year, in commemoration, my children will speak the words of remembrance. I honored my mother when she died because I had no doubt that she was my mother. Now I will teach all in our band to honor their fathers.

"And best of all, my son and daughter, knowing that I am the only man who gave my essence to them, will continue to ask my council after I am gone. They will believe that I can still hear their prayers. Who knows for sure? Per-haps I will be able to visit with them from the other side. I shall exert an influ-ence on the living long after the flesh has fallen from my bones."

And so our previously depressed, forty-thousand-year-old *Homo sapi-ens* dances with joy at his discovery. So many troubling questions answered by just cranking up his new-fangled *if* = *then* algorithm device. He returns to the center of the camp to catch another glimpse of the little boy he is now certain is his son. He watches him proudly from a distance, not sure yet what to say to the mother or how to behave toward the little boy. Nevertheless, he beams. "Imagine that. . . . A son. *My* son. How delightful! How significant!"

A sense of profound wonderment washes over Adam as he continues to reflect on his discovery of the secret behind paternity. From this insight, a subsidiary but equally momentous thought begins to form in his mind, one that will as profoundly affect future relations between men and women as Adam's initial understanding of fatherhood.

"If my son was the result of one pleasurable episode between the furs with his mother, then what about all those other times we made love and Eve's belly did not swell?" He thinks about one woman who has never had a child, and all the men who have taken pleasure with her. "Apparently she does not have what it takes to make a baby," he thinks. "How sad for the men who mate with her."

Then, a more disturbing thought creeps into his mind. "Suppose it is the men's essences that are not strong enough to make a baby? What if their fluids are not potent enough to pull off this magnificent feat?" Adam's chest begins to swell with pride as he realizes that his essence had a quality that made it feistier than other men's. Masculine egos are easily subject to self-puffery, and Adam's is no exception. He experiences a feeling of manly superiority.

And then it strikes him: "My manhood is somehow tied to whether my essence can produce a child—no, not just a child, a boy child. A son. Up till this moment, I had thought that the two things that most proved a man's virility were the number of animals he killed and the number of women he convinced to let him take pleasure." Adam's thoughts drift to the other powerful male animals he has observed. Fighting and mating have been the two major criteria by which he and the other men judged the strength, courage, and heart of male animals. But Adam's new insight about the potency of his essence forces him to consider another criterion. "A man's virility can be measured when he proves to all that his essence is powerful enough to make a son!"

Adam continues along this line of reasoning. "Why, a man unable to prove that he could make children would be humiliated before the other members of the tribe. If he couldn't demonstrate beyond a reasonable doubt that the child a woman bore was as much his as hers, then everyone would whisper behind his back that his essence was not very manly!" He luxuriates in the thought that, once he explains to everyone the significance of his relationship to Eve's little boy, everyone will look at him in a new way. And he thinks, "If one son will raise my stature among the other men and women, how much more they would admire me if I could demonstrate my virility by having many, many sons and daughters."

Again his thoughts return to the redheaded boy's robustness. "But it is not enough that Eve and I have produced a son. The boy's vigor and strength also reflect on the vigor and strength of my essence. I will be judged by the kind of son I produce. It is important, therefore, that my son grow up to be a strong, skilled, leader among men."

These thoughts create a kernel of responsibility for his son that goes beyond his desire to achieve a sort of immortality. Although he, like the other men, is generally protective toward all the children in the tribe, he has never before felt so *personally* responsible for one *particular* child. The more he mulls these new ideas over, the more determined he becomes to see to it that his son has every advantage.

A strong maternal instinct resides in the heart of every female mammal. Very few male mammals possess the neurocircuitry for a strong paternal instinct. And for those few in possession of one, concern for the welfare of their young rarely extends beyond infancy. Stemming from Adam's facility of foresight, he begins to feel deep in his solar plexus a paternal instinct unrivaled by any other living creature.

The *concept* of paternity, in distinction to the *instinct* for paternity, is an idea incomprehensible, I suspect, to the nonhuman male. A powerful shaping influence on human relationships, it forced each man to face the prospect that sex brought with it a passel of heavy responsibilities. Some men denied them outright, some chafed at the idea, some men, frightened, ran away from them; but most men, considering the onerous nature of these responsibilities, unexpectedly welcomed these added burdens with open arms.

Adam is one such man. The idea of immortality, mixed with the realization that his little boy is a diminutive version of himself, engenders in Adam tender thoughts that are the beginning of an entirely new kind of love.

The only way to stack the odds in the little boy's favor so that he can accomplish all the things Adam dreams for him is for Adam to stay close to aid and protect him and his mother. He feels a new sense of urgency to see to it that Eve has everything she needs to nurture his son so that the little boy can resist disease, injury, or death. Adam realizes that he will have to establish an entirely new kind of long-term relationship with his son's mother, but is not exactly sure how to go about it.

Beads of sweat form on his forehead as he contemplates the complex negotiations that he will have to enter into with Eve in order to settle all the details resulting from his newly acquired paternal determination. Finally, all these new thoughts, each demanding attention from his overworked brain, exhaust Adam. He sits down on a log and mops his brow, exclaiming out loud to no one in particular, "Being *personally* responsible for bringing a child into the world is far more complicated than first I thought it was."

Further confusing Adam is the emergence of another unfamiliar emotion. Mixed with his anxiety is a warm feeling toward Eve that has been increasing in intensity ever since he understood that the two of them are tightly connected in the person of the little red-headed boy. "Hmmm," Adam wonders, "how will the boy's mother, Eve, react to all my new interest in her . . . um, er . . . I mean my—no, wait!—*our* son? I think I know her. After all, I had sex with her. That means I know her, right? Or does it? We did talk a lot. Let me see, what can I remember of the things she said?" Adam realizes that he must get to know her better. Energized by all the thoughts whizzing around his mind, he abruptly turns on his heel and heads in the direction of the forest. He wants to be away from the others so he can savor all of the startling insights he has had on this most momentous evening.

Another happy thought enters his head. "If pleasure results in children who have my essence, then pleasure with lots of women will give me lots of children. I will have as many different women as I can, and accumulate many sons and daughters. Each son and daughter will possess my essence. . . . Lots and lots and lots of my children, all running around looking just like me—this is a very good thing!" Loping along, a new spring in his step, he suddenly stops short and frowns. "But wait! How will I know for sure if a new child is mine? Suppose he doesn't look exactly like me? What if another man puts his essence into the same woman? How will I know that this child is mine and not his? Hmmmm."

Resuming his walk but now at a much slower pace, he entwines his fingers behind his back. By bringing to bear all his powers of concentration on this problem, he comes up with a solution. He will force the woman to take pleasure only with him. Immediately he discards that idea. Shaking his head involuntarily, he mutters, "Forcing a female to do anything is a very bad idea. And," he thinks grimly, "nearly impossible." Gradually, another solution occurs to him.

It occurs to Adam that if he understood this amazing fact about how a man and a woman made a specific child, then Eve, being more attuned to the subject, surely would have been aware of this knowledge for some time. He strikes his forehead with his palm. "I am so stupid! Why hadn't I figured this out much sooner?"

He carefully reasons through his options and finally comes to what he is sure is the perfect answer. "I know! I will ask the mother if the child she bears is mine. She would certainly know." A bit of doubt creeps in. "Or would she? I will simply have to trust her. But suppose she doesn't tell me the truth about whether she has had sex with other men? We hunters often talk among ourselves about how differently women think about things from men. Especially when the subject is either sex or children. Suppose she has sex with lots of the men, as some women in the band are prone to do. Could *she* even know for sure who the father was?" Putting his hands to his head Adam laments, "Aiee, all this thinking hurts my head. I cannot keep this to myself. After I have thought more about this great secret, I will tell the other men. We should talk this over around the fire. Perhaps they might have some suggestions."

For the *Homo sapiens* male who first had the paternity insight, sex passed from being great fun to being deadly-serious business, all in a single afternoon. What the others who had not yet figured out this connection did not know is that relations between men and women were on the verge of a convulsion. Once everybody *understood* the function of sex, *sapients'* mating patterns would never be the same again.

Let us leave the befuddled, trailblazing ancestor of all dads and return to Haeckel's Biogenetic Law, "Ontogeny recapitulates phylogeny."* Previously, I conjectured that a critical mass of humans most likely discovered death approximately forty thousand years ago. I based that assertion on the fact that this was the time frame when *sapients* first began to fashion what appeared to be lunar calendars. It was also the period when they began to bury their dead consistently in what appear to be elaborate funerary rituals. I compared these insights in the life of our species to similar insights that occur to a seven-year-old child. Earlier, at about the age of five, children learn how to tell time in terms of multiple months. Before that, they had lengthened their time horizon from first a day, to a week, to a month; but at around the age of five, the significance of their birthday hits home, and they grasp that one year hence on the same day they will enjoy another special day. This milestone signifies that a child is capable of anticipating an event that is a year in the future. But it is not until one or two years later that a child finally understands the personal implications of death.

Making an analogy to Haeckel's law again, I set the date of men's discovery of paternity also at about forty thousand years ago. Children usually comprehend death coincident with the moment they grasp that storks do not bring babies. Both insights depend on a well-developed sense of the future. The archeological record of modern humans reveals a tediously slow accretion of cultural advances for the first 110,000 years without any gross

---

*This principle, remember, states that each human fetus revisits in stages, over a nine-month period, each of the stages of evolution that took 3.8 billion years to complete. Borrowing Haeckel's principle, we can make an analogy between the cultural maturation of a single child and the entire prehistory of the human species. I propose that each child's development recapitulates the progression of key insights occurring during the human species' 150,000-year cultural maturation.

discontinuities. Throughout this period, which covers more than two-thirds of our existence as a distinct species, there is no consistent record that *Homo sapiens* created art, music, funerals, or grave goods, or routinely adorned their bodies.* I shall advance the idea, endorsed by others, that the first 110,000 years of the species *Homo sapiens sapiens* were analogous to the early childhood of a single individual. My emphasis will be on the timing of three major human conceptualizations.

If the insights about sex, death, and paternity were not behind the extraordinary Creative Explosion, what was? No significant physical changes occurred in our ancestor's brain or body forty thousand years ago. Nomadic hunting and gathering in small bands of a little over one hundred individuals seems to have remained the norm during the transition from pre–Creative Explosion to post–Creative Explosion. Nothing in the environment seems to have altered so drastically that it could explain *Homo sapiens'* traverse across that mysterious forty-thousand-year-old date line.

Some have posited that human language fully emerged at the advent of the Creative Explosion. There are strong counterarguments to this hypothesis. The human vocal tract and its complex supporting apparatus were in place at the outset of our species. It is doubtful that these sophisticated adaptations would have sat idle in the preceding 110,000 years. Endocasts of early *Homo sapiens'* brain cases reveal the same degree of left-hemisphere language-center indentation as those present in a contemporary man's skull. Most linguists would agree that humans of forty thousand years ago were sophisticated symbol-makers and most likely would have been using an advanced form of vocal language.

If the Creative Explosion was not due to a physical change in our body or brain, and there were no sharp shifts in how our species responded to environmental challenges, then one must consider a third option. Through the marvel of human language, the few early consciousness explorers were able to communicate what they had discovered to others nearby. Rapidly disseminated crucial insights concerning the human condition would have brought about a revolution in human culture. Like a spark struck from a hard flint, a culture-wide mental conflagration ignited.

Some might argue that archeologists have not looked hard enough or long enough. Perhaps the record prior to forty thousand years ago is too de-

---

*A recent archeological find at Blombus, in Southern Africa, of finely chiseled bones with ochre-inscribed abstract designs on them was dated to 70,000 years ago. Consistent early finds of art, however, continue to be rare compared with the number catalogued after the 40,000-year watershed mark. Anthropologist Richard Klein recently proposed that the Creative Explosion was due to a mutation in a 50,000-year-old *Homo* or *Gyna sapiens'* brain. See note 5 (chapter 20) on page 383 for a discussion of the pros and cons of his argument.

graded to supply evidence that would point to a more gentle transition. But this is unlikely, because caves and mounds can faithfully preserve evidence of art and burials for exceedingly long periods.[4] Some researchers propose, however, that the Creative Explosion is an artifact. Their counterarguments notwithstanding, the evidence accumulating from numerous excavations of earlier *Homo sapiens'* living sites points to something quite remarkable occurring to our species around forty thousand years ago. Though it is still possible that a magnificent cache of sophisticated art or bones that predate the Creative Explosion will be discovered, it becomes increasingly unlikely as teams of archeologists fan out across the world.[5]

Another puzzle: The Creative Explosion seems to have developed nearly synchronously in areas extremely distant from each other. There is no feasible way Aborigines in Australia could have learned from Cro-Magnons in the Dordogne to bury their dead or create art (or vice versa). Yet both groups seemed to have made these cultural markers commonplace nearly simultaneously.

Since the work of Piaget, psychologists have accepted that human brains mature according to a remarkably stable schedule. A parent using Piaget's developmental guidelines can predict with reasonable accuracy when a child will sit, walk, and talk. The sum of all individuals' maturation schedules traces out a bell-shaped curve.

I would assign a similar schedule to the mental maturity of the *sapient* species as a whole. Within a few years after children learn how to tell time, they understand that they are going to die. Soon afterward or simultaneously, they roughly understand how sex has something to do with how babies are made. Analogously, after ancestral humans learned how to track the moon's phases, they comprehended death. Soon after these two insights, they deduced the male's contribution to birth. Suggestive evidence that the male's involvement in procreation was the last of the three culture-changing insights that early *sapients* comprehended has emerged from anthropological field studies, most notably of those few cultures that had not yet made the connection between fathers and their offspring.

In the early years of the twentieth century, anthropologist Bronislaw Malinowski lived among the Trobrianders, a Melanesian people who inhabited several isolated South Pacific islands. Trobrianders believed that a mischievous water spirit, the *balima,* gained entrance into an unsuspecting woman's vagina when she waded out to the point where the water reached her groin, and caused conception. Trobriander society allowed their young people considerable sexual latitude. Invoking the *balima* was a creative way for the tribe to accept a new child into the community without ever having to know who the father was.[6] Other indigenous peoples—the Arctic inland

Chukchi, the Australian Aborigines, and the Sumatran Bataks—had also not yet discovered the secret of impregnation when they first came in contact with explorers.[7]

The Siriono, a remote tribe in the Amazon, believe in *partible* paternity.[8] They hold that every man who has sex with a woman prior to her evident signs of pregnancy contributes a small part of his essence to the making of her child. The man having the most sex with her contributes the most. Each child begat under the concept of partible paternity can have many fathers. Partible paternity is a clever way to involve multiple men in the care of a child. In the event that the actual father is killed or dies, other men will remain engaged with the mother and her child, as they remain convinced that part of their spirit lives in the child.

These few isolated cultures are rare exceptions. The overwhelming majority of people living in major contemporary and historical cultures throughout the world understand and understood paternity as we do. All cultures understand the succession of the seasons far into the future, comprehend personal death and care for their deceased, and personally adorn their bodies. Since there remained a few cultures that had not yet uncovered the secret of paternity by the early part of this century, it would be a fair presupposition, then, that, of the three, paternity was the last and most difficult insight for *sapients* to comprehend.

A woman would have been far more likely than a man to be the first to grasp the connection. How could she have not? Having been the first to make the connection between sex and pregnancy, a woman was in a better position to make the next logical leap—that sex with a specific man led to the birth of his specific child.

Women's close association with their offspring would have prompted them to recognize the similarities in both features and character of each of their children and those of the child's father. In an age without mirrors and few chances to see one's reflection accurately, a woman looking at a man she suspected of being the father and then at her children would have had an easier time ferreting out the links of paternity than a man looking at a woman and then at her children. Few men would disagree with the statement that women in general are more astute at discerning the slightest nuances of a child's expressions, and that a woman's intuition is more finely tuned than a man's. This is especially true when it concerns subtle matters relating to interpersonal relationships.

The mother, most likely, would have known the father intimately, and would be familiar with his distinguishing physical characteristics. She also would have had the opportunity to observe his idiosyncrasies. Since children inherit many physical traits and personal quirks from their fathers (as

Knowledge of fatherhood gave men a new sense of purpose.

well as their mothers), it's reasonable to assume women made the connection sooner than men.

We will probably never know whether paternity was an independent discovery by a man, or whether a woman taught this life-changing connection to her lover. Identifying which of the two sexes first made the discovery is not nearly so important as assessing how each sex was affected by the knowledge. Unquestionably, it was for the male of the species a profound revelation. His response to this news then deeply affected all females.

Knowledge of paternity forced men to make major adjustments in their attitudes and behavior toward children and women. These, in turn, greatly affected child-rearing and the relations between the sexes. *The structure of human society and its dramatic divergence from the mating and parental-investment strategies of every other of the three million sexually reproducing species has come about largely as the result of men's recognition of the paramount importance to them of paternity.*

Norman O. Brown describes man's longing for immortality as "the wish to be father to oneself."[9] Nietzsche believed that men's fear of death underlies both the religions of immortality and the economic institution of hereditary property. He encapsulated this urgent desire when he wrote, "I want 'heirs' thus speaks all that suffers; 'I want children, I do not want *myself.*'"[10]

For women, understanding paternity was far less revolutionary. As children literally come from and through a woman, she never is in doubt as to who her children are. Establishing her "lineage" is never a problem. A mother *knows.* Lord Kilbrandon put it thus: "Maternity is a matter of fact, paternity one of mere inference." Matrilineal inheritance customs were once the norm, because it was far easier to establish who was the child's mother than the father. In Homer's *Odyssey,* a disguised Athena approaches Telemachus and asks him if he is the son of Ulysses. Telemachus replies, "Can any man knoweth who hath begotten him?"* Herodotus, reporting on his travels in 500 B.C., wrote, "Ask a Lycian who he is and he answers by giving his own name, then that of his mother and then his grandmother's and great-grandmother's and so on."[12] Among Etruscan married couples' grave inscriptions, only the wife's name was written on their sarcophagus. (Later, the Romans reversed this custom.)[13]

The prospect of death holds less dread for a woman, so having an "heir" to carry on her name is less important to her. The male's *need* to know beyond a reasonable doubt that his children were his became the force behind many draconian new customs men imposed on women, sometimes by force. Other customs strengthened a woman's hand in the increasingly tangled negotiations between the sexes. For the most part, however, the radical reconfiguring of culture was deleterious to women and significantly eroded much of the power *Gyna sapiens* had acquired when she learned how to exercise control over a man's access to both his sexual pleasure and his legacy.

---

*Telemachus was prescient. In a study conducted by zoologists Robin Baker and Mark Bellis on a block of flats in Liverpool and now repeated in many other cities, DNA testing of fathers and their offspring indicated that the DNA of the man listed as the father on the birth certificate differed from his ostensible child in these studies 10 to 20 percent of the time![11]

Men, now wide-awake, sought to restructure society so that their heirs could carry on their name.

Chapter 21

Incest/Dowries

The political system is a breeding system. When we apply the word lust to both power and sex, we are nearer the truth than we imagine.
—Lionel Tiger and Robin Fox[1]

Upon the chastity of women all property in the world depends.
—Samuel Johnson

*I*magine a group of men sitting around a campfire late at night, sipping fermented juice from hollowed gourds, joking, bragging, arguing, and gossiping. They are engaged in the gentle verbal tug-of-war that has characterized male bonding in pubs, clubs, guilds, and agoras since time immemorial. This particular gathering took place sometime around forty thousand years ago in southern France. Instead of sports and elections, the men were chewing over the highlights of last week's bison hunt and the latest intrigues concerning who might replace the band's aging chieftain. At some point, a man named Adam cleared his throat and announced that he required their undivided attention.

Let us listen in on this important moment that would be repeated around campfires circling the globe. Allow me to use poetic license to telescope into one evening's conversation a complex set of interrelated issues that were resolved over thousands of years through trial and error. The men in my story will speak with a level of sophistication that was most assuredly unavailable to them. They are men with many rough edges, on the threshold of making a series of major adjustments concerning their relations with women. What

Adam has to say will have a profound effect and impel them to change their mating behavior radically. Let's eavesdrop on their concerns and try not to judge them too harshly.

"Fellow hunters," Adam began, "I think I have discovered one thing that will change everything."

Eager to learn what could be the subject of such a pronouncement, and aware that Adam was known as a particularly skilled hunter, the others quieted.

One of the men piped up, "What is it? A new weapon?"

Another broke in before the first had finished: "Have you discovered a new hunting tactic?"

"Perhaps a new tool?" another speculated.

"No," Adam answered, "it has to do with sex."

With that, every head turned toward him, and an expectant silence fell over the group, broken only by the crackling of the logs on the fire, which illuminated the tall figure of Adam standing with arms folded before the assembled men.

"I believe . . ." Here Adam paused for emphasis. ". . . that each of us contributes his essence to the creation of children. Each child is a unique combination of only one woman and one man. I have created the word 'father' to describe the relationship a man has to his special child. We must all learn this new kin word and place it alongside other familiar ones such as 'brother,' 'sister,' and 'mother.'"

"How could that be?" one of the men objected. "Everyone knows it is the woman that bears children."

Others chimed in with their opinions, as they all began to talk excitedly at once.

"You are not telling us something we that we don't already know," spoke one hunter sarcastically. "Everyone here understands that sex has something to do with birth. We have all witnessed the great massing of reindeer and bison during their rutting times and observed that in the following season the females deliver their young. We have watched how the stags and bulls fight among themselves to decide which one will have the most sex. Not unlike us, eh, boys?" he added, looking about the group.

The men laughed, nodding their heads in agreement.

"No, no, that's not what I have discovered," Adam interrupted. "I am sure that each child has a specific father. Each of us can be the special father of a child that is part of us. No other man has anything to do with the birth of that child." He let his pronouncement sink in.

"How did you discover this? What is your proof?" one man demanded.

"I observed Eve's little boy," Adam replied. "There can be no mistaking that the boy strongly resembles me. When I recalled how much sex Eve and I had several snows ago, I put the two observations together. I am certain that I am the one and only father of her little boy."

A lively discussion broke out, for each man had been more or less on the verge of making the same discovery. They could not have missed what was right before their eyes daily, either. Some children born to a woman with whom a particular man had had sex bore a striking resemblance to him. Each man had also noted that some of the children born to a woman with whom they had not had sex were spitting images of the woman's particular lover.

Intermittently crowding in at the edges of their consciousness was the dimly formulated thought that perhaps one man and one woman made each baby. But none of the men had ever linked the observations with their thoughts the way Adam had.

After Adam shared his revelations concerning a man's potential to live on through his heirs and the pleasure, benefits, and responsibilities that would accrue to a man who knew who his children were and they him, he began to tackle more thorny issues he knew would be much more difficult for the men to accept.

"But, men, I must warn you. Before we can enjoy the advantages of fatherhood, there is a tricky matter that will sorely challenge us. Since the child does not come through us like it does a woman, and we are separated by three seasons from our pleasure with a woman, how can any one of us be sure that a child is his and does not belong to another man?" At this pronouncement, a collective look of puzzlement came over the intently listening men's faces. Adam continued, "I have given the matter considerable thought. First I will tell you the problem we face, and then I will present what I think is the solution.

"The first problem is of our own making," Adam said. "We men rarely pass up an opportunity to have pleasure with a woman—any woman." Several of the men grinned and nodded knowingly. "But if we continue to behave in this manner, then we will never be able to sort out whose child belongs to which father. So the first step we must take is to learn to exert self-control over our desire to have sex at any time, any place, with any woman we want. Then we must make sure everybody else is doing the same. Together we must outlaw free sex, and we must invent a means to get everyone to obey these new rules. We've learned to hunt together for the good of the band. Now we have to learn how to deal with sex in the same way." Several of the men exchanged nervous glances.

"Henceforth, each man will have to limit himself to sleeping with only one woman. This may be the most difficult task we have ever attempted, but our best chance to achieve immortality depends on our mutual cooperation."

"That's the craziest thing I have ever heard!" exploded one man, as others furrowed their brows at Adam's suggestion. Many shook their heads in disagreement.

Adam spoke forcefully: "Hear me out! I am not finished. We must also somehow gain control over women's sexual freedom. If we allow a woman to choose the men with whom she has pleasure, none of us will know which of her children are ours. Remember, a dead man without an heir will quickly be forgotten."

The men began to warm to Adam's declarations, brightened by the prospect of forcing women to obey their wishes. One puzzled man asked, "Why would any woman allow us to tell her who she can sleep with and who she may not?"

Adam smiled knowingly. "Because, in return, we will agree among ourselves to protect all women from men who try to force themselves on them. Once the women realize that we will not let them be raped, I believe they will be more inclined to take a single partner for pleasure. Besides, there will be other benefits to them besides protection. I believe we can get them to go along with us."

"So how are we going to do this thing you propose?" asked another one of the men.

"I put before you, this night, that we will punish any man among us who violates a woman. If we do not agree on this point, then we will never know for certain which children are whose. If a man puts his essence into a woman without her agreement and they both know there will be no punishment for his transgression, she will have little incentive to bring the matter to the rest of us so that we can avenge her.

One man shook his head in wonderment. "You are saying that we can't use force anymore and that we men are going to be the ones to enforce this new rule. Adam, you are talking nonsense! What you propose will never happen."

Adam continued: "Giving up sex with any woman we want and not allowing forced sex aren't nonsense. They are just new ideas. But it does mean that we will have to make radical changes in the way we live."

"Wait a minute," one man interrupted. "Suppose she sneaks off and has sex with another man?"

"That, too, will be an everpresent danger to us men," Adam replied. "I haven't yet figured out how to completely prevent something like that from happening. Somehow we must monitor the sexual activities of the woman of our choice extremely closely. A better way would be to treat her so well, and keep her so sexually satisfied, she wouldn't even want to stray."

Adam went on: "We'll have to set up strict rules to prevent either a man or woman from breaking the new rules. This means that each man must stake a sexual claim to one woman. He will have to win her agreement for this union, and to do this he must prove to her that he is worthy. Once each man and woman agree, I think it would be a good idea to have a public ceremony combining great solemnity with a festive party. Before all the others in the band, the new pair will vow to protect, honor, and be true to each other. It will be a difficult vow to keep, but the stakes are high."

Adam continued: "I will call the ceremony 'marriage.' We will recognize a new relationship between a man and a woman who are not of the same blood. The man will be called a 'husband,' and the woman, 'wife.' Only through the use of this oath can we begin to eliminate the problem of men having sex with multiple women, and women having sex with many men." He began to chuckle. "The women will believe that all the attention showered on them on this special day is about them. They will only be vaguely aware that our real purpose for this new living arrangement is to make sure that we men know who is the father of each woman's children. It will be our secret that this is the reason we men invented marriage, an institution whose primary purpose will be to control with whom women have pleasure."

Adam grew pensive and stared deeply into the flames. "We must do this," he stressed, "because, if we don't, then for each of us death will be final. Unless we have done extraordinary deeds, no one will care to remember us, and our memory will disappear like so much smoke. Men, we *must* know who our children are."

"What about the woman to whom we are partnered, the one who will bear our children? We will have to stay very close to her. How are we supposed to treat her?" The man who asked the question kicked a coal back into the fire.

"That," Adam answered, "is a most complex and delicate problem. But, first, allow me to finish.

"Now, where was I? Oh yes, one other very important issue. Each of us would prefer that our special child be a boy. I believe that, as men, we feel more at ease with sons. However, there is always the possibility that your wife will present you with a girl. Daughters, despite what you may initially think, are very desirable. They will treat their fathers kindly once they understand the part we play in giving them life. As fathers of daughters, we must learn entirely new ways to behave. We must treat a daughter with love and tenderness, so as to teach her to love and respect us. The more she loves her father, the more she will be inclined to keep his memory alive."

"And the more inclined to give us pleasure," added another man.

The mood of the whole group brightened. Older men always sought to have sex with the younger women. Establishing a close relationship with a young, pretty woman who loved and cared for you, each man thought, would be a wonderfully convenient arrangement.

"We must as a community absolutely forbid that," Adam declared sternly.

A ripple of hostility circled the fire. "Why would you prevent us from taking pleasure with our daughters?" a man yelled angrily. "Didn't you just tell us that they belong to us?"

Adam looked grave. "It is very important that we never have sex with those girls we know share our essence. If we are to have them care for us as their fathers, we cannot also be their lovers. A sexual union with our daughter would create much disharmony among the women." The men in the group grunted grudgingly, having learned the hard way how stressful life in the band was when the women were arguing among themselves. The battles between mothers and daughters had the potential to be the most contentious.

Adam could not foresee the genetic implications of incest, with its elevated risks of birth defects. Even as late as pharaonic Egypt, the link between deformed offspring and incestuous marriage remained obscure. However, assume that Adam intuited that sex between fathers and daughters would be highly disruptive to the survival of the band. The biological implications would become apparent many thousands of years later, when domestication of animals permitted humans to observe firsthand the results of inbreeding. With the advent of modern science, the exact perils of consanguinity would be quantified. In one study, matings between first cousins increased infant mortality by 60 percent.[2] In another, many of the infants born of father-daughter and brother-sister matings either died within the first six months or manifested a disproportionately higher percentage of serious birth defects (including severe mental retardation) than the control group.[3]

"We must distinguish between sex with a daughter and sex with another young woman. I will name the former behavior 'incest.' Do you not see that incest would destroy the bonds of trust that we must have if our band is to survive? It is essential that we forbid taking pleasure with our daughters. If we discover that a man has done this shameful thing, we must all condemn and punish him. We will call our agreement on this matter a 'taboo.'"

"Aren't those other rules that you gave us earlier also taboos?" a man asked.

"Yes, but some are more crucial than others. We must enforce the incest taboo above all others."

"What about mothers and sons?" asked one of the men. "If we cannot take our pleasure with our daughters, then shouldn't we forbid the women to take pleasure with their sons."

"Ah," said Adam, "that is where women have it easier. They are not inclined to mate with their sons. Surely you've noticed that no animal mothers have sex with their sons. They do not need a taboo. But we men do."

One of the men asked, "How, then, can we protect ourselves from our desire to take pleasure with a young woman, especially one who lives with us?"

It was then that Adam made his most shocking pronouncement: "We must rid our camp of our daughters as soon as they begin to bleed."

"How are we to do that?" one of the men sputtered incredulously.

"I have given considerable thought to this problem, and here is my solution. We must offer to give our daughters to the men of other bands."

At this novel notion, the group displayed genuine astonishment. They all began to object at once. One growled indignantly, "Why would we give away our own pretty young women? Would you prevent us from having sex next?"

"How will we survive if we do not have young women to become mothers? There will be no more hunters! We will all starve!" another yelled.

"No sex and no food. Have you gone mad?!" cried yet another.

"Hold on and hear me out," cautioned Adam. "We will offer our young women to the men of other bands in exchange for *their* young women."*

Now the others looked at him as if he had surely lost his mind.

"Why would other bands give us their young women?"

"Because," Adam continued, "we will teach them that bartering their young women for ours will benefit both bands for generations to come."

What Adam was talking about—what we call "long-term interest"—is a concept foreign to other animals. Confined to live for the most part in the present, they cannot imagine initiating an action that may be detrimental in the moment yet might benefit them in the next generation. Only an animal possessed of a keen sense of foresight can behave in this manner.

As the men grappled with Adam's novel bride-barter proposal, they never contemplated exchanging their sons. Given the importance of hunting and defense to ancestral bands, this idea would have been out of the question.

"Must a girl *always* be traded to another band?" asked a man who had his eye on one of the pretty young females who was nearly of bleeding age.

"Well, not always. Sometimes we can bond a young woman with a young man within our own band, as long as we know that they do not closely share any of the same essence."

"Who will decide which women will be exchanged?" another queried.

"We will leave that to the elder males, for whom the passions of lust have cooled. They already know how to make alliances for hunting rights and how to trade for goods. We know that barter is the safest and most peaceful way to obtain ochre, ivory, and seashells from strangers, so why not women? We will entrust them to trade with the elder males of other bands for females as well as goods. They will do what is best for our band as a whole."

---

*What follows is a brief summary of the basic premise of Claude Lévi-Strauss's work on kinship concerning bride barter: "The men who own and the women who are owned . . . wives who are acquired and sisters and daughters who are given away."[4]

A glint, not from the fire, shone in Adam's eye. "All of us here would probably agree that the thing we want the most in life is regular sexual access to a young woman who cannot refuse us, right? If there were a way we could have this without shedding blood among ourselves and at the same time ensure that we would have strong, healthy children of our own, then why would we *not* exchange our young women?"

It was at this moment in history that a man began to conceive of a woman as a commodity. By moving her into the category of objects that could be traded, men began an invidious trend that would sour relations between the sexes far into the future. Adam could not have foreseen just how thoroughly the custom of fathers' giving daughters away would become embedded in subsequent human cultures. He could not have imagined that, forty thousand years later, a father would still exercise his right to "give" his daughter away to the groom of a stranger's family at a wedding ceremony.

In virtually all cultures in the world there is an unspoken assumption that something of value must be traded to obtain a woman for marriage. Originally, men had to pay a bride-price to a woman's family to convince them to part with their daughter. As patriarchy became the norm and women became devalued, this tradition reversed, and at present it is the bride's family that must add value to the marriage transaction in the form of a dowry.

Adam continued: "But before we can initiate the first exchange of daughters, we will have to convince the other bands of its value. We will have to learn to trust each other. It isn't going to be easy. There may be seasons in which they do not have an equal number of daughters to exchange with us."

"Then we will be cheated!" cried one man.

"No," said Adam. "In some seasons we may have to give them more women than we receive. But the next season, when they have more girls than we do, we will have to trust them to turn over those women. Over many seasons, I believe, it will all be fair. Should the situation become too unbalanced, we can always offer something of value in exchange for their women."

Thus, one of the earliest and most important balance-of-trade issues in history concerned sex. With the active barter of their young women, Homo sapiens shaded into Homo economicus. What Adam could not have known was that his band's willingness to exchange their most precious possession with a group of strangers would form the foundation upon which all future civilizations would rest.

Mutual trust among strangers had to begin somewhere. In general, social predators, the class of hunting animals to which Homo sapiens belongs, behave with territorial hostility toward strange conspecifics. The exchange of women would begin only after men understood the implications of paternity.

Bride barter would be the only solution that they would have to eventually arrive at that could mute men's naturally lustful tendencies.

Bride barter demands an extraordinary degree of faith in the integrity of strangers. Wars between peoples have seemingly occurred constantly throughout history, but the majority of people have based their interactions with neighboring communities on mutual trust. The custom of fathers' trading unmarried daughters with fathers they may barely know continues to be a common practice throughout the world.

Many of the men were shaking their heads at the inconceivable idea of trusting hunters from another band. "I am not sure," one finally said. "You are asking us to cooperate with hunters we don't even know."

Adam could see that the men were having a difficult time grasping this complex issue. He crouched as he patiently explained the far-reaching ramifications of his ideas. "Don't you see, once we establish a working relationship with another band by exchanging our women with them, we will no longer want to fight with them. They will have our daughters. And they will no longer want to fight with us because we have their daughters. What would be the point? If we fight and capture their women, all that will happen is that we will be bringing our own daughters home."

The last speaker began to take Adam's side. "Suppose, after we have exchanged women, we propose to them that our two groups form an alliance? There is strength in numbers. That would be a good thing."

Adam smiled, pleased that the men were beginning to understand the benefits of bride barter. "We would not really be strangers to each other anymore. When our daughters have children, there will be young ones who share our essence once removed who live in another band. Naturally, the exchanged women will long to visit their mothers and their fathers. They will insist on more and more peaceful contacts between our two peoples."

"But those children will not wear the same clothes or put the same designs on their bodies," objected one man. "They will not be of our band."

"No, not of our band, but we will be connected nonetheless. I can see a time when many bands will be linked together by our children and our children's children. Men, I tell you, things are going to be very different. If we need help from a band that hunts over that mountain and we have some of their children's children living with us because we have exchanged women, those warriors are bound by blood to help us."

Now Adam became really animated. "In fact, I can envision a time when bands we don't even know about now will be linked to us through our children's children's children! Imagine what it would be like if all the bands got together." He stood up and expansively spread his arms wide to create a better picture for the men around him. "Once a year we could organize a great meeting of all the clans, and we could have contests to see who among us is the fastest and who can throw a spear the most accurately. And we could dance, feast, and exchange valuable information with each other around one huge bonfire!

"We could also look over the women we would be getting, and our elders could make shrewd bargains so that we will get the best women for ourselves. Ah, what wonderful things will happen to us when we give our daughters away!"

The first major trade in something of immense value was bride barter. Repugnant as it is to modern sensibilities, commerce between strangers received an important boost when men began trading their young women. It ultimately fostered friendship, understanding, and the critical exchange of skills, goods, and knowledge.*

Many nonhuman primate societies have social arrangements that resemble the human mating system in this regard, but theirs are instinctual, not planned. Young females leave their natal group and establish themselves in a strange troop. Natural selection programmed this behavior to prevent inbreeding. (There are also many primate species in which the young males must leave their natal group to find another troop. There are no species of nonhuman primates in which both sons and daughters remain within the same community.)

Primate social groups easily absorb strange females from other communities. Strange males have a more difficult time and must first prove their worth by patrolling the perimeter of their adopted troop, a most dangerous position. After a time, they begin to challenge other males and, by testing their mettle, aspire to become the alpha male through competition and alliances. Despite all this moving about among nonhuman primates, the presence of new females or males from a neighboring troop does not reduce the hostility that each troop's males display toward males of another when two groups confront each other in the forest.† Only by comprehending the concepts of birth, death, and paternity did the males of the human species begin to overcome this hobbling but natural enmity.‡

The prospect of receiving a fresh supply of nubile women from other

---

*Economist and anthropologist Elias Canetti proposed that all human barter stems from the ability of primates to grasp one vine while letting go of another—a necessary prerequisite to swinging through the tree's canopy. This complex maneuver of releasing and acquiring simultaneously is the essence of all economic exchange, and Canetti traces it back to the human's opposable thumb.[5]

†Among chimps and bonobos, if food is plentiful neighboring troops can sometimes peaceably coexist.

‡The Iroquois Nation comprised over 300 tribes spread over a million square miles of the American Northeast. Arranged in a loose federation with some central authority, they maintained what they called the "Great Peace" for over 300 years. So pacific were the dealings among the Iroquois that Benjamin Franklin recommended to Thomas Jefferson that America's framers incorporate many of the Iroquois's governing principles into the United States Constitution. The Iroquois concept of governance was one of many strains that went into the United States' republican model of states' rights and a federal government.[6]

bands pleased the men and compensated them, somewhat, for the idea that they would have to trade away their own daughters.

But they still had questions. "Suppose, as you suggest, the men of another tribe are willing to offer us their young women in exchange for ours. How shall we distribute these girls among ourselves?"

"You may not like my answer, I fear," Adam replied. "Few among us would possess these girls. We must pledge them to marry our young sons."

"What?! First we give away our young women, and now you say we do not get anything in return! This is a very bad idea. I say we do not agree," one of the older men said, beginning to get truculent.

"Wait, wait," Adam cautioned. "We will gain as a band even if we don't gain as individuals. The mating of our sons with another band's daughters will produce healthy hunters who will replenish our ranks. If the young couple has daughters, they will have a better chance to survive to reach an age when they will give birth to another generation of hunters.

"But there are other benefits for us, too. This new way of living will eliminate a lot of fighting among ourselves over women." Adam smiled to himself. Bride barter, as strange a concept as there ever was, would actually solve a host of irksome problems.

A short man who had not spoken now stood up and, lowering his voice conspiratorially, began to address the others about another benefit of bride barter that Adam had not mentioned. "We all know how wily women are. They always seem to be a step ahead of us. We are stronger, but they can easily outsmart us, especially when the matter concerns relations between us. If we give away their daughters to strangers, we will reduce their power."

One of the men next to him furrowed his brow in puzzlement. "Huh? What do you mean?"

The new convert to Adam's idea continued: "Although a man is stronger than a woman, two women can mob one man in a fight and the women may actually win because they are two against one. By separating a woman from her closest ally, her daughter, we will diminish women's collective power. You see, a woman becomes isolated and less powerful when her daughters leave home to live far away. Of course, she still will have her son to protect her and look out for her interests. But he is a man and he owes loyalty to both the other men and his father.

"On the other hand, the girl who comes to live with us will be severely handicapped. Her husband has remained on familiar ground, but she will be bereft of friends and family. I predict that she will be very submissive in order to gain acceptance into her new family. We will encourage her to be meek because it will serve our interests. In time, she will teach her daughters to defer to men.

"To reinforce her submission, we will make each new wife live at the hearth of her husband's mother. Generally, mothers are possessive toward their sons. The new wife will be in competition for her husband's loyalty. This rivalry will serve our purposes well because the new wife must learn to live with not one, but two new people. She will soon discover just how hard that can be. We men have learned to employ the principle of 'divide and conquer' when hunting. Why should we not use it on our women to reduce the power they hold over us?

"We will tell the women that we have to exchange their daughters for the good of the band. But among ourselves, we will know that this arrangement will give us a major advantage. Once we make such an important set of rules, women will begin to believe that it is our natural right to make the rules by which they live. By asserting control over such a vital issue as the future of their daughters, we demonstrate to the women who is boss. We will begin to convince them that they never really had any rights in the first place."

The new speaker sat back down and the others mulled over what he had said. Another man raised a question, redirecting the conversation back to Adam.

"But will the mothers of the girls let us go through with the exchanges? Why would a mother permit us to give her daughter, whom she loves, to a strange man in a faraway tribe?"

"That will always pose a problem," Adam admitted, "but the women will consent for the following reasons. Mothers want to protect their daughters from pain. They love their daughters and realize that a nubile young woman living in the same quarters as their randy old husband might tempt him beyond the limits of his control. A woman does not want her husband to impregnate her child. Putting the welfare of her daughters before her own, a mother will want them to have healthy babies more than she wants them to stay near. She will be willing to give up her daughters precisely because she loves them so much."

Remember that, at the outset of these new living arrangements, men did not possess the degree of self-mastery they would acquire much later. It would require thousands of years of acculturation before men could begin to gain control over an urge as overpowering as their sex drive. The ineradicable incidence of incest in our society today is a vestige of a much greater problem that most likely would have plagued an ancestral community.

Adam went on: "Also, once daughters begin to exhibit signs of womanhood, some mothers become anxious that their daughters are becoming more attractive or more desirable than they are. A few may actually be relieved to have their daughters leave the family quarters.

"But the final reason the mothers will let their daughters go is that they want grandchildren. The women understand, better than men, that they will increase their chances for healthy grandchildren if they let us trade their daughters with other bands. I believe they will accede to these new arrangements."

"I have had enough," cried one man. "Grandmothers, fathers, wives, husbands, sons, and daughters—who can keep track of all these new relatives? Adam, you have raised many troubling issues tonight. I, for one, am going to turn in."

There was a general agreement among the men on this last sentiment. The fire had turned to glowing embers, and the men slowly stood up and began drifting away. Adam's pronouncement regarding his exciting but extremely thought-provoking concept of paternity had put each of them into a strange mood: a combination of elation, reflection, anxiety, and concern.

All of the issues raised in the above discourse took thousands of years to resolve. But of one thing there can be no doubt—the third major insight, the one concerning paternity, like the two major ones preceding it, profoundly affected relations between men and women.

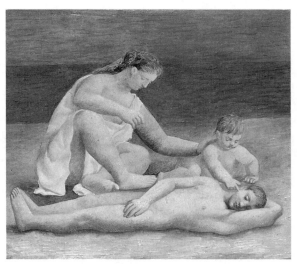

Three major insights—about time, death, and paternity—led humans to form stable families.

Chapter 22

Wife/Husband

I will not allow either boyfriend or husband
To approach me in an erect condition
And I will live at home without any sexual activity
Wearing my best makeup and my most seductive dresses
To inflame my husband's ardor.

> —Oath that Lysistrata makes her followers
> repeat as they stage a sex strike against
> the men, in Aristophanes' *Lysistrata*

Males are a breeding experiment run by females—a proving ground from which females cull winning genes.　　　—John Hartung[1]

Love, and the lack of it, change the brain forever.

> —Thomas Lewis, Fari Amini, and
> Richard Lannon[2]

*T*he band's women sat alongside the stream washing hides and weaving grass mats. While they were engaged in lively banter about the issues that interested them most, a radiantly healthy young woman named Eve brought the group's conversation to a halt by announcing:

"I think Adam finally figured it out."

"What do you mean?" one of the women asked as the others turned in her direction and quieted.

"Well, for weeks now, I have purposely let Abel run past Adam. Each time, Adam barely looked at the boy. I was beginning to despair. The big lug can be so dense! Yesterday, however, when he was in one of those black moods he gets into lately, Abel was making a ruckus. Suddenly, he started looking at Abel very long and hard, as if something was happening inside his head. I saw the precise moment when he finally got it! You should have seen the expression on his face!" Eve threw her head back and laughed with delight.

"Tell us how he looked," one of the other women gleefully cried.

"Well, it was as if he had been hit squarely between the eyes with a bison hipbone."

"What did he do next?" the same woman asked.

"After the initial shock wore off, he began to pace back and forth, all the while looking at Abel and then looking at me and then back to Abel. It was clear he finally realized that he was Abel's father. He then turned on his heel and disappeared into the woods."

"What did he do after that?" asked one of the women.

"After a while, he came back and just kept staring at Abel."

"Didn't he say anything?"

Eve shrugged. "Nothing. He seemed quite agitated. I kept expecting him to say something, because he looked as if he would explode if he didn't, but he appeared to be completely tongue-tied. He understood, though. I know he did."

"Well, it's about time!" an older woman exclaimed. "We women have known for many seasons that each of our children comes from a particular man. Getting these dolts to make the connection, however, has been almost impossible. Men are so thickheaded!"

"What do you think will happen now that Adam knows?" another woman asked.

Eve grew thoughtful. "I think that it will be better for everyone after he has explained this concept to the other men. I can't help feeling a little sorry for them. They're going to have to change their ways and give up some of their precious freedom. They will have to learn how to live with us instead of just trying to take their pleasure and leave when they want. If they desire to know who their children are, they must treat us better."

Another woman chimed in, "You mean a man will no longer seek sex and, once he satisfied himself, abruptly leave." They all laughed, and a young girl who had not yet been initiated into the ways of a woman tittered nervously.

"But I fear that Adam's discovery will bring us sorrow as well," Eve added.

"How so? What makes you think that?" one of the woman asked, puzzled. "What ill can come of a man's knowing his own children?"

Eve sighed as the other women formed a circle, their wet mats and hides forgotten. "Once a man finally understands that he can be a father to a special child, he will try to control the woman he has chosen to have his children. He will want her to take pleasure with him and only him."

An older woman snorted. "How is a man going to do that? As all of us here know, a woman can have any man, anytime, anywhere she wants. All she has to do is give the slightest look that she is interested and every man in the band lines up in a hurry. They are so easy!"

The other women, laughing, all nodded in agreement.

"I don't get it. What does Adam's discovery about children have to do with how we women have sex?" the young girl persisted.

Eve drew in a deep breath before continuing. "As we all know, men don't handle uncertainty well. They fear death more than we do. Just look at all these strange new stories, rituals, and elaborate burials they have invented to convince themselves that death is not the end of life. For such brave hunters, they can act like such frightened children when the subject is death. Knowing who their children are, especially their sons, is going to convince them that they can continue to live on even after they are dead. Men will call this their 'lineage.'" Eve spoke with a conviction that impressed the other women. Evidently, she had thought long and hard about these issues. "The beneficial part of his belief for us is that, in order to protect his lineage, he will have to make sure his children live to have children of their own. Meaning"—here Eve paused for emphasis before continuing—"that a man will have to make some drastic adjustments in the way he conducts himself." She grinned mischievously and added, "We will make sure of that.

"You see, he can't have it both ways. He can't be free to come and go as he pleases, visiting us just long enough to get us with child, and then be off. If he does, he will never know for sure which children of the band are his, and even if he thinks he knows, unless he invests time and effort in their upbringing, his certainty will mean nothing, because his children will not know him and will not honor him in life or in death."

Eve continued: "A man must be willing to settle down with one woman. Oh, they're going to try to convince us that there are other ways, but in the end, it will all come to this: A man will have to trust his woman. He will have to face the harsh fact that, if he wants to be as sure as he can be that he is a father, he must keep a woman from taking pleasure with any other man. And the best way for him to achieve his goal is to treat his mate so well that she won't be looking around for another man.

"You can be sure that a man will have mixed feelings about being a father," Eve explained further. "He won't be able to take his pleasure with a lot of other women, either. If he wants us to be true, then he is going to have to stop sleeping with anyone who takes his fancy. If he can't control his desires, then we'll take other partners. What is good for the buck is good for the doe."

"Eve, you are talking nonsense," one woman interjected. "There is no way that what you propose will work."

Eve bristled. "I didn't say it was going to be easy. The few men and women who understand what is at stake must convince everyone else to consent to these new living arrangements. They will need our cooperation in reorganizing the entire band so that we are all paired off. If one couple try to be faithful to each other while the other men and women in our band continue to seek sex based on their whims, then the mated couple's good intentions will most assuredly fail. The temptations to join in on the fun with the others would simply be too great.

"Therefore, *all* the men and *all* the women must join in the pact. We must all agree that this new way of life will be best for all concerned, especially our children." She looked around the group. Several of the women were scowling.

"What's to stop a man from making us be true to him through threats and punishments?" asked one woman, who bore a bruise on her arm from an ugly fight with her lover.

"That will always pose a problem," Eve replied sadly. "A man might be able to get away with such behavior some of the time, but how many of us are going to put up with it all the time? Unfortunately, there will always be men who try it," she added as the woman with the bruise hung her head. "I suspect the men may make it work in a few groups; we all know what it's like in the band across the valley." The women nodded again. A bully who mistreated the women ruled the neighboring band, and the other men took their cue from him. That tribe was one of the least successful bands in the area.

Eve continued: "I believe most men will come to love and care for their partner. Cooperation and mutual respect benefit everyone. Men are smarter than most of you think. They will learn that love is more powerful than brute force."

One of the younger women said, "I don't know if I like the sound of this. I don't want to give up my freedom to stick with one man. That would give him too much power over me. Why would I consent to your new arrangement when I could have lots of men competing with each other to bring me meat?"

"Well," replied Eve, "you may change your mind when you learn what a man must do to get you to let only him and no one else monopolize you. He will have to persuade you that he is exceptionally deserving. He must prove to you that he is trustworthy and reliable and that he won't leave you with a bunch of his children the minute his desire for another woman tickles him. It won't be easy, though. Men will fight tooth and nail against the idea."

"A woman is not made of stone, either," the same young woman persisted. "Why would we give up our pleasure with many to have pleasure with only one?"

"Because," Eve continued, "although we love our freedom, after we have borne children we will love them even more than our pleasure. There are many footsteps between a 'woman' and a 'mother.' Having one committed man promise to help you raise your children who he is sure are his is better than depending on the whims of many."

The young woman, still unsatisfied, raised yet another concern. "How can we be sure he will follow through? How do we know that a man is even interested in having children, or that he is willing to help us raise them? After all, we women know that men will tell us just about anything to get us between the furs. After they get what they want, they often lose interest and look for another conquest."

"To keep that from happening," Eve answered, "we women must agree to lay down some very strict rules. First and foremost: absolutely, positively no granting of sexual favors until a man passes a test." She gazed across the river. "Let us call this test 'courtship.' We must make this test so difficult for him that by the end we will have a pretty good idea whether he has what it takes to satisfy our needs."

"Unlike that of all the other animals, our courtship will not just be for a few days. We all know how many turns of the moon it takes before a child can walk or find food. So we must take our time in choosing our mates. All the while, we will judge their health, their ability to bring us meat, and what kind of lovers, companions, and fathers they have the potential to become.

"We will prevent a man from having his way until he is mad with desire." She laughed a little. "Then we will demand that he stay at our side. If we arrange it just so, by then he will want to have pleasure with us so badly he will agree to almost anything."

She paused for dramatic effect. "Many of you may have wondered why I kept shoving Abel under Adam's nose. I had a hunch that when Adam finally grasped that he was Abel's father he would willingly consent to this new ultimatum."

"Ultimatum?" asked the youngest girl.

"Yes. I will tell him: If you want sex from me, then you must assist me with the children that will result from our union and favor no other woman, ever again."

Eve went on: "Once a man decides that he wants to be with only one woman, I suggest that we have a large public ceremony and have the man stand next to his intended before all his family and friends. In front of such an assembly, he will have to swear publicly, by a solemn oath, that he will honor, cherish, and provision his bride with a supply of resources beyond many seasons, so that she can concentrate on raising their children."

Eve smiled, and then lowered her voice conspiratorially. "The man will bask in all the attention on him on this special day, and he will only be vaguely aware that this ceremony is the women's idea. We will insist on it, because, once he makes this proclamation before all the people he cares about, he will be less likely to break his oath.

"As we women know," Eve continued, "it is one thing when a man whispers sweet promises in a woman's ear, and quite another when he has to say it out loud in front of both his mother and father and her parents, too. He will be more inclined to honor his commitment, because he will not want to lose face by appearing untrustworthy, especially when he has sworn his oath in front of those people whose respect he most wants to retain."

The women were gradually beginning to understand Eve's line of reasoning. Then, as they exchanged glances and talked excitedly among themselves, it became clear to Eve that they were coming around to her perceptive assessments.

But one woman remained unconvinced. "You point out that we play an important role in choosing a man. But isn't he making a similar judgment about us? If he wants us for the children we can give him, I can see that he will look us over carefully to figure out if we are young enough, pretty enough, fertile enough, and motherly enough for him."

"Are you kidding?" another woman interrupted, laughing. "All a man wants is sex. He's not going to be thinking about those other things."

"Not entirely true," Eve responded. "As men begin to realize the importance of knowing who their children are, they will increasingly care about how we mother. But there is another factor you are overlooking. Men fall in love. For the majority of young men, all common sense disappears when they are in love."

A fit of laughter momentarily stopped the discussion as one woman cut in: "We've all heard men brag about how they are more logical and reasonable than us girls. But when it comes to love, men are much bigger fools than women."

"What *is* this thing called love?" the teenage girl asked wistfully.

Eve smiled and looked at her. "Someday, you will know love yourself. It is an emotion every human yearns to experience. I cannot explain the experience to you, because it is beyond words. But I do know that men have a more difficult time figuring out what they are feeling than we do. When they first experience love, it seems a very foreign feeling to them. I've seen it beginning in Adam. If other men are like him, love will so addle them that they won't be able to think of anything else. A man will do almost anything to convince us that he is the only one we should choose to father our babies." Eve laughed with the other women. "A man in love will do many silly things that he wouldn't ordinarily do."

"But won't we be as susceptible to the same sort of delicious madness?" asked one of the women.

"Of course we will. Some of us, at least. But I don't think lovesickness will disorient women as much as men. After all, love, as we all know, is more our province."

Much has been said of the superiority of one human trait after another when compared with other animals. Language, intelligence, abstraction, and foresight are commonly trotted out to justify our dominion. Strange, is it not, that we are also the one animal that can love the longest, deepest, and best of all creatures. Our capacity to love one another, to care for one another, to nurture, protect, minister to, and grieve for one another, also sets us apart from the other animals. We alone among the creatures can crawl into the mind's cave belonging to another person and understand him or her so completely that we can become one with the other. Although love can also be twisted and grotesque, it is our capacity to love one another and the power to change each other through that love that makes the bond between two people in love so exceptional.

Many biologists and ethologists have observed what appears to be behavior consistent with the human definition of love between other mated pairs of animals. This is especially in evidence during their courtship phase. In a few species, love appears to be the glue binding a couple for a lifetime. In general, the intense, white-hot phase of love between two animals seems to be instinctually coordinated to peak at the height of the female's ovulation. Human love differs from that expressed by a few other species in that its acute phase can persist over a much longer duration and then shade into a love that endures. Dostoyevsky, moved by the human need to love, wrote, "Fathers and Teachers, I ponder the question, 'What is Hell?' I maintain it is the suffering of being unable to love."[3]

"Come!" Eve clapped her hands, wanting to change the subject. "Let us make a list of what we want in a mate."

"Any man I'd choose has to be strong and brave," said one of the women.

"I agree," said Eve. "But what else? What other things should we look for in a mate?"

One older woman who had been silent spoke up: "Before I let a fellow court me, I insist that he be clean."

"He should be cute!" one young woman added.

Another piped up, "He should be witty and clever, too! He must make me laugh."

"Don't leave off the list a hairy chest and a tight behind."

"Definitely!" Eve agreed. "We'd all agree that cleanliness, looks, and laughter are important. Anything else?"

"Has anyone mentioned kindness?" the old woman of the band interjected. "How considerate a man is foretells what kind of father he will make. And mark my words, the best way to learn how a man will treat you is to pay attention to how he treats his mother. Beware of men who dislike their mothers. And also avoid those men that are *too* attached to their mothers."

"Thank you," Eve said to this last woman. "You're very wise."

And then, addressing the younger women, Eve began: "All of the traits we've talked about are important for a good relationship between a man and a woman. But we must also be able to gauge a man's potential to be a father; that is more important than the way he looks or how strong, brave, or funny he is. The land is full of good-looking, strong hunters who would make sorry husbands and terrible fathers. Any woman who chooses one of them for her mate will be filled with regret."

"How are we to choose the best?" the youngest girl asked.

"You are forgetting 'women's wisdom'—our special knowing," Eve answered. "Men are not so adept at this knowing as we are, and most men will never understand it. It is our birthright as women. But alas," she said sadly, "some women do not have it. Women who do not use their intuition will most likely choose a mate foolishly. Their poor decision will ultimately endanger their babies. Therefore, it is important that we pay attention to the voice that speaks to us from within."

The other women grew silent and thoughtful.

Their sense of deeptime separated Gyna sapiens *from the females of other species. Not only did she know the connection between sex and pregnancy, she was also able to anticipate her long-term needs. This knowledge would make the process of mating, for both men and women, considerably more complex. No other female animal found it necessary to take into account so many different factors when choosing a mate. Absent among the other sexually reproducing creatures, complex, protracted negotiations over sex became the norm in the human line.*

One of the women piped up, "I think I've understood you so far, but exactly what is it that you are saying we should expect from a man in exchange for promising to sleep only with him?"

Eve turned thoughtful. Slowly, a wry smile crossed her face. "It is," she answered, "a promise. A promise we shall call from this day forth 'commitment.' The word will send shivers down the spine of the boldest hunter whenever we drop it into a conversation. We should require it of him before we let him kiss us, before we let him have sex with us, and before we agree that we will not have sex with any other man but him."

"What exactly is it that we want him to commit to?" the same woman pursued.

"That will be our prerogative. We will get to decide—and we can change our minds. It will drive the men crazy trying to figure it out. But, in general, the kind of commitment we require for a stolen kiss will be considerably different from what we will expect from a man before we promise to bear his children. That will be the big one. He's going to have to promise to keep his manhood in check. And," she added, "he is going to have to promise to provide us with whatever we need to raise the children we have together. He must swear that he will love, honor, and cherish us." Eve grinned. "I suspect 'commitment' is going to cause the boys a great deal of distress, indecision, and anguish. They are going to be faced with a whole series of choices they would just as soon not have to make."

Had Eve paid a visit through time travel to the present, she would find her suspicions amply confirmed. The current version of Homo sapiens *has had many millennia to evolve and mature gradually. During these years, he has slowly come to terms with the kind of compromises he would have to make before a woman would agree to have his heirs. (Just as she has had to make her peace with the compromises she would have to make to ensure his ongoing commitment.) Many a contemporary young man, if asked why he wants children, will say it is because he loves children. But love was not the precipitating motive that originally initiated an ancestral man's quest for offspring. At the outset there was an awful dread—the fear of death, the loss of identity, the erasure of his ego. These fears stirred him more than his love of toddlers.*

Homo sapiens *is the only father among the millions of species that reproduce sexually who cares enough to arrange marriages for his daughters and to teach his sons how to throw a baseball. That ancestral man wanted heirs because he feared death more than he loved children is evidenced in anthropological studies of diverse present cultures. There exist many examples in which a father shows minimal or no affection for his children. His inexplicable neglect exists alongside his paranoid concern regarding the paternity of "his" children.*

The shadows began to lengthen, and some of the women started to stir. "Before we return to camp," Eve said, "I have one last thing I want to share. It's something I have observed since Adam figured out that Abel is his son. Now that he knows this, he will want us to have more children together. He will be especially concerned that I might have a child with another man: Then I might not devote enough time to *his* child. And he's right. Infants take a lot of work, and we all know that we have to spend more time with the new ones that come along.

"But Adam will fear that I might prefer the child of another man over his son. Now, I'm not sure he has thought this completely through yet, but, believe me, he will. He is smart enough to figure it out eventually, and he will not want to be snared into providing for some other man's child. This is the main reason he will want me to be faithful to him, and why he will sacrifice his freedom so that I won't want someone else's baby.

"I'm not interested in having any other man's children, but Adam doesn't know that, and I'm not going to tell him. I'm just going to let him think that there is always the possibility that I might be interested in someone else. That way, I can be sure he will take extra-special care of both me and little Abel, as well as any additional children we might have together."

"Isn't that a little, well, devious?" asked one of the women.

Eve smiled enigmatically. "Men are stronger than we are. We must use the control we have over their desire for both pleasure and heirs in such a way that we achieve equality. Remember, our ability to refuse a man's advances is the source of our power, and it is for his own good. Entwining a man's legs with little children will soften him. A man will gradually learn over time the joy that comes from loving his children—and their mothers."

Over the years, as the male's investment of time and effort in his children increased, so, too, did his concern for the welfare of their mother. Men began to realize that if their mates died their young children would not likely survive. Increasingly, men came to understand that their fate and that of their future heirs, both born and unborn, were intertwined with their wives. Increasingly, the man pledged what was his to her. His successes were her successes; her travails or illnesses affected him as if they were happening to him. Knowledge of paternity markedly diminished men's innate selfishness. Men ceased thinking only of themselves, and began to assume a greater responsibility in ensuring the welfare of their mates.

The same thinking process shaped his relationship to his children. What he accumulated in his life, whether it was skill, knowledge, or goods, he wanted to share with them. He bequeathed these things first to his wife and then to their children when he passed on. He learned that he must protect his little ones, even if it involved risking life and limb, because if his children died then a part of him and his claim on the future died, too. He thrilled at their successes as if they were his and anguished over their failures. In short, the destiny of his children gradually came to hold his heart hostage. Fathers, like mothers before them, realized the truth in the saying, "A parent can only be as happy as his or her unhappiest child." More than compensating for his concern, a father learned, somewhat to his own surprise, that the multitextured relationship he could develop with his son or daughter could exceed in complexity the relationship he had formed with their mother.

A father watching his growing family swell in size and number experienced pride in his manhood. Besides hunting, men had a new reason for living . . .

and loving. With few exceptions, men eagerly longed to pair up with a woman and participate in the magnificent adventure called "creating a family." A man's stature among the entire band grew as he submitted proof positive of his potency. Other men treated him with respect when they saw that he could keep a woman happy enough not to stray.

A man's fear of death began to ease as he watched his children grow. He learned the secret that mothers have always known: Despite the finality of his own inevitable demise, a piece of him would live on. Increasingly, he was disposed to develop a closer relationship with the mother of his children. This benefited his children, for, as an old proverb goes, "The greatest gift a man can give his children is to love their mother."

As a couple incrementally built a shared history of significant events, those that were uplifting as well as those that seared the soul, they constructed an edifice, stone by stone, that became stronger with age and better able to weather future turmoil. The mental and emotional boundaries that separated them began to blur with the scouring of age. The two members of a couple began to speak the same thoughts at identical moments, as if their minds were functioning as a single unit. The constant proximity of their two souls began to meld into each other until the two were as one.

As each came to know the other better, a man gradually realized that intimacy enabled him to become a better man, a more nurturing father, and a more complete human being. His love for his wife awakened in him his dormant feminine side, and this new awareness made him a more balanced person. This, in turn, encouraged him to be a more loving husband. And all these changes occurring in the male psyche benefited his mate. The changes in his character as a result of intimacy with a woman more profoundly altered him than her. The three psychiatrists Thomas Lewis, Fari Amini, and Richard Lannon put it succinctly: "Who we are and who we become depends, in part, on whom we love."[4]

Before men gained the crushing knowledge of death's inevitability and grasped the mechanism of paternity, Homo sapiens were just male mammals belonging to the order of primates in the family of apes and the species of hominids. After these two insights reconfigured their psyches, men rose above their taxonomic classification to become husbands and fathers.

Men and Women

Part V

A little boy begins life loving his mother. Why, then, are virtually all societies steeped in misogyny and patriarchy?

Chapter 23

Misogyny/Patriarchy

... a central paradox of the human condition is that our species possesses the capacity to carry out sexual inequality to its greatest known extremes.

—Sarah Blaffer Hrdy[1]

No woman is completely free unless she has control over her own reproductive system.

—Margaret Sanger[2]

I never yet touched a fig leaf that didn't turn into a price tag.

—Saul Bellow[3]

Adam and Eve must now sit down with each other and work out the details of their new mutual child-rearing life together. As they try to mesh their conflicting everyday needs and wants, they will become painfully aware of how much their two agendas vary. Complicating matters will also be the very different ways the two of them think about things. But they will succeed through persistence, because, at last, they also both realize that each of them needs the other for reasons beyond sex, death, and DNA—to satisfy longing, to love deeply, and to enjoy each other's companionship. As in all negotiations, they will eventually settle their differences through compromise. The framework that the first couple finally hammers out will serve as the template for the man-woman relationships of the most successful subsequent human cultures. Let us leave the first couple's initially prickly but heartfelt "relationship" discussion so that we may review where we have been.

My aim has been to focus attention on a series of elements that I believe have been either overlooked or not given their proper due in the telling of the evolutionary story of our young species. The untangling of the cat's cradle of

factors that initiates the emergence of a new species must, by the very complexity of the task, remain in the realm of speculation. Nevertheless, I hope
that my endeavor has added to the readers' understanding of relations between the sexes and has predisposed them to conceive of human mating behaviors in a new way.

The key factor shaping human sexual relations was the fallout that occurred following *Gyna sapiens'* acquisition of Free Will. Her ability to *choose*
a course of action different from the commands of her potent sex hormones
was the result of genetic mutations. The impetus for the installation of Original Choice in her genome was Natural Selection's imperative to grant women
veto power over impregnation. This became necessary because birth carried
with it—for the first time in any complex animal—the mordant threat of maternal death. She evolved cognitively more quickly than he did, because the
sword of Damocles* hung over her sex, not his.

To acquire the resolve necessary to refuse sex when she was ovulating,
this one breakaway female primate had to undergo a major overhaul in the
design of her brain coincident with a gear-grinding resetting of the major
timers within her reproductive system. A perplexed *Homo sapiens* discovered to his chagrin that he had to respond to the challenge she posed to him
or lose the opportunity to pass on his genes.

Compounding his problem were changes simultaneously occurring in
his genome, particularly his increased eagerness to have sex with a woman,
any woman, *all the time*. The psychosexual emergency precipitated by this
dissonance of desire between the sexes set the stage for a battle, the tocsin
sounds of which have reverberated down through all the generations ever
since African Eve proudly held up her firstborn for all the other members of
her pressed band to behold.

Obscured by her flamboyant reproductive changes, *Gyna sapiens'* internal milieu began to exhibit a puzzling inability to retain iron atoms. Leaking
from multiple avenues throughout her fertile life, the metal's loss imperiled
her health and destiny. The homeostatic warning light on the gauge tracking
her serum-hemoglobin levels blinked ominously during the time she was
gestating, birthing, or breast-feeding her baby. Some other species also manifest a male-female hemoglobin disparity, but they do not have to answer to a
greedy, demanding brain that appropriates up to over one-fourth of all the
hemoglobin's precious oxygen cargo. Another factor adding to her potential
peril: The breast-fed infants of other mammals do not have iron-hungry
brains that more than double in size during the first year of life.

---

*In an ancient Greek myth, the sword of Damocles was a sword hanging over someone's head suspended
by a thin thread. The threat of its falling was ever present.

This, then, was the crux of the problem: By exercising the power to post-pone sex, women gained a significant advantage over men; men, however, un-burdened of child-rearing could more easily obtain the essential item, iron (along with a raft of other precious, hard-to-obtain foodstuffs), that women and their babies needed. For the first time among the animals, Mother Nature required that the members of the two sexes enter into complicated negotia-tions in order to agree mutually on the terms and conditions under which they would engage in consensual sex.

Men enthusiastically increased their interest in the hunting life to ac-quire the nuptial gift most desired by ancestral women—meat. Diets rich in animal products rapidly shrank their primate vegetarian-designed gut, re-leasing more oxygen to build out multistoried, elaborately gabled brains. The discovery of cooking facilitated the absorption of difficult-to-digest plant and animal foods (which further accelerated the process), but also burned enzymatic digestive bridges, making it difficult for this one primate to revert to subsisting on a diet of raw plant foods.

Hoping to impress women, men set out on their quest for ever-larger and more dangerous quarry. Hindering them in their endeavor was their unpre-possessing appearance. Large prey would likely not even have bothered to raise their snouts upon picking up the scent of this newest predator on the block. Natural Selection would call upon *Gyna sapiens* to come to his assis-tance and properly arm him.

Gyna sapiens underwent a major overhaul of the key features of her re-productive life cycle, the purpose of which was to give her access to the won-der weapon, foresight. Upon acquiring the key that would open the gates leading to the future, she used the information to make the connection first between sex and pregnancy, then between sex and possible death during la-bor, and finally between sex and a lifelong commitment to her child's welfare. She was shaken, apprehensive, and changed as a result of these three purely mental extensions. Nevertheless, unselfishly, she used her newfound knowl-edge to gird the loins of her man with the invisible shield that would ensure his success first on the savanna and later in the snows and dunes of more in-hospitable climes. Returning triumphant from his forays, he provisioned her with the iron-rich meat she needed to maintain her health and raise smart, vigorous babies.

Mother Nature captured *Gyna sapiens'* attention by aligning her menses with the moon, allowing this one creature to grasp the meaning of a month and discover the advantages of maneuvering in deeptime. This single evolu-tionary advance increased the opportunities open to the human species more than any other. A major insight occurring to *sapiens* as a result of breaking through to the future was the realization that, among the many diseases that harried them, the only one that carried a 100 percent mortality rate was life. Everyone, no matter how young, strong, or ebullient, was doomed to die.

This epiphany more deeply haunted men than it did women and initiated a sea-change in consciousness second only to the connection a woman had previously made between sex and pregnancy. The awareness of a *future death* changed the outlook, demeanor, and aspirations of the human species, separating individuals further from their relatives on life's extensively branching tree, both close and distant. Increasingly complex rituals, beliefs, and customs emerged that were attempts to soften the blow.

Soon after these discoveries, the facility to think ahead led to the third transformative insight. A man finally discovered his role in the sex-birth process and realized he fathered specific children. This knowledge alleviated some, but not all, of his limited-life anxiety. Awareness of paternity spawned in him an intense interest in the fate of his offspring, never witnessed in the males of any other species. For the first time, a male wanted to keep track of the exact whereabouts and fortunes of his genetic legacy. He hoped to live on through the memory of his children, and he committed himself to being involved in their upbringing.

A powerful emotion, hitherto never experienced to the same degree by the males of any other species, also moved him. Sustainable over an extended period of time, love greatly aided his comity with both his offspring and the mother who bore them. His burning desire to know his children required establishing an entirely novel relationship with their mother. And so men needed the routing equipment installed in their nervous system that could lead them easily to the kingdom of love. These cables and modules had to coexist alongside others that sounded the clarion call to arms or the frenetic cries of the hunt. Men had to be killers and lovers simultaneously—a merging of opposites that has never completely been successful.

Men eventually concluded that the only way they could be confident that a man's "begats" were his and his alone was to plot together to restructure societies' sexual relationships. To accomplish their long-term goals, men set out to achieve the impossible—control women's sexuality and reproductive abilities. Severely hampering their efforts was their own need to gain a semblance of control over their sexual urges.

These three insights concerning birth, death, and paternity, all of which are intimately bound to sex and indirectly connected to iron metabolism, molded every culture in the world into its present shape. Women were at a significant disadvantage if they tried to resist these male impositions, because they were weaker physically, and their babies' enormous needs required that they constantly seek male support. Despite these evident drawbacks, women were not entirely without resources. Many a woman was able to influence men by subtler means and could, not infrequently, wrap a husky man around her dainty little finger. Still the evidence remains undeniable: Men have craftily sought ways to blunt the power of women's Original Choice.

Our species is 150,000 years old. In the last ten thousand years, we have experienced a series of technological revolutions that have markedly affected the relationship between men and women. Yet all of us are walking around with a nervous system designed to work optimally within a small band of hunter-gatherers. The substantial creation of wealth and the social upheaval attendant upon the rise of first agriculture, then industrialization, and most recently technology, tend to obscure this important fact.

The most critical legacy of those bygone times was a sharp division of labor between men and women—a necessary condition for the survival of the individual and the progression of the generations. Then, abruptly, the key three insights ignited synapses within the recently enlarged brains of *Gyna* and *Homo sapiens*. Throwing the switch that illuminated these three lightbulbs one after the other triggered a massive reconfiguration of society. Many of the sexist biases and social institutions that persist in the world came into being as a result. The two most pervasive that affect relations between men and women are misogyny and patriarchy.

Misogyny is a disdain for women and denigration of the values commonly associated with the feminine. Patriarchy is a set of institutionalized social rules put in place by men to control the sexual and reproductive rights of women. An investigation into the root causes of these two will throw light on how we have come to be the way we are today.

When asked, many men will gallantly express their admiration for women in general and profess a profuse love for their mates in particular. Despite these touching personal testimonials, society is rife with misogyny and patriarchy. A cursory glance at the current newspapers or television news reveals a global society in which the majority of men disdain women. While some cultures are more egalitarian than others, men's actions suggest that they believe firmly in their superiority over women.

The historical record presents an even bleaker picture. The Western canon consists of many brilliant tomes written by white males, the overwhelming majority of whom were unapologetic misogynists. Isaiah, Jeremiah, Plato, Aristotle, Paul, Pliny, Jerome, Augustine, Aquinas, Bacon, Luther, Calvin, Nietzsche, Schopenhauer, Marx, Hegel, and Freud ranged from outright woman-haters to those who weren't quite so blatant but nevertheless strongly promoted the patriarchal agenda.

Even during the majestic periods in which human dignity flowered—classical Greece, Renaissance Europe, parliamentary England, revolutionary

America, and Enlightenment France—such champions of human rights as Socrates, Pico della Mirandola, Erasmus, Locke, Jefferson, and Voltaire did not consider the need to elevate women from their second-class status an item high on their political agendas. Western culture had to wait until the nineteenth century before a prominent male philosopher, John Stuart Mill, was willing to stand up and speak out for women's equality.

The record in non-Western cultures is also dismal. Both Confucius and Buddha were misogynists. Of the few leaders who seemed friendly to women, Jesus, Muhammad, and Lao Tzu stand out, but what they really had to say has been so filtered by subsequent patriarchal commentators that it is difficult now to know exactly what were their true attitudes toward women during their lives. The history of Christianity, Islam, and Taoism darkly demonstrates that the religions that flowed from the teachings of these three influential leaders have been most unkind to women.* In every case, after the death of the founder, men with harsh patriarchal leanings seized the reins of power and revised whatever gentle counsel the originators of these traditions may have had to impart about women.

Something is awry if the majority of individual men profess their love for women and many of them proclaim they are for equal rights, but the greater society, dominated by these same men, continues to denigrate the feminine and suppress women's rights. With few exceptions, the social institutions of commerce, religion, the military, education, and government are rife with overt misogyny and unrelenting patriarchy. These stances, pervasive in global policies and culture, are puzzling to comprehend when one considers the attitudes of each man toward women when he begins life.

The very first person filling a little boy's world is his mother. His love for her is boundless. His close attentiveness to, dependence on, and affection for her are unmistakable. Watch the eagerness and love that light up a little boy's face when he catches sight of his mother after a temporary absence. And anyone who observes his mother's countenance as she beholds her son can recognize that their feelings, with few exceptions, are satisfyingly mutual. The bond in their intense dyad is one of life's most uplifting events to witness and is a world unto itself.†

Surely, girls, too, love their mothers and vice versa, but because the sub-

*Male Taoist priests began to venerate Lao Tzu as a god in the tenth century A.D. and banned women from holding any major institutional positions.
†For a discussion of how the image of the Madonna and her child has played out in art, see page 384.⁴

ject of this inquiry is misogyny, let us focus on the special relationship between a mother and her son. There is much truth in the saying, "The most important event in a woman's life is the birth of her son; in a man's life, it is the death of his father." One might modify this statement by adding that the most important person in a little boy's life is his mother. Dying men on a battlefield cry out for their mothers, not their fathers. The last word they utter is likely to be the same as their first—"Mama."

Stemming from the remarkable love that a boy has for his mother and the warm reciprocity that flows back over him from her, one would expect a lifelong male inclination to be kindly disposed to the feminine principle in general and to the women in his life in particular. As a rule, no such transfer occurs. Instead, a pervasive, sometimes barely concealed disdain of and unease with the feminine tinges the jokes and conversations among men out of earshot of women. How could this be? If all boys start out loving their mothers and women in general, how is it that every major historical and contemporary society is founded on antithetical principles?

Psychoanalytic theory answers the question by proposing that, in order for boys to make the transition to men, they must reject the ways of their mothers. Girls aspire to be like their mothers and therefore do not have to undergo this wrenching shift in affiliation. To support their thesis, psychoanalytic writers often submit as evidence the rigorous initiation rites elder men design specifically to separate the men from the boys. Psychoanalytic literature, however, has not thoroughly addressed the question why women would not equally dislike men. Although young girls, in general, passionately love their fathers, they do not seem to have to reject them in order to become complete women.

Anthropologists have put forward an alternative explanation. Having gained control over the symbol-making authority of language, men became the progenitors of laws, religions, and other key social institutions. By exercising this potent power, men, more than women, set the parameters and directions that dictated the evolution of culture. This came about because of the extreme division of labor between the sexes. At the dawn of our species, *private* space was under a woman's sphere of influence but *public* space fell under the purview of men. Home, hearth, and children were a woman's responsibility; men retained control over pow-wows, peace pipes, and policy.

Simone de Beauvoir proposed that hunting made men *transcendant* because it was an exciting, dangerous occupation. A heroic endeavor, it filled men with resolve. The taking of life paradoxically gave meaning and purpose to a man's life. In contrast, de Beauvoir observed, women remained *immanent*. Their daily round of washing, cleaning, and cooking was a repetitive grind and not particularly ennobling, even though men tacitly recognized that the band could not survive without women's contributions.[5] A mother's primary responsibility to her children denied her the freedom to engage in structuring the society to the degree that men did. Another factor was that meat has always, in all cultures, had a higher value than plant foodstuffs, although most hunter-gatherer societies rely more on women's gathering than men's hunting.[6]

The crux of the anthropologists' argument can be distilled in the question raised in an influential article by anthropologist Sherry Ortner in 1974: "Is the Female to Male as Nature Is to Culture?"[7] Men, according to Ortner, are traditionally associated with *culture*, and women with *nature*. Women's bodies discharge a variety of fluids during menses, birth, and breast-feeding, connecting them inextricably to the elements of earth and water. Combining a woman's close link to the cycles of nature and her inseparable association with the mysteries of womb and tomb, men blurred the distinctions between them and in their subconscious conflated women with a force of nature.

Men's relationship to nature, for the most part, has always been antagonistic. As protector and provider, men from an early age have dedicated themselves to conquering and subduing nature. At the same time, elders have inculcated in young men the need to uphold and honor their culture and, when called upon by old men who deem it necessary, to die for it.

In Ortner's view, men transfer the hostility they harbor toward ice, fang, claw, and storm to women. Culture is a man's defense against nature. Whether it is the crude lean-to he has fashioned out of saplings, or sturdy hurricane shelters built of concrete and cinder block, the trappings of culture are the ramparts and weapons men have used first to repulse and then to overcome the forces of nature. Using cultural knowledge held in the net of language, they have converted nature's raw material into tools and weapons to exploit Her. Arising in the breast of men as a result of the culture-nature distinction was a feeling of male superiority. This, along with the other factors discussed above, has fostered an attitude of misogyny that clings tenaciously to contemporary social institutions.

Another factor contributing to this culture-nature dichotomy is the fear men often have of women's power. Menses mystifies men, and the potency they have superstitiously attributed to menstrual blood is all out of proportion to this relatively innocuous mix of discarded proteins and expendable red cells. Men ascribe the most astonishing and, in many instances, malevolent attributes to menstrual blood.*

---

*Lest one assume that these attitudes affect only men who run around in loincloths hefting spears, consider this. Until very recently, physicians routinely employed bloodletting as the most efficacious treatment for a wide variety of disorders. Using leeches or simply opening veins in the patient's arm, the most distinguished scientifically trained doctors, including the revered father of American medicine, William

Menstruation, combined with a woman's mysterious ability to nurture life within her belly, made men fear and, in some cases, envy her close connection to nature. Her uncanny relationship with the moon could have only added to man's sense of unease. Further confounding his mind: His irrational fears of women dimly flicker against the backdrop of his repeated hosannas of relief that he was not born a woman.

Some physiologists place the blame for misogyny and patriarchy squarely at the pedestal of male biochemistry. *Homo sapiens'* extraordinarily high levels of testosterone, they point out, lie behind his domineering and sometimes cruel treatment of women. Testosterone builds muscle mass, increases strength and stamina, and fuels aggression.[10] Men have as much as ten times more circulating testosterone than women do.

Testosterone shapes men's attitudes toward the weak, timid, and cautious, because it makes them want to dominate those around them. This hormone weakens the bonds of attachment and love. Married men have lower levels than single men, and when a man divorces, it rises sharply. Men with high baseline levels of testosterone marry less frequently, are more likely to be abusive when they do, and are more likely to divorce.[11] Some physiologists contend that misogyny is an insoluble problem because of this harsh biochemical fact of life.

Supporting the claim is the correlation between testosterone levels and a man's attitude toward women. Men are the most misogynistic immediately following puberty. Generally, as men age and their testosterone levels fall, their disposition toward women improves. Some might say it is due to a man's gaining experience; others would claim that his change of heart has more to do with his failing testicles. I doubt the physiological explanation can account fully for the rampant misogyny present within cultures, because many other animals have significant disparities between male and female testosterone levels and, with a few exceptions, males do not interrelate as malevolently toward females as humans do.[12]

The list of attributes *Gyna sapiens* seeks in a prospective mate is considerably longer than *Homo sapiens'* list (if he even has one). Because the standards by which a woman judges a man are more stringent than among any

---

Osler, ardently believed that draining blood from a man's body to rid it of toxins would mysteriously increase his health. George Washington's misguided physicians caused his death by bleeding him excessively after a non-life-threatening injury.[8] This nutty, pervasive idea stems from the men's irrational belief that menses was key to women's power. If she could maintain her health while bleeding every month, then bloodletting should be just what the doctor ordered to make men heal from a wide range of illnesses.

Among a diverse cross-section of cultures around the world, men frequently include bleeding in their purification ceremonies. For example, the Californian Yurok men would retire to their sweat lodge to periodically renew their spirituality. Bathing, fasting, and gashing their legs with quartz crystals until blood freely flowed down their legs were central to their ritual.[9]

other female animals, men who do not measure up receive an extra dose of frustration. Cross-cultural surveys reveal that a woman seeks a mate higher on the social and intellectual scale than she is. This is true even among women independently successful and wealthy.[13] Men, in contrast, are often attracted to women whose physical charms outweigh their wit or standing in the community.[14]

The above discussion represents a brief list of some of the theories that have been proposed to account for what Frederick Engels called "the world historic defeat of the feminine."[15] I shall present an alternative to the sketches catalogued above, which I believe may be at the root of the problem.

In my view, testosterone plays a role, but not quite in the manner that others have envisioned. Misogyny takes root and grows immediately after puberty, when iron and sex become the central issue in every boy's and girl's life. During the halcyon days of childhood, boys' attitude toward girls resembles the apathy that male mammals express toward females that are not in heat. Girls ignore boys nearly as much. And then the rolling wall of puberty, like a giant tsunami wave, slams into young bodies. This is the moment when the political agendas of the sexes sharply diverge, when economic negotiations over sex become paramount, and when men confront the "world historic defeat of the masculine" at the hands of mere girls.

Prior to the onset of puberty, a boy's testosterone levels are nearly undetectable. Then the juice of Eros jolts his nervous system with a twenty-to-fortyfold increase. Testosterone floods the young male's incompletely myelinated brain, creating a dangerous and unstable situation. Among its protean manifestations, the androgenic hormone boosts the level of the male's sexual tension to unbearable levels. Though frequent masturbation offers him a partial solution, a teenager realizes for the first time in his life that he requires the intimate cooperation of a willing female. Therein lies the problem.

Boys enjoy playing games. By the time he discovers that he can mime a gun by cocking his thumb above his index finger, a boy seeks out other boys for group play. Put several of them together, and within a few moments they will be engrossed in a cooperative form of play distinguished by clear winners and losers. "Pickup ball," "choose-up," "marbles," and "cops and robbers" require minimal negotiation at the outset. (The arguments start shortly thereafter.) Rarely will a boy refuse an invitation from another boy to join a game. Imagine, then, a postpubertal boy's surprise and consternation when he first invites a girl to play with him the game that will instantly become the most fascinating and intriguing sport of his young life—the one in which the closer he gets to her bare skin and external openings, the more points he scores.

It is like no game he has ever played before. The other player is, from all outward appearances, a member of the same species, but the boy soon perceives that he is dealing with an alien. She seems to speak a different language from the one with which he is familiar. Worse, he must play by her rules, and, much to his surprise, despite her smaller stature, she is extremely effective at tenaciously guarding her goal. Initially, he may be convinced that it is he who is in control of the situation; after all, he is the one who initiated the game in the first place. Over time, however, he dimly perceives that she is setting the pace and conditions and, more often than not, that he is acceding to her requests in order to please her.

In those contemporary cultures in which harsh patriarchy has not yet destroyed a young woman's exercise of Original Choice, a young man's initial foray into this unfamiliar arena most likely occurs at a middle-school sock hop. (There is no reason to doubt that some variant of this ritual also occurred in the Pleistocene.) Perhaps the first girl he asks refuses his offer to dance (a skill at which she excels and he has barely learned the rudiments, since dancing was not something terribly interesting to him before his testosteronal transformation). She confounds him by her inexplicable rejection. Rarely in his life has a potential playmate refused to engage in a game. If a conspiratorial peal of giggles among her close group of friends follows her refusal, then, in addition to his feelings of perplexity, his nascent masculinity suffers a crushing blow.

Undeterred, sometime afterward he will attempt to discover firsthand the consistency of the two fascinating protuberances growing so invitingly on the chest of the student with whom he shares homeroom. Repulsed again, he will quickly learn that, although he considers this a harmless exploration, it, too, is verboten. Her resistance further baffles him, because he thought he had made it clear to her that he did not intend to hurt her but only wanted to satisfy his curiosity. Our young explorer soon comes to the realization that his quest is going to be far more difficult than he had initially anticipated. Perhaps he is fortunate enough to be rich, suave, handsome, strong, athletic, and never at a loss for words. Even if he has all those qualities, convincing a girl to join him in the play called sex will remain a pointed-elbows contest of wills at this stage in their lives.

Throughout the animal world, females are a scarce resource and males compete among themselves for the right to impregnate them. This struggle may be the invisible death swim of schools of sperm racing toward a prize only one of them can win, or the very visible and resounding clash of locked horns on the field of battle. Once a male has won the contest among other males for mating rights, he meets virtually no resistance from a female in heat. She desires sexual union as eagerly as he does, because her hormones are firmly in control of her frontal lobe's Executor (if she even has one). Through impeccable timing, the height of her lustful frenzy coincides with the departure of her eggs from her ovaries.

The irresistible siren song of her sexual instinct compels a female's willingness. The vibrations that shake her frame are set in motion by a cocktail of different hormones around the time of her internal ovulation. Estrus, or its equivalent in other female mammals, initiates a distinctive set of behaviors *driving* a female primate urgently to seek sexual union.

A biologist appropriately coined the word "estrus" by borrowing the Greek name for the stinging gadfly that harasses cattle. This large insect deposits its microscopic eggs under the beast's tough hide. When the gadfly's eggs mature into larvae, their squirming drives the host animal mad with itching. Sexual desire's most apt metaphor is an itch that *must* be scratched. (The intense frictioning of human coitus is an exceedingly complicated form of scratching an itch.) The human female is the only mammalian species who we know for sure has lost estrus (or its equivalent). However, what she lost, he seems to have gained; a young male of the human species exhibits ample behavioral indicators signifying that he is in a state of full-blown "estrus" *all the time.*

Behind every expectation is a frustration waiting to happen. Sadly, a young man learns, to his eternal disappointment, that the person of the opposite sex upon whom he has just recently focused his attention is balky, recalcitrant, and uncooperative. It is his incredible frustration with his sexual counterpart's unwillingness to comply with even his simplest sexual advances that begins the baleful dirge of misogyny. Cultural convention and media may reinforce his feelings later, but this is the fount from which it springs. That is exceedingly unfortunate, because his testosterone fog prevents him from seeing the major difference between his agenda and hers. At his young age, he is blithely unaware of the stakes involved—while he indulges himself thinking he is just playing a game, she is playing for keeps. For her, this is not a game but the opening skirmish in a campaign she cannot afford to lose.

Elder women train girls very early for the mission they must accomplish postmenarche. During her initial contacts with the opposite sex, a young woman must learn to recognize the key landmarks of the male mind's strange terrain and to assess quickly the outlines of his character. She must often make snap judgments so as not to waste valuable time that could be better used birthing babies with Mr. Right. The considerable advantages afforded her by her youth and beauty during this early period of her life serve her well.

Her ultimate goal is to negotiate his willing surrender—preferably "unconditional." She wants him to declare publicly that he will no longer seek the attention of any other woman and will devote his time, resources, love, and help to raising any children that may result from their passionate embraces. She knows that she cannot claim victory until she hears it in his voice, sees it in his eyes, reads it in his gestures, and feels it in her marrow. From a young woman's perspective, this contest is a matter of life and death. At some tellurian level, she understands that she will soon be called upon to put her life on the line. If she survives the dangerous delivery of her offspring, she requires the impregnating male to make a very long and very intense commitment in order to maintain her health and ensure her ability to continue to birth intelligent babies.

Anovulatory cycles grant her some leeway to experiment, but not much. Parents, older women, and her culture's conventions constantly remind her that if she fails to secure a future reliable source of resources the result will be catastrophic. And it doesn't matter how badly she herself wants to have sex— she *must* exercise restraint over her hormonal urges. She knows that her ability to withhold that which a man so hungers for is her most potent bargaining chip. Her agenda differs so markedly from his at this stage of life that his extraordinary frustration is unavoidable. *Her veto over sex is the primary source of her power and becomes the root of his anger.*

Many men resist the pull of misogyny and remain favorably disposed to gender equality. To a large extent, their mothers' love undergirds their equipoise. Reciprocating that love allows them to translate the deep feelings they have for their mothers into loving relationships with their lovers, wives, sisters, and daughters. Maternal adoration is the balm that can offset the bane a man experiences when confronted by a woman's repeated failure to cooperate with his sexual agenda.

Unfortunately, war, loss, illness, and catastrophe may prevent a boy from properly bonding with a loving mother. Surrogate mothers, sisters, aunts, grandmothers, teachers, and even fathers and grandfathers can often substitute for this crucial missing ingredient of a man's character, but a subset of emotionally damaged men exists for whom the art of loving women is a difficult and daunting challenge. Moreover, not all mothers love their sons equally, and some even irreparably wound them.

In general, without the unguent of a mother's love to salve the erupting adolescent antagonism toward women, men may evolve a suspicious and angry attitude toward women, making it nearly impossible for them to love any particular woman. Many a woman married to a man lugging this lumber around in his unconscious storehouse has discovered that, despite her love for him, the structural damage stemming from his childhood remains too extensive to repair. The most critical choice of her life often turns into a disaster, for her mate will behave maladaptively toward her and disappoint as a father to their children. Of course, many exceptions to these observations abound.

Through persistence, a woman gradually convinces a man to compromise his primary reproductive goal of unfettered access to a variety of women—a dream he relinquishes reluctantly, and one for the loss of which he never really forgives her. When he is young and his hormones are spiking, he is inclined to bargain away the right to be sexually free in exchange for exclusive sexual access. In the intense negotiation that occurs between the sexes, she wants him to make this key promise to her. The likelihood that he is in love vastly increases his willingness to please her by agreeing.

Male leaders of religion, government, and culture further women's agenda, because they understand that restraining the male's sexual urges reduces conflict, aids babies, and provides all fathers with increased confidence concerning their genetic heritage. The immense social pressure of his culture is brought to bear on him. The weight of tradition, ritual, and taboo must be marshaled against his towering natural impulse to pursue Master Testosterone's goal—*all women, all the time.*

Initially, his concession does not seem all that important to him. His volatile mixture of love and lust persuades him that he is obtaining his heart's desire. And then passing time erodes the fortitude underpinning his promise. Familiarity inexorably dims the intensity of his ardor for his mate. His hormones begin to back up against the door to the paddock and kick against the thin barrier that culture has constructed to hem them in. Neighing insistently, testosterone urges his faulty Executor to follow the evolutionary imperative adhered to by 99.9 percent of male creatures—to spread his seed far, wide, and often.

A moment arrives in every husband's life when he begins to question the fairness of his decision to pledge himself to one woman forever. (The same doubt also occurs to his wife, but she is less likely to act on it, because once she has children, deciding to leave her mate is a more hazardous and difficult course of action for her than it is for him.)

Many men experience a sense of guilt for their sexual fantasies and/or transgressions, because their paleomammalian limbic brains rebel against their culture's onerous restrictions. One part of a man resents women for preventing him from having what he wants. On the other side of this divide, a husband's failure to adhere to his "I do" oath in particular, and the male gender's failure to live up to women's expectations in general, induces women to harbor a deep resentment toward the male sex. This mutual rancor abets the war between the sexes and renews itself in every generation.

Further exacerbating his sense of pusillanimity when dealing with women is his stunning epiphany that his legacy depends on knowing for sure who his children are. Fatherhood is the least taxing route open to him to satisfy his longing to soften the terror of death. Knowing his children means he can deposit a part of himself—especially his name—in the next generation. Although he also has the alternative option of performing a memorable heroic deed, a sort of "memic" immortality, the more reliable method open to the average man would be to father children who adore and respect him.* And so a young man learns not only that a young woman can control his access to pleasure but that, further diminishing his negotiating stance, she is also the guarantor of his lineage.

Besides their irritation with the hoops that they must jump through to convince a woman to say *Yes!*, men's inability to maintain constancy to the degree that women expect fosters additional male frustration. Chafing under their load of "should"s and "ought"s, men crave to establish a social structure that will allow them to have their cake and eat it, too.† A man wants one woman to be exclusively his, and at the same time he wants to philander. He demands that all children "his" woman bears are exclusively his, yet he resists any restrictions on his sexual escapades even if they result in his illicit partner's pregnancy. Men see no paradox in the contradictory goals of the double standard that grow directly from the three grand transformative insights—sex causes babies, everyone's going to die, and "begats" ensure immortality. The resentment men experienced upon realizing their situation led them to resort to the powerful weapon they had at their disposal that they could wield against women—their ability to make public symbols. *The institution of patriarchy came into existence because men needed, first, to control women's sexuality and, second, to control women's reproductive rights. A man's control of the former ensured that he could relieve his intolerable itch on terms favorable to his sex; control of the second assured him his place in posterity.*

---

*Richard Dawkins, in his influential 1976 book, *The Selfish Gene*, introduced the concept of the meme. A meme is a purely mental idea that spreads from one individual to another, passed along by fashion or language, until it takes on a life of its own. Certain rites, ideas, trends, customs, insights, superstitions, and linguistic expressions mysteriously become woven into the fabric of culture.

†Sociobiologist Robin Fox hypothesizes that the human male's need to abide by stringent rules concerning his naturally aggressive sexuality produced "a creature who was capable of becoming extremely guilty about his sexuality."[16]

Male frustration over a female's veto power over sex is the primary source of misogyny.

Chapter 24

Unknown Mother/African Eve/ Modern Woman

And perhaps the sexes are more related than we think, and the great renewal of the world will perhaps consist of this, that man and maid, freed from all the false telling and aversion, will seek each other out not as opposites, but as brother and sister, as neighbors, and will come together as human beings.

            —Rainer Maria Rilke[1]

Our species is at a crux in its earthly history, aggressively mastering techniques that will let us guide and change our own nature.... The gate is ajar to a realm of wonders.   —Melvin Konner[2]

*T*wo features of contemporary human society—pornography and cosmetics—encapsulate the dilemma confronting each sex. An odd couple, the two complement each other in many subtle ways. Both are estimated to be ten-billion-dollar industries. Pornography is very nearly the furtive habit of postpubertal boys and men. Cosmetics are very nearly the exclusive province of postpubertal girls and women. Both result from stresses unique to the respective gender. Let us first examine the *animus* side of this couple.

  Pornography is fantasy; the pornography industry is not. Revolving around the nuts and bolts of sex, pornography never involves romance.* The women in the male fantasy are nearly always strangers with whom he has not, and will not, develop any kind of emotional attachment. Pornography fulfills wishes that cannot be readily, if ever, attained in a man's real life. Present in

---

*A woman shown a picture of male hunk will gaze first on his face, then his upper torso, and then his entire trunk. A man shown a picture of a nude woman rarely bothers to glance above the neck.[3] Also in this discussion I distinguish between pornography and erotica. The latter does involve elements of romance.

many variations, it plays out in three main imaginary scenarios. The first, nymphomaniacal pornography, depicts a man surrounded by multiple women whose sexual appetite is equal to or more voracious than his. Pedophilic pornography involves the fantasizer's having sex with a child. The third kind, bondage pornography, portrays men physically restraining women, rendering them defenseless.

In nymphomaniacal pornography, insatiable women demand as much sex as the male could possibly deliver. In pedophilic pornography, the preference for children ensures that sex is nonconsensual. The dominance relationship that normally exists between an adult and a child, combined with the trust that children always place in adults, assures that the level of the child's resistance will be minimal. In bondage pornography, a restrained woman cannot prevent a man from doing to her whatever he wishes to do to her. Central to all three types of pornography: The man *does not need to ask permission* to engage in sex. Moreover, he does not need to bring the woman something in exchange for her acquiescence, there is no negotiation, and he can leave anytime he is finished.

Men purchase pornography because in real life they have discovered it is nearly impossible to find a willing woman who does not expect something in return for sex. And because men seek young, beautiful women, most men can never hope to attract women as young and beautiful as those they can see in porno magazines.

Barely concealed beneath the veneer of civilized society, pornography's ineradicable appeal exposes the deep well of resentment that men harbor toward women. *Pornography would disappear tomorrow if women were as eager to have sex and behaved sexually as indiscriminately as men.*

On the *anima* side of this odd couple lies the effluvium of the vanity table. Women apply cosmetics to their face and body to increase their sexual attractiveness to men. No doubt exists that the application of cosmetics is a form of art. When performed artfully, the act bestows upon the artist a sense of accomplishment. In general, however, the fundamental motivation behind the purchase of cosmetics is to increase a woman's sexual desirability. The Great Game cannot begin *until* a man expresses interest in a woman. The leverage inherent in a woman's veto power is rendered useless unless a woman can attract a man. Therefore, most women firmly believe that they must accentuate their physical charms prior to advertising them.

Women often seek to satisfy their own urges and, on many occasions, desire sex more than a man; this situation will occur more frequently over the passage of time, as a man's sex drive wanes and hers waxes. Nevertheless, the threat of pregnancy, with its enormous implications, must give every young woman pause. This was especially true in the long period when effective birth control was nearly nonexistent.

Natural Selection's drastic reconfiguration of *Gyna sapiens'* reproductive system resulted in many unexpected consequences. None would prove as rapturous, yet as anxiety-provoking, as the unique signal human females used to attract men. To understand better why this unusual adaptation was such a radical departure from the norm, a review of the standard that Mother Nature installed in females to signal the other three million or so coupling species' males is necessary.

In the majority of species, this is an olfactory alert in the form of pheromones. Among primates, a combination of sight, smell, and body language alerts males to the proximity of an estral female. Missing from the extensive species catalogue of sexual semaphores is the one that draws *Homo sapiens* to *Gyna sapiens.* Humans call this signal "beauty."* Nowhere among the multitude is there another species in which a female's overall appearance singularly activates male lust.

No other female animal has to "dress to the nines." Lionesses never experience a bad hair day. Young bucks fight over old hinds as fiercely as they do over young does. Ailing or disfigured female chimpanzees never worry that they may become wallflowers. An ovulatory female's age, disability, complexion, figure, or hair tint virtually *never* interferes with a rutting male's sexual fixity. Nonhuman males nearly *always* respond to a female when she wafts her ovulatory signal.

Consider, then, the predicament of the *sapiens. Gyna* does not manifest a consciously identifiable signal to alert her *Homo.* Lacking this staple of sexual attraction, Natural Selection had to invent a substitute rapidly to ensure that *Homo sapiens* continued to express interest in mating. In Her haste, the Red Queen picked a doozy. The unusual alternatives upon which She settled were youth, health, and beauty. Farther down the list were character, wisdom, intelligence, industry, creativity, compassion, and nurturance, all of which ran distantly behind the sexual troika.[4]

The circuits that involuntarily swivel a man's head to follow the gently swaying backside of a passing young woman are firmly soldered into his nervous system—a fact of life that has ever been thus and will always be so. *Homo sapiens* lust after *Gyna sapiens* who are young, beautiful, and shapely.

Replacing periodic, swollen, flaming red vulvas and monthly titillating aromas with a never-used-before signal created a unique quandary for *Gyna*

---

*"Was this the face that launch'd a thousand ships / And burnt the towers of Ilium?" asked poet Christopher Marlowe about Helen of Troy's beauty. Whimsical scientists jokingly quantify beauty in units called "helens." A millihelen is just enough beauty to launch one ship.

sapiens. Before she can exercise Original Choice, a woman must first project the image of being youthful, healthy, and beautiful. Enticed by one or more of these elements, a man's courting behavior involuntarily flares, resulting in his lingering long enough for her to have the opportunity, if she is so inclined, to ignite the kindling necessary to encourage further exploration between them. Only with time will he discover that a woman may possess additional worthy attributes. Should a woman observe the erosion of Mother Nature's paramount "attractiveness index" during her daily mirror inspection, she will, to a lesser or greater degree, grow anxious, for she intuits that her ability to beguile a man will also diminish.

An example of women's desire to appear youthful is their generous use of hair dyes. Throughout their reproductive lives, women stay keenly attuned to the attributes that attract men. It would not have escaped women's perceptive notice that Caucasian children commonly begin life with a lighter shade of hair than they will exhibit as adults. Men unconsciously associate dark hair with maturity, gray hair with old age, and blond hair with youth. Because men constitutionally prefer youthful-looking females, women dye their hair blond more than any other color. All women, including non-Caucasians, have intuited that gentlemen prefer blondes.[5]

A further feature investing a woman's hair with sexual overtones is the curious fact pointed out by Alison Jolly that, unlike the hair, fur, feathers, and scales of other animals, a human's is not self-maintaining. No other female animals feel compelled to wash, shampoo, curl, perm, or style their crown tufts. Hair's centrality as a sexual signaling device can be measured in the amount of time and energy humans, particularly women, expend fashioning their scalp hair.

Underscoring Mother Nature's bizarre choice of beauty as the human female's primary sexual signal was the substance She chose to boost a woman's charm most in a man's eye. Underpinning the full lips, high cheekbones, firm breasts, smooth skin, beguiling shoulders, and taut behinds of a young *Gyna sapiens* lies Mother Nature's secret enticer. Instead of gorgeous colors, fantastic markings, sleek hides, sumptuous fur, bright feathers, or polychromatic scales, Mother Nature decided to use fat—common, greasy, yellow, subcutaneous fat. Positioned alluringly beneath the skin in just the right proportions and in just the right places, oleaginous avoirdupois can drive a man wild with intoxicating desire.

No other female creature uses adipose tissue as the infrastructure of its primary sexual signaling device. Few animals, other than sea mammals, even *possess* subcutaneous fat deposits. Humans sport *ten times* more fat cells under their skin than other land animals, easily surpassing even rotund pigs in this category.[6]

The human male is a consummate hunter. His success depends on his constant alertness for the slightest movement occurring in the periphery of his field of vision. The motion sensors out in the extremities of the male retina can detect a nearly subliminal shudder at the edges of his line of sight. "Critical flicker-fusion threshold" is the point at which a human can distinguish that a light is being turned on and off, just before it alternates too rapidly to be noticed. Men surpass women at this skill.[7] This attribute, so essential to hunting, was shanghaied into the service of Natural Selection. The Red Queen engaged a man's hunting rods to attract him to a woman. When a man says, "She caught my eye," he means it both literally and figuratively, whether it is the slight jiggle to a woman's breasts when she moves, the captivating commotion of her buttocks when she walks, or the curve of her ankle, calf, and thigh. The jazz-era pianist Fats Waller best captured men's fascination with a woman's subcutaneous fat deposits when he crooned, "It must be jelly, 'cause jam don't shake like that." Paradoxically, fat plays a lean role in a male's attractiveness to a female. Chiseled muscular definition and washboard abdomens devoid of subcutaneous fat are the physical features most likely to "catch the eye" of a woman.

Mother Nature's peculiar choice of fat to entice men is not so strange as it seems. Minus a clear signal from her that she was ovulating, the male of the human species must make an assessment as to the potential fertility of a female based on her appearance. The metabolic needs to gestate a human fetus are so great that Mother Nature will not allow menarche to begin until a prospective mother has enough of the yellow energy stuff in reserve. A girl cannot menstruate until a certain critical mass of fat has accumulated on her frame.* These fat reserves could have been stored deep inside the abdominal or chest cavities (where they reside in most other animals). Instead, fat is strategically arranged just under the skin to round out the human female form.†

Natural Selection put the human species in a precarious bind when *Gyna sapiens* abandoned estrus signaling. The problem was somewhat ameliorated by the installation in the male psyche of an attraction to a female exhibiting

---

*A leading authority in this field, Rose Frisch, calculated that before menses can begin, a girl must increase the fat on her frame until it constitutes at least 17 percent of her total body weight. Whenever she falls below 17 percent, her menses will cease. Her fat cells produce a hormone called leptin that, upon reaching a critical level, signals the pituitary to initiate menarche and maintain menses.[8]

†"Cellulite," the dimpling of fat that occurs as a woman ages, may be Natural Selection's method of signaling to a man a woman's approximate age, so that at an unconscious level he may determine how many more babies she has the potential to bear. Plastic surgeons and dermatologists, discovering that they can enhance a woman's appearance of youthfulness, health, and beauty by rearranging her body's distribution of fat, confirm the oily substance's importance as the substrate of male desire. Liposuction, silicone implants, and fat-cell transfers to eyes, hands, and lips reinforce my observation that fat plays the paramount role in human sexual signaling.

soft curves in all the right places.* Feminine beauty and potential fertility are inextricably grounded in subcutaneous fat.

*   *   *

To highlight the basic difference between human and nonhuman mating systems, revisiting the coupling habits of our closest relative, the chimpanzee, will be illustrative. Besides sharing the most genes with us, chimpanzees are representative of the mating pattern of primates in particular and mammals in general. Upon entering her period of estrus, a female chimpanzee begins to ascend toward the acme of popularity among the males. At her first visible and aromatic estral signs, males begin to jockey for position to mate with her.

Despite the crowd of overeager males always surrounding an estral female at her peak, jealousy from nonovulating females is relatively nonexistent. When a female experiences her state of estrus, she has absolutely no inhibitions about soliciting sex from a male. This cycle, which in a female chimpanzee recurs for a little over a week every thirty-seven days, will continue throughout her reproductive life.

Unlike human males, chimp males express little interest in juvenile estral females and favor those that have successfully mothered multiple offspring.† This pattern is common among other nonhuman primates. For example, primatologist Barbara Smuts observed that some savanna baboons formed friendships and sexual preferences for certain older females.[10]

Now consider how the reproductive life history of a typical *Gyna sapiens* differs from *Gyna all-the-others*. Instead of a season of desirability that recurs approximately once every month, she has a season of maximal desirability that comes (and goes) but once in a lifetime, tracing out a parabolic curve whose apogee occurs in her late teens. And it flames out at a time in her life when she is not likely to understand its function fully. Instead of knowing she is fertile right up to the time she dies, she becomes acutely aware that her

---

*Adolescent eating disorders most commonly revolve around a girl's self-image. Culture plays an important role in determining standards for beauty. Although fat may have been alluring in ancestral populations (witness the amplitude of the numerous sculpted fertility figures from the last Ice Age), at present the opposite perceptions concerning fat hold true. Thinness is in, obesity is out. Disordered perceptions of subcutaneous fat are central to anorexia nervosa, bulimia, and obesity.

†Among the males of one troop of chimpanzees in the Gombe Reserve, Jane Goodall observed that some males exercised a modicum of discernment. Males rarely refused *any* copulatory opportunity, but the majority seemed to prefer to mate with one particular female. Dr. Goodall named this chimp *femme fatale* Flo. Basing their opinions on human standards of beauty, Goodall and her co-workers all found Flo old, scrawny, and relatively unattractive. Yet males would sometimes postpone offers from younger females to line up patiently behind Flo, waiting their turn. Goodall hypothesized that the reason males preferred Flo was her superior mothering skills.[9]

biological clock will stop ticking at not the stroke of midnight but more like six-thirty or seven o'clock in the evening. Just when the party is really getting interesting, and she has finally figured out how all the participants interact, the pumpkin coach arrives. This, then, is the cruel double bind for women.

Adding to her dissatisfaction is the gross mismatch of her internal peak of sexual desire and her external peak of sexual desirability to males. During the early years of her postpubescence, a woman's libido tends to be low (and if it isn't, the heavy hand of cultural admonitions serves to restrain it). Yet this corresponds to the moment in her life when all men lust after her. When her libido rises to its maximum, in her late thirties and early forties, the majority of men are no longer wildly interested in her. As her sexual urges intensify, the pool of men she can attract shrinks. The inconsistencies in the human male-female sex drive are numerous and vexatious. Their emergence suggests that Mother Nature responded to an evolutionary sexual crisis with what, to many mere mortals of both sexes, seems to be a hastily thrown-together design that, in many men's and women's opinion, contains considerable room for improvement.

Imagine polling women and asking them to choose between two lifestyles. The first represents the status quo, containing every one of the familiar idiosynchronicities that exist between contemporary men and women. In the alternative life-style, a woman, upon attaining menarche, would experience a distinct one-week period out of every month during which every man, young and old, rich and poor, handsome or homely, would think she is the most alluring creature in the world. Her state of health, age, or appearance would not matter in the least to prospective suitors.

She would not compete with other women for any man's attention, because other women would not begrudge her the shower of masculine interest. During her monthly reign as Queen of the May, she would exist in a sublime bubble, unconcerned about the consequences of her promiscuity. Best of all, she would never care how men view her, spending not a nickel on cosmetics, because they are completely superfluous.

For the remainder of her month, she would go about her business assured that no man would take the slightest sexual interest in her—no harassment, no demands, no unwanted wolf whistles or disquieting looks from strangers. She would be free to speak to any man secure in the knowledge that he would not misinterpret her stance, dress, words, looks, or gestures as a sexual come-on. She could go where she pleased and do what she wanted. For those weeks, she would brook no interference from any man who deigned to tell her how to raise her children. Bolstering her independence would be her confidence that she could provision both herself and her children without any male assistance. Nothing would be required from men other than that they provide her and her children with protection, and they would willingly grant her unspoken request without asking for anything in return.

She could forgo her daily mirror inspection of how her face and figure were faring with the encroachment of age. Her self-esteem would not depend on how a man responded to her "appearance." She could dress as comfortably as she pleased and could spend her days in the company of other women, who were generally supportive. Or, if she was so inclined, she could hang out with the guys, who would treat her as one of the boys. There would be no fear of rape by a strange male, murder by an angry one, or stalking by an obsessive-possessive. During this time when she was not sexually available, no male would dominate her.

Were pollsters to conduct such a canvass, I suspect that the majority of women would eagerly ask where they could sign up for such a feminine utopia. Yet such is the state in which virtually all other female mammals find themselves—with the sole exception of *Gyna sapiens*.

Now let us give the same choice to a man. His alternative world: He could join a society where only one-fourth of the women would be possible candidates to engage in sex with him in any given month. Lest he be crestfallen at his reduced odds of achieving his sex drive's delight, this quarter of the female population would make up in hedonistic enthusiasm for any shortfall in numbers. Further, a constant rotation of eager fresh faces would assure him new opportunities every week. Women would approach him urgently demanding that he have intercourse with them. They would ask nothing in return, have no expectations that he bring them something of value. There would be no pounding of one fist on the other palm, declaring it was time for him to decide whether or not he was going to commit. No entreaties that he restrict his sexual life. And no responsibility to work to provide for a wife and family. He would be free to come and go as he pleased. I would venture that the line for the men in this sexual paradise would be as long as it was for the women.

Oh, there is one other thing a woman should know before signing on and packing her bags. Should she enter this parallel sexual universe, she must surrender any notions of love. She could continue to love her children and bask in the love that they return, but the heart-fluttering, dreamy, pink-cotton-candy, hazy love that can enmesh a man and a woman would not be a part of this world—lust, yes; love, no. This caveat would tend to stay the writing hand of most women.

The fine print in the contract for the men would contain similar clauses, warning them that they would never know the joy of walking through the door and being greeted by several small ones, faces all lit up, exclaiming

"Daaaaddy!" while running toward him with arms outstretched. Nor would a man experience the oceanic feeling of losing the boundaries of his body while lying in the dark, arms and legs entwined with a woman he loves. And when a man grows old, the companionate love that builds between him and his mate, with whom he has shared a lifetime of successes and defeats, would also be absent.

Many might take issue with my presentation of the pornography-and-cosmetics complementarity, or my notions about the causes of misogyny and patriarchy, arguing instead that both are more dependant on the processes of acculturation. I am aware that culture plays a part in shaping behaviors, but I believe that these conditions have firm roots in evolutionary adaptations. They appeared first inside the twists and turns of chromosomes handed down from distant relatives who lived and died in a remote area of Africa's Great Rift Valley.

Recorded history, remember, only began approximately five thousand years ago. In this narrow 3 percent slice of the life of our species, patriarchy and misogyny have been the dominant norm in most major civilizations and throughout nearly every historical period. Despite the tantalizing suggestion that there might have been a time before recorded history in which relations between the sexes were more pacific, the historical record speaks for itself.

Perusing the daily news confirms that patriarchy and misogyny persist in every major contemporary society. From the Taliban in Afghanistan to Japanese men's repression of women, from male pro-life fanatics in the United States to the practice of bride burning in India, evidence for these invidious twins abounds. Wide swaths of Africa and the Islamic world still practice female genital mutilation. Forced prostitution is rife among the former Soviet republics, and sex slavery is an ugly fact of life from Saudi Arabia to Thailand.

Humanity staggers on like a person who has suffered a stroke that left half of the body paralyzed. The masculine half of the body politic resists acknowledging the obvious: Disdaining, ignoring, and dismissing its distaff half is extremely counterproductive. Until individuals, couples, and cultures can facilitate and appreciate the contributions of both halves of the human psyche, the human species will continue to be hobbled by this serious handicap.

What can we do to redirect culture's gender relationships and set them flowing in a more congenial direction? The many framers of this question most often posit their solutions in the context of education: If only we could somehow change the content and context of what we teach children, then we could begin to eradicate this poison that prevents harmony between the sexes.

Certainly education has a role to play, but in this work I have tried to move the conversation away from an emphasis on culture and more toward the realm of genetics and evolution. Some might protest, claiming that this approach is but a warmed-over version of Freud's determinist "anatomy is destiny" argument, which men have used to legitimize their dominance over women throughout the ages. That is not my intention. I do not seek to *justify* why things are the way they are but, rather, to *understand* why they came to be.

To many, gaining an understanding of a human behavior that is unalterable may seem akin to winning a pathetic consolation prize. "It is but sorrow," Tiresias told Oedipus, sighing, "to be wise when wisdom profits not." And yet this must be the first step in the process to initiate meaningful change in our cultural institutions. Chiseled into the lintel above the entrance to the Oracle of Delphi in ancient Greece was Apollo's first commandment to all mortals—*Gnosti Seautum*—Know Thyself.* This Socratic imperative remains very relevant today. Sigmund Freud gave it new meaning when he hypothesized that an individual who failed to appreciate his or her primal shaping influences was doomed to repeat destructive patterns of behavior throughout his or her life.

Discoveries in the biological sciences have superseded many of Freud's original ideas, but his guiding principle holds true, especially as applied to evolutionary theory. To understand and change the present condition of our species, we must gain insight into the past. If we do not, we cannot exert a lasting influence on the future. Grasping the evolutionary reasons behind the many quirks in the human mating system will lead, I believe, to an improvement in the relations between men and women. This overriding of the "anatomy is destiny" conundrum is possible because we have arrived at a critical juncture in the life of our species.

Homo sapiens sapiens—the Doubly Wise Human, as we used to be called—is at present undergoing a metamorphosis. We humans are in the process of changing into something else. A new species is being born right before our eyes, but because we are so close to the cataclysmic event, we cannot appreciate its full import.

An apt example of the process of metamorphosis is the life cycle of a caterpillar, a notoriously rapacious and greedy insect. Resembling an invading army living off the land, caterpillars ravage any greenery in their path, leaving a wake of destruction. Yet the caterpillar's insatiable hunger is driven by an evolutionary imperative—it must convert the available nutrients in its surroundings into the surplus essential to fuel its metamorphosis. Once it

---

*Just below it was his second rule to live by, "Nothing in Excess."

spins its cocoon and sequesters itself inside, its gargantuan appetite ceases. The changeling-in-process lives off the stored food it so hungrily consumed earlier so that it has an energy source sufficient to allow it to morph into an entirely new organism.

Upon its maiden flight from its chrysalis, a butterfly is utterly transformed from its former state. Not only is it inexplicably far more beautiful than it had been, but also it performs a function for nature that is the opposite of that of its caterpillar incarnation. Instead of munching its way mindlessly through vegetation, the butterfly flits from flower to flower, partaking of very little, destroying nothing, and inadvertently serves as a pollinator, enabling more flowers to bloom. Humans seem to have an instinctual aversion for insects—except butterflies. Somehow, humans have developed a strange affinity with these colorful creatures and are soothed and delighted by their presence.

In Buddhist parables, an ordinary man may behave in a selfish and self-serving manner until he becomes "awakened." To achieve this state, a man must withdraw temporarily from the world and turn inward. The transformation that is occurring within his soul is not visible to anyone observing him from the outside. Upon achieving enlightenment, the person is utterly changed and is said to have attained the state of *satori*. One so enlightened is then entitled to *nirvana*, a Sanskrit word meaning "extinguished." He is freed from the obligation to return reincarnated to the world on another turn of the karmic wheel of fate.

In Buddhist tradition, many enlightened souls elect to return to this world and serve as bodhisattvas. They teach and give aid to alleviate the suffering of the unawakened. The Buddhists frequently use the metaphor of the metamorphosis of a caterpillar into a butterfly to describe poetically this transformation of the individual. Could it be that the entire human species began its existence as a collective caterpillar? As we enter our most environmentally rapacious stage, are we on the verge of transforming into the metaphorical equivalent of a butterfly? Perhaps, as a result of our transformation, individuals within our species will increasingly behave like bodhisattvas and act to heal our relationship with nature and with each other. I realize that my optimistic stance will be considered by many as hopelessly naïve. But I submit the following historical evidence to buttress the above appraisal concerning the current status of the human condition.

Two and a half million years ago, the evolving hominid line differentiated away from its apelike predecessor by fashioning the first stone tool. Imagine the moment. A slow-witted, little-more-than-one-pint-brained *Homo habilis* held a cobble in his hand and, through a concerted effort of sustained mental concentration, conceived a tool residing inside the rock. The imprisoned tool could not be released from the stone unless the first Handy Man (or Woman) sat still and laboriously chipped away at one side of the rock. Employing patience and persistence, the first toolmaker was rewarded at last with a working stone tool that sported a crude cutting edge. *Homo habilis* continued to make this identical simple tool with very few modifications for almost the next one million years!

Then the next hominid version, bigger-brained *Homo erectus*, figured out that a stone's cutting edge could be sharpened considerably if he or she simply turned the rock over and began knapping each side alternately. The bifacial hand ax fashioned in this manner sliced through hides and cut bone from joint much more cleanly than those previously fashioned by *Homo habilis*.

Here is the astonishing part. Though Acheulian bifacial hand axes, the signature tool of two-pint-brained *Homo erectus*, have been discovered nearly everywhere our predecessor settled in Africa, Europe, and Asia, these artifacts remained mysteriously unchanged across enormous distances and over incredibly long stretches of time. Despite a rapidly enlarging brain, *Homo erectus* could not *imagine* a way to improve on his technique for the next eight hundred thousand years. Glynn Isaac expressed disappointment at this "shuffling of the same essential ingredients" for an extraordinarily lengthy period of time "in a minor directionless change."[11]

And then, 150,000 years ago, along came *Homo sapiens*. At the start of the *sapiens'* adventure, some novel toolmaking techniques appear in the archeological record. But despite the *sapiens'* having acquired a three-pint brain, the pace of innovation advanced with the speed of spreading molasses. And then, rather suddenly, *sapiens* crossed some sort of an invisible barrier forty thousand years ago. Art, tombs, and artifacts began to appear in profusion. The pace of novelty accelerated with the agricultural revolution ten thousand years ago and went into hyperdrive in the last century. At present, it has switched to a breathtaking perpendicular ascent. The amazing shift in the celerity with which humans embraced innovation has necessitated converting the 2.5-million-year-old curve from a geometric representation to a logarithmic scale for the last 1.5 percent of its length. In an hour-long film chronicling the hominid toolmaking epoch, the industrial age of the machine would flash by in the last few seconds.

Whenever a measurement curve switches from an incrementally slowly rising horizontal one to a steeply ascending vertical one, conditions are favorable for a major transformation. Typically, this is the moment when the object being measured or observed changes from one state into another.* This alteration in states occurs in a relative instant, without any transitional gradations. In physics this process is called a phase change; in biology, metamorphosis; and in evolution, punctuated change.

Technological innovation has advanced so rapidly in our lifetime that it has become the primary environmental stimulus responsible for refashioning the animal that began life as *Homo* and *Gyna sapiens*. Linguist Derek Bickerton noted "the two most shocking facts of human evolution: that our ancestors stagnated so long despite their ever-growing brains and that human culture grew exponentially only after the brain had ceased to grow."[12] Using newly discovered technology, humans have brought about intended and unintended changes in both their external surroundings and each individual's internal milieu. Warnings about the impact of the negative consequences have been widely disseminated and have induced appropriate levels of anxiety in thoughtful people; the positive consequences, however, possess the power to transform us as a species. This metamorphosis will affect all aspects of the human condition. In the following discussion, I wish to focus on the realignment of the relationships between men and women as a result of the impact of technological innovation.

For most women in the First World, the threat of maternal mortality and iron-deficiency anemia, the two key factors that I have proposed molded the interactions of ancestral *Homo* and *Gyna sapiens*, verge on irrelevancy. Advances in obstetrics, including asepsis, anesthesia, and blood replacement, have made the passage of the big head through the small opening such a routine event that women have all but forgotten just how dangerous sex with its common consequence, pregnancy, was in the past. Cesarean sections are imperceptibly boosting the intelligence of our species as babies possessed of such large brains that they would have killed their mothers (and themselves) are now routinely surviving delivery and passing on their big-brain genes.†

Effective birth control has brought the decision to become pregnant and carry a child to term more firmly under the personal control of most women living in technologically advanced societies. Prenatal testing for birth defects, widespread vaccinations, and neonatal care have drastically reduced infant mortality.

---

*A familiar example of this phenomenon is the three states of water: solid (ice), liquid (water), and gas (steam). All three states remain in effect even though the temperature slowly rises. When the temperature advances one degree above 32°F, ice changes state and turns into water. It remains in this liquid state until the temperature exceeds 212°F, and then the liquid transforms into vapor.

†In 2001 in the United States, 24 percent of babies were cesarean births.[13] On the same day that this was reported in the *Washington Post*, a grim complementary statistic was announced on NPR radio. Twenty-four percent of mothers delivering their babies in rural Afghanistan died of complications of childbirth.

Changes in the structure of modern societies, economics, and food distribution, combined with advances in nutrition, have markedly reduced a woman's dependence on a man to supplement her diet with iron, vitamins, essential amino acids, and essential fatty acids. Recent discoveries in reproductive physiology and gene therapy stagger the imagination and challenge bioethicists with their realities and possibilities. And medical advances continue to extend our life span, enhancing the quality of old age. Increasing numbers of us live up to the potential that was encoded into each of our chromosomes at the moment of our conception. There can be little doubt that all these drastic alterations in our environment are collectively functioning as transformative agents fueling the human species' metamorphosis.

The long-lived intergalactic anthropologists present at the birth of our species express wonderment at the changes they observe happening right before their eyes. Remembering the demise of the Unknown Mother 150,000 years earlier, they can recall the many dire prognostications most members of the expedition floated over the future prospects of the hominid line. And there would be, most likely, not a few pessimists among them who would indulge in similar bouts of hand-wringing as they watch African Eve's descendants muck up the planet they inherited.

The more perceptive aliens, however, would discern the outlines of an emergent brave new species that was undergoing an unheralded form of metamorphosis. They would excitedly notify the mother ship that they were witnessing the birth throes of something that none of them could have anticipated. They would report that the old reliable parameters of physical attributes, such as brain size or bone length, were inadequate to distinguish the new species from the old, but one distinctive marker was the dramatic changes in the way some men and women had begun to relate to each other. *Sapients*, it was clear to the observers, were experiencing a period of punctuated change.

Humbled, perhaps, by the knowledge of how wrong they were when predicting African Eve's survival chances 150,000 years earlier, the aliens would most likely add the qualifier that their report on the mating patterns of this new species was a work-in-progress. They promise to keep sending regular updates. They hedge their bets, because the transformation they are observing is entirely without precedent when compared with anything that has occurred in the previous 3.8 billion years of life on Planet Earth.

Knowledge of time, death, and paternity molded men into husbands and fathers. To a lesser extent, these insights reinforced women's roles as wives and mothers. In humans, the capacity to love, which is amply present in both sexes, exceeds in intensity and duration that of any species.

❧

Epilogue

I felt compelled to write this book because, after the publication of my earlier work, *The Alphabet Versus the Goddess: The Conflict Between Word and Image,* I had the feeling that I had incompletely answered the question I had originally posed. To wit: Why was global society so shot through with misogyny and patriarchy? Although I remain proud of my earlier work and stand by its premises and conclusions, something seemed missing.

To review briefly: In *AVG,* I questioned the nature of the historical event that could have been so immense and so pervasive that it changed the sex of God. Indisputable evidence exists in the often meticulous, sometimes fragmentary, record of the ancient world that there had been a historical time when both men and women venerated a female deity. Rome, Egypt, Japan, China, India, Greece, and Mesopotamia provide ample evidence that a goddess once reigned supreme in the hearts of both sexes. In classical Athens, the male citizens voted whom they wanted as their patron deity. Athena won over Poseidon. The famous city we know as Athens might have been called Poseids had men been inclined to want a male deity rather than a female one to watch over them.

The period during which goddess worship occurred spans the beginning of recorded history, five thousand years ago (and most likely a much longer period before that, of which little can be confirmed despite many tantalizing clues). It seems to have lasted approximately two millennia. Women's rights and prestige in culture appear to have paralleled the reign of goddess worship.

And then a wrenching change occurred. Particularly in the West, three monotheistic religions arose—Judaism, Christianity, and Islam—each of which had as its core premise that there existed only a singular deity and He was indisputably male. All three Western religions emphatically insisted that goddesses did not exist. Eve, Mary, and Fatima were mortals, and although some may have worshipped them, these figures never possessed the power to resurrect the dead, a quite routine function for goddesses of old.

I hypothesized in that book that the invention of writing, particularly alphabetic writing, reconfigured the brain of anyone who learned the new skill in such a way as to reinforce the masculine *animus* at the expense of the feminine *anima*. Reading and writing superseded speaking and listening, and strengthened the power of the already dominant left brain and right hand over the right brain and left hand.

This internal shift in neurocircuitry convulsed culture and manifested in the disappearance of goddesses, the suppression of women's rights, and a general abhorrence of image information. Alphabetic sacred texts became glorified to the point where masses of people were willing to kill each other over minute doctrinal differences. And they still do. Zealots destroyed and/ or banned iconic information (principally perceived by the right hemisphere) beginning with graven images. They then extended their draconian prohibitions to include every form of representative art. Literate information, the written word, was elevated to an exalted position until it congealed into dogma. The Old Testament, New Testament, and the Koran became for their respective believers the final arbiters for sacred truth. In the beginning was the Word. Denying the pre-eminence of the images of Lascaux, Luxor, and Nineveh, the word superseded the image.

I had based the hypothesis of *AVG* on the media theorist Marshall McLuhan's aphorism, "The medium is the message." The *process* people use to absorb and generate information is a more important factor shaping culture than the *content* of the information that they are absorbing or generating.

Though *AVG* obviously struck a chord with many readers, judging by its popularity and many favorable reviews, there were those who criticized it, protesting that men's malevolent attitudes toward women were more deeply rooted. I remain firmly convinced that the central thesis of *AVG* is essentially correct. Too many historical correlations existed for critics to dismiss my research as just a series of mere coincidences. Yet, at the same time, I felt a need to respond to these critics. They had a point: There are deeper currents feeding misogyny and patriarchy.

Another subject that has always intrigued me was the reasons we humans diverged so far away from the reproductive life cycles of the other three million sexually reproducing species. If it worked so well for so long, why did *sapients* abandon it for a system that seems, to so many thoughtful observers of the human condition, so dysfunctional? Curious about the roots of human sexuality, I felt compelled to trace the origins of misogyny and patriarchy back further than the beginning of recorded history. I suspected that there were subterranean primordial undercurrents feeding this malevolence that must have something to do with features unique to our species.

In a parallel mystery intriguing me, I puzzled over the reason why human females menstruated. I had long ago concluded, on the basis of what I had learned about it in my medical training, that it was a trait that seemed to defy common sense. No one has ever been able to demonstrate clearly a single incontestable benefit, and many have amplified the dreary list of the negative impacts it has on women's health, outlook, and metabolism. If the other 3,999 mammals could get along just fine without having to experience a menses as vexatious as the human one, what on earth was it doing so stubbornly resisting ejection from the human gene pool?

These, then, were the various strands of thoughts gently cobwebbing my mind when I found myself forced to take an overnight drive through the Columbia River canyon, traveling between Eugene, Oregon, and Missoula, Montana. It was in the dead of night on an empty highway that I had the multiple insights that formed the woof and warp of this book.

Let me explain how I came to be in that magnificent, silent space at such a strange hour. *AVG* had recently been published, and I had embarked on the obligatory author's tour. In the hectic pace of scheduling, I had inadvertently agreed to speak at the university in Eugene, Oregon, at 7:00 P.M. and also to keynote the Montana State Teachers Convention in Missoula, Montana, the following morning. Failing to recognize properly that I had left precious little time to get from Point A to Point B, I belatedly called my travel agent and naïvely asked her to arrange a late flight from Eugene to Missoula. She replied, somewhat incredulous at my ignorance of airline schedules and geography, that no connections existed that could possibly accommodate such an itinerary within my time constraints. Only after I consulted a map and surveyed the vast reach of Western emptiness in between those two cities did I fully appreciate the problem.

I considered canceling one of the engagements, but both sponsoring organizations had distributed flyers and expressed dismay at any suggestion I might be a no-show. I decided that the only feasible way to cover the distance was to rent a comfortable car and hire a reliable driver to put in the ten hours required to cover the six hundred miles, so that I would arrive an hour before my talk in Missoula the next morning. I planned to sleep in the back seat and emerge reasonably rested just before my presentation. My driver, Steve Nolan, and I started out on our trip immediately after my talk at the university.

Although I had planned to sleep during the bulk of the drive, I was concerned that this long stretch might be too exhausting for Steve. Besides, I found trying to sleep in the back seat nearly impossible. At about 2:00 A.M. I switched places with him.

As I steered the car, gently wending my way alongside the river, a nearly full moon rose majestically over the canyon rim, suddenly flooding the landscape with illumination. Moonlight reflecting off the river allowed me to see more clearly the beautiful scenery I had been missing without the moon's beneficence.

I had been driving in silence and was reflecting on the question-and-answer period that had followed my earlier talk at the university. I had been asked to explain how my theory of brain reconfiguration due to alphabet literacy could account for the presence of so many preliterate cultures in which men repressed women. As I switched back and forth between admiring the moon and grappling with the questioner's criticism, the idea for this book quite suddenly popped up within my mind.

The great epiphany that waylaid me and laid claim to the greater part of my waking hours for the next few years was that menses was a necessary evolutionary trait in humans. Entraining a woman's bleeding with the moon's orbit taught humans how to tell time. The ability to range into the future was so tremendous an asset to our species that whatever the price we had to pay would be worth it. That one sex of the species, but not the other, bore the cost, was a detail that would shape all subsequent relations between the sexes.

Menses had to be copious enough to attract the attention of women, so that they would recognize its connection to the moon's monthly periodicity. An unfortunate side effect, however, was the increased danger to the health of a woman and her offspring from the pernicious effects of iron loss. Now wide awake with creative possibilities, I continued to follow the meander of the river, while periodically glancing at the moon. My stream of consciousness about this subject began to swell into ideas that gradually took on the semblance of coherent flow.

Many readers of my earlier books have asked me to explain the process by which I arrive at the ideas that propel a relatively unformed concept through the tooling of refinement until it sits in your hand as a bound, illustrated, typeset, four-hundred-page book. A writer and his or her reader form an unusual couple. One does his or her work in isolation, without any input from the other, and then, years later, the other gets to evaluate whether or not the trees were worth the felling. Now that my task is over it is your turn to respond. Feel free to let me know what you think by visiting my Web site bulletin board, www.sextimepower.com, or contact me at Lshlain@aol.com. Thank you for exploring the bulrushes to find the basket and then being willing to lavish so much of your time and attention on my child.

—Leonard Shlain

Notes

Preface: Iron/Sex

1. Quoted in Pinker, 1997, p. 461.
2. Gould, 1993.
3. Blum, 1997, p. 101.
4. Hrdy, 1999, p. 136.
5. Kaplan, 1994, pp. 753–91.
6. Van Valen, 1973, pp. 1–30.

Chapter 1: Unknown Mother/African Eve

1. Santayana, 1988, p. 41.
2. Washburn, 1973.
3. Trevathan, 1987, p 108.
4. Gould and Eldrege, 1993; Gould and Lewontin, 1979.
5. First reported by Rebecca Cann, John Wilson, and their associates in 1987. These researchers exploded the field of evolutionary history by introducing molecular biology into it as an investigative tool (Cann, Stoneking, and Wilson, 1987, pp. 32–36). Disagreement continues between paleontologists and molecular biologists as to the precise date of birth of our species. Some place it 200,000 years ago; others claim it was a mere 100,000 years ago; the majority of opinion lies somewhere between at around 150,000 years (Stringer and Andrews, 1988, pp. 1263–68). Many other innovative researchers are adding to this burgeoning field. Luigi Luca Cavalli-Sforza, a linguist, has compared the genetic movements of populations and compared them with the ethnic makeup of the speakers (Cavalli-Sforza, 2000). He built upon the earlier work of Joseph Greenberg, who classified the world's languages. The work of Douglas C. Wallace and his colleagues has also detailed the human family tree. Peter Underhill and Peter Oefner have traced the Y chromosome and constructed a detailed history of the human male (Underhill, Oefner, et al., 2000, pp. 358–61).
6. Mitochondria are the tiny energy modules that exist outside the nucleus in the cell's cytoplasm. Curiously, they have a DNA code different from the cell's nuclear DNA. Scientists conjecture that the mitochondria are remnants of an archaic virus that invaded cells at the dawn of life on the planet and then found the neighborhood so inviting they stayed on and developed a symbiotic relationship with their host cells. The discovery of mitochondrial DNA was fortuitous in presenting molecular biologists with a tool to track the lineage of a species, because mitochondrial DNA, unlike nuclear DNA, does not alter with sexual reproduction. Charles Sibley and Jon Ahlquist began applying this new technology in 1980 to bird family trees, and then began applying the knowledge they had acquired about birds to the hominid family tree (Sibley and Ahlquist, 1984, pp. 99–121).
7. Washburn, 1960, pp. 60–75; Rosenberg and Trevathan, 1996, pp. 161–68.

Chapter 2: Big Brain/Narrow Pelvis

1. Ellison, 2001, p. 22.
2. Bowlby, 1972.
3. Furuichi, 1987, pp. 309–18.
4. Jolly, 1999, p. 155.
5. Avers, 1974.
6. Trevathan, 1987, pp. 72–88.

Chapter 3: Red Blood/White Milk

1. Quoted in Ardrey, 1976, p. 75.
2. Zuckerman, 1932, p. 98.
3. Angier, 1999, p. 117.
4. Dunbar, 1996, p. 3.
5. Labrecque, Eason, and Marcoux, 2001, pp. 1753–54; Myers-Helfgott and Helfgott, 1999, pp. 305–25.
6. Hrdy, 1999, p. 434.
7. Trevathan, 1987, pp. 104–6.
8. Angier, 1999, p. 248.
9. Jacobs, Butler, and Blanche, 1965.
10. Strassman, 1998, pp. 167–84.
11. Strassman, 1996, pp. 181–220.
12. Norwitz, Schust, and Fisher, 2001, pp. 1400–1408.
13. Hrdy, 1999, p. 7.

Chapter 4: Plant Iron/Meat Iron

1. Daly and Wilson, 1978, p. 321.
2. Grahn, 1993, p. 132.
3. Crawford and Marsh, 1989, p. 88.
4. Aiello and Wheeler, 1995, pp. 199–221; Leonard and Robertson, 1994, pp. 77–88.

Chapter 5: *Gyna Sapiens / Gyna All-the-Others*

1. Morgan, 1972, p. 92.
2. Comfort, 1967.
3. Zahavi and Zahavi, 1997.
4. Profet, 1993, p. 352.
5. Diamond, 1997, p. 71; Alexander and Noonan, 1979, pp. 436–53; Alexander, 1990.
6. Buss, 1994, p. 187.
7. Burley, 1977, pp. 3476–79.
8. Hrdy, 1999, p. 34.
9. Daly and Wilson, 1988.
10. Symons, 1979.
11. Benshoof and Thornhill, 1979, pp. 95–106.

12. Morris, 1967, p. 65.
13. McClintock, 1971, pp. 244–45.
14. Batten, 1992, p. 117.
15. Wrangham, 2001, p. 140.
16. Dunbar, 1988.
17. Turke, 1984, pp. 33–44.
18. Jolly, 1999, p. 1.
19. Knight, 1991, p. 215.

Chapter 6: Periods/Perils
1. Delaney, Lupton, and Toth, 1979, p. 1.
2. Profet, 1993, p. 336.
3. Grahn, 1993, p. 52.
4. March, 1980, pp. 125–27.
5. Diamond, 1997, p. 122.
6. Konner, 2002, p. 104.
7. Countinho, 1999, p. 97.
8. Ibid., p. 60.
9. Grahn, 1993, p. 98.
10. Loebenstein, trans., 1983, p. 8.
11. Pliny, 1963, chapter 23.
12. Dunbar, Knight, and Power, eds., 1999, p. 98.
13. Nicholson, 1995, pp. 779–84.
14. Profet, 1993, pp. 335–85.
15. Strassman, 1996, pp. 181–220.
16. Dalton, 1964.
17. Delaney, Lupton, and Toth, 1979, p. 19.

Chapter 7: Her Climax/His Climax
1. Quoted in Symons, 1979, p. 219.
2. Twain, 1938, p. 43.
3. Sagan and Druyan, 1992, p. 335.
4. Konner, 2002, p. 293.
5. Sherfey, 1966, pp. 88–92.
6. Baker and Bellis, 1993, pp. 861–85.
7. Konner, 2002, p. 293.
8. Bermant, 1976, pp. 76–103.
9. Quoted in Symons, 1979, p. 208.
10. Masters and Johnson, 1966.
11. Hines, 2001, pp. 359–62.
12. Angier, 1999, p. 63.
13. Bodger, 2000, p. 227.

Chapter 8: Grandmothers/Circumcision

1. Quoted in Fisher, 1992, p. 308.
2. Spock, 1989, p. 43.
3. Quoted in Fisher, 1999, p. 204.
4. Sherman, 1998, pp. 759–61.
5. Hawkes, 1997, pp. 551–77.
6. Angier, 1999, p. 250.
7. Gardiner, 1949, pp. 1433–37.
8. Manniche, 1987, p. 8.
9. Wilson, 1997, pp. 7–8.
10. Modern perceptions of circumcision's origins are a confusing mix of myth, religion, and science, so it would be helpful to review their place in history. At present, circumcision is associated with the demands of a stern Old Testament patriarchal god.

The ancient Hebrews declared that it was one of the four (and only four) conditions of the covenant between Yahweh and His Chosen People. In this ancient contract that begins with a conversation between Abraham and Yahweh, God demands that Abraham (1) foreswear fealty to any other god except Yahweh and (2), to demonstrate his resolve, sacrifice the last several millimeters of the loose skin covering his and all subsequent converts' penises. (The latter must rank among the strangest requests made by a god of his followers in history.) In exchange for these commands, Yahweh promises Abraham that (3) he will become the leader of a great and generously populated nation and (4) his new nomadic nation will finally have a homeland to call its own (Genesis 17:1–21). Because of the centrality of this covenant to their religion, Jews consider the practice of circumcision essential to their identity.

After the death and resurrection of Jesus, Paul faced the daunting task of converting the inhabitants of the Roman Empire to Christianity. He had the political acumen to recognize that a major stumbling block in his mission was the onerous Biblical requirement that each new convert must undergo circumcision. A skilled negotiator, Paul realized that circumcision was a real deal killer and he simply dispensed with it. Paul replaced a potentially hazardous surgical procedure with the much more palatable requirement that a convert simply declare that he believed in Christ.

A few centuries later, Muhammad faced the same conundrum as Paul had, but was proselytizing to a vastly different audience. Instead of the literate, sophisticated inhabitants of cosmopolitan Roman centers, the Prophet needed to convince a desert-hardened constituency of the need to adhere to a new strict set of rules markedly at variance from what had passed for religion before his received revelation from the angel Gabriel in A.D. 610. To promote literacy among the illiterate and to keep his converts focused on the gravity of a sacred written text, Muhammad maintained a close connection to the Old Testament in order to link his teachings with ancient scripture. A central tenet of his new religion was the lineage connecting Arabs to Abraham through his son Ishmael. Circumcision was therefore imperative, and its practice remains unquestioned among Muslims. Lost among the precepts of the world's three Western religions is that circumcision long antedated the Old Testament, as evidenced by the Saqqara mural, and most likely was a ritual performed in prehistoric times.

The cosmologies of hunter-gatherers contain a plethora of spirits. Supernatural genders can be masculine, feminine, or hermaphroditic. In general, if a tribe exalts the spoils of the *hunt,* the spirits tend to be more virile than fertile. If the fruits of *gathering* are the main staple of the tribe's diet, then feminine spirits tend to be more influential. With the advent of agriculture, a major shift occurred in every culture. The power previously invested among multiple spirits became concentrated in one all-powerful mother deity known as the Earth Goddess. She had a different name in virtually every society studied, but no one has yet identified a major ancient culture in which both the men and women failed to venerate this omnipotent Creatrix. The Earth Goddess presided over life, death, and the fertility of animals and crops.

In many early agricultural societies, it was the custom to crown a vigorous young male specimen king for a season, during which time he was encouraged to mate with many healthy, beautiful young women. At the end of his yearlong reign, he was sacrificed on orders of the high priestesses of the Goddess and his blood was sown into the freshly turned soil so that his vitality would ensure the fertility of the season's plantings (Frazer, 1960, pp. 559–88). (Many feminist writers question the veracity of these accounts, claiming that patriarchal religious leaders, attempting to discredit the Goddess in the eyes of the people, spread these stories that were tantamount to propaganda.)

In many goddess religions of old, men not uncommonly voluntarily castrated themselves to serve Her. As late as the fifth century Roman Empire, Augustine decried the contemporary practice among men who mutilated themselves in the name of Cybele, the reigning Roman mother goddess (Augustine, 1972, p. 286).

If a priestess or a priest of the Goddess could convince men to submit to such extreme sacrifices, surely those in charge of the welfare of the tribe could manage to institute the mild (in comparison) practice of male circumcision. Perhaps the following speculative scenario can explain the reasons behind its adoption by the Hebrews.

When the Hebrews began their Reformation of the Egyptian religion, they set out to reject all aspects of Egyptian ways. Their spare religion would embrace an invisible god—no images, no goddesses—and a sacred alphabetic text. They so thoroughly rejected the Egyptian obsession with mummies, pyramids, and the Land of the Dead that the Hebrews initiated the only world religion whose founding document, the Torah, makes *no* mention of an afterlife.

Yet it is rare for a people rebelling against a parent culture (or a child rebelling against a parent) not to carry over some traditions from the preceding society. The Buddha fulminated against Hindu religious practices, yet he incorporated the Hindu idea of reincarnation, making it central to his new doctrine, destined to become Buddhism. After breaking with the Jewish religion that spawned it, Christianity maintained its connection to the Old Testament, and Islam continued to venerate the Kaaba, the black stone monolith in Mecca left over from the pre-Muslim worship.

Therefore, it would not be surprising that the Israelites might have retained the Egyptian practice of circumcising males. To expunge any idea that this practice was originally the directive of the Goddess, the early Israelites claimed that a stern male deity now decreed it, not the formerly powerful female deity.

Among the many ancient civilizations inclined to honor feminine values, Egyptians held goddesses in the highest esteem, especially in the earlier years of their exceedingly long history. Isis, Maat, Hathor, Nepthys, and many others were venerated with the same degree of fervor as were male deities. The Egyptians' oldest creation myth posited that two goddesses, without the assistance of any god, created the world (Larrington, 1992, pp. 24–25). Egyptians practiced matrilineal inheritance customs, and women exerted considerable influence at all levels of society. They often wielded major political power and controlled financial resources. As acknowledged in surrounding ancient cultures and in their own historical accounts, Egyptians were also experts in the art of love.

The name of the drug belladonna means "beautiful woman" in Italian. Cleopatra was the first woman in history that we know of to use the drug to dilate the pupils of her eyes. Egyptian women understood that a man is more sexually attracted at an unconscious level to a wide-eyed female than he is to one whose pupils are of normal diameter. I conjecture that the women in a culture this sophisticated, having discovered so subtle a secret of sensuality, would most likely have discerned circumcision's elusive benefits. The Israelites carried the practice over into their new religion and three thousand years later it remains firmly embedded in modern culture.

11. Maimonides, 1956, pt. 3, ch. 49.
12. Fisher, 1999, p. 171.
13. Sherfey, 1966, p. 87.
14. Richards, 1999, pp. 308–14, 508.
15. Konner, 2002, pp. 102–5.
16. Quoted in Fisher, 1999, p. 182.

Chapter 9: Prey/Predator

1. Sahlins, 1960, p. 82.
2. Schreiner, 1978, p. 176.
3. Leakey and Lewin, 1978, p. 105.
4. Stanford, 1999, p. 151.
5. Wrangham and Peterson, 1996.
6. Quoted in Donald, 1991, p. 104.
7. Quoted in Johanson and Edey, 1981, p. 309.
8. Quoted in Morgan, 1990, p. 41.
9. Laporte and Zihlman, 1983, pp. 96–110.
10. Wheeler, 1994, pp. 339–50.

11. Tattersall, 1998, p. 117.
12. Hardy, 1960, p. 642.
13. Daly and Wilson, 1978, p. 337.
14. Stanford, 1999, pp. 40, 70.
15. Donald, 1991, p. 111.
16. Symons, 1979, pp. 158–62.
17. Ridley, 1993, p. 228.
18. Sagan and Druyan, 1992.
19. Quoted in Leakey and Lewin, 1992, p. 181.
20. Hanbury-Tenison, 1982, p. 95.
21. Zerries, 1968, p. 272.
22. Harrington, 1933, p. 179.
23. Biesele, 1993, p. 84.
24. Hawkes, 1991, pp. 29–54.
25. Stanford, 1999, p. 157.
26. Landes, 1938, p. 131.
27. Lee, 1988, p. 266.
28. Siskind, 1973, p. 103.
29. Goodall, 1986, p. 484.
30. Fisher, 1982.
31. Richards, 1987, pp. 166–67.

Chapter 10: Carnivory/Vegetarianism

1. Quoted in Pinker, 1997, p. 197.
2. Kingdon, 1993, p. 124.
3. Batten, 1992, p. 158.
4. Milton, 1984, pp. 249–79; Stahl, 1984, pp. 151–68.
5. Fisher, 1992, p. 147.
6. Corballis, 1991, p. 135.
7. Wrangham, 2001, pp. 123–43.
8. Rodwell-Wilton, 1985, p. 93.
9. Ibid.
10. Calloway and Kunzer, 1982, p. 356.
11. Morgan, 1972, p. 174.
12. Crawford and Marsh, 1989.
13. Crawford and Crawford, 1972, p. 17.
14. Ridley, 1999, p. 169.
15. Ibid., p. 170.
16. Ibid., p. 89.
17. Speth, 1987, pp. 13–29.
18. Halterman, 2001, pp. 1381–86.
19. Bruner et al., 1996, pp. 992–96.
20. Brabin, 1999, pp. 690–91.

Chapter 11: Menarche/Mustaches

1. Quoted in Symons, 1979, p. 96.
2. Low, 2000, p. 83.
3. Symons, 1979, p. 253.
4. Grahn, 1993, p. 35.
5. Weinstock, 1947.
6. Goodall, 1986, p. 443.
7. Money and Eberhardt, 1972.
8. Martin and Bumpass, 1989, pp. 37–51.
9. Bogin, 1988.
10. Herodotus, 1954, p. 121.
11. Marshall, 1971, pp. 103–62.
12. Bettelheim, 1954.
13. MacLean, 1990.
14. Sowell, 1999, p. 861.
15. Schiffer, 1998, p. 69.
16. Thatcher, Walken, and Guidice, 1987, pp. 1100–33.
17. Damasio, 1994; LeDoux, 1998; Coolidge and Wynn, 2002.

Chapter 12: Premenstrual Tension/Masturbatory Tension

1. Ridley, 1993, p. 133.
2. Small, 1995, p. 61.
3. Gross, 1975, pp. 526–49.
4. Rosenblum, 1976.
5. Hrdy, 1999, p. 221.
6. *New York Times,* June 24, 1999, p. A20.
7. Sagan and Druyen, 1992, p. 229.
8. Hill and Wenzel, 1981.
9. Benedek and Rubinstein, 1939, pp. 245–70, 461–85.
10. Dalton, 1964; Dalton, 1969, pp. 242–47.
11. Markee, 1940, pp. 221–308.
12. Dalton, 1964, 1969, 1975.
13. Janowsky, 1966a.
14. Baker and Bellis, 1988, pp. 937–80.
15. Hall and Devore, 1965.
16. Tutin, 1979, pp. 29–38. See also Wallis, 1997, pp. 297–307.
17. Hrdy, 1999, p. 85.

Chapter 13: Moon/Menses

1. Stanford, 1999, p. 43.
2. Grahn, 1993, p. 157.
3. Ardrey, 1961, p. 104.
4. In the ancient myths of the major cultures, people credited goddesses with having taught mortals the meaning of time. Because of women's close association

with moon, menses, and measurement, the first calendars fell under the goddesses' aegis. Linear time, however, is a left-hemispheric function commonly associated with the masculine. It was not long afterward in the development of these same cultures that a god usurped the claim of primacy from goddesses for both the notion of time and the invention of calendars. In the real world, timekeeping functions presided over by priestesses were taken over by priests.

Beginning in 3000 B.C., Mesopotamians believed that the goddess Inanna taught mortals how to track time. Around her neck she wore the tablets that determined the fate of each person. Inanna's totem was the owl, and she was the Mesopotamian goddess of wisdom.

About 1750 B.C., the fierce god Marduk gained power by murdering his grandmother, the Great Goddess Tiamat, and wrested control of the calendar from Inanna. *The Seven Tablets of Creation,* written in the Mesopotamian city of Babylon, coincided with Marduk's triumph and the historical reign of King Hammurabi. The myths associated with Marduk soon displaced the myths of earlier times, and it was Marduk around whose neck hung the tablets of fate. Thereafter, a male priest oversaw the important task of marking the days of the calendar. So well did these Chaldean priests who replaced priestesses perform their duties that they became renowned throughout the ancient world as astute astronomers and mathematicians.

The Rig-Veda, the oldest Sanskrit account of Hindu myths, credits the goddess Sarasvati with teaching the secrets of the calendar to the people. Later, the god Shiva became the one who dispensed this gift.

In Greek mythology, an archaic version of the goddess Athena taught men about the periodicity of the months, equinoxes, and how to keep track of a year. Later, it was the Titan Prometheus who supposedly gave the gift to humankind. When classical Greece was at its zenith, in the fourth century B.C., the sun god Apollo was in charge of the function of time and calendars. The influence of women on the notion of time, however, can be discerned in the Greek psyche by the persistence of women's roles over such matters.

The three female Fates, the Greek Moirae, played a paramount role in the pantheon of archaic Greece. *Moira* is the word from which our English words "mortuary," "morbid," "mortician," and "mortal" are derived. In French, the word for "death," *mort,* comes from the same derivative. *Destino* is the Latin word from which "destiny" stems, and means "woven." The myth of the three Fates uses a metaphor of weaving a long thread to delineate the span of a man's life. Each Fate plays a different role. Clotho the Spinner begins spinning the thread (she represents a man's birth). Lachesis the Measurer determines how long the thread will be (she represents life). Atropos the Cutter snips the thread and ends each mortal's life (she represents his death). A long thread captures poetically the essence of linear sequence and is an appropriate metaphor for that of passing time.

When archaic Greece passed into classical Greece, the Fates came under the control of the god Zeus, who assumed the ultimate authority over the timing of a mortal's life and death. Zeus, who was not a weaver, became the master of destiny.

The Gregorian calendar, adopted in A.D. 1177, represented a major realignment. The sun, not the moon, now reckoned a year. To fit the scheme of twelve months into 365.25 days in a year, months were assigned arbitrary lengths that (with the exception of February) were no longer in synchrony with the actual lunar month. Sun time eclipsed moon time as the new improved method to keep time. Despite this increase in the accuracy of computing a year, many cultures continue to reckon time by the moon (Walker, 1983, p. 303).

5. Menaker and Menaker, 1959, pp. 905–14; Law, 1986, pp. 45–48; McClintock, 1971, pp. 244–45; Vollman, 1968, pp. 1171–75; Cutler, Garcia, and Krieger, 1980, pp. 163–72.

6. Menaker and Menaker, 1959, pp. 905–14.

7. Knight, 1991, p. 215.

8. Goldsmith, 1990, pp. 14–15.

9. Briffault, 1927.

10. Brown, 1959, p. 247.

11. Walker, 1983, pp. 636–43.

12. Shuttle and Redgrove, 1999, p. 149.

13. Marshack, 1964, pp. 743–45; Marshack, 1972, pp. 445–77.

14. White, 1989, pp. 211–31.

Chapter 14: Woo/I Do

1. Huxley, 1943, p. 3.

2. Rich, 1978.

3. Eliot, 1986 (1871), p. 9.

4. Holloway, 1983, pp. 105–14; Geschwind and Galabruda, 1985, pp. 428–59.

5. Frisch, 1967.

6. Dunbar, 1996.

7. Byrne and Whiten, 1988.

8. Dunbar, 1996, p. 172.

9. Ibid., p. 173.

10. Leakey and Lewin, 1992, p. 181.

11. Chris Knight, personal communication.

12. Pinker and Bloom, 1990, pp. 707–84.

13. Jesperson, 1922, p. 436.

14. Calvin, 1983, pp. 121–35.

15. Kimura, 1979; Lieberman, 1984.

16. Donald, 1991.

17. Bickerton, 1990, pp. 341–58.

18. Deacon, 1997.

19. Corballis, 2002.

20. Christiansen, 1994, p. 187; Christiansen, unpublished manuscript.

21. Miller, 2001, p. 357.

22. Fisher, 1999, pp. 58–59.

23. Miller, 2001, pp. 369–75.

24. Walker and Leakey, 1978, pp. 54–66.
25. Fisher, 1999, p. 59.
26. Hrdy, 1999, p. 235.
27. Fitch, 2002. See also *Proceedings of the Royal Society,* vol. 268 (Mar. 1999), p. 1669.
28. Pinker, 1997, p. 192.

Chapter 15: Anima/Animus

1. Quoted in Brown, 1959, p. 134.
2. Jung, 1964.
3. Kruuk, 1972.
4. Thornhill and Thornhill, 1990a, 1990b.
5. Brownmiller, 1976.
6. Low, 2000.

Chapter 16: Gay/Lesbian

1. Potts and Short, 1999, p. 74.
2. Symons, 1979, p. 300.
3. Reyes, Winter, and Faiman, 1978, pp. 74–78.
4. Smuts and Wantanabe, 1990, pp. 147–72.
5. Sagan and Druyan, 1992, p. 299.
6. Hamer, Hu, Magnuson, and Pattatucci, 1993, pp. 1405–9.
7. Blanchard, 1997, pp. 27–67.
8. Ridley, 1993, p. 264.
9. Small, 1995, p. 178.
10. Kruijver, Zhou, Pool, et al., 2000, pp. 2034–41.
11. LeVay, 1991, pp. 1034–37.
12. Winter, trans. and ed., 1961, pp. 88–90.
13. Laumann, Gagnon, Michael, and Michaels, 1994.
14. Wilson, 1978.
15. Mondimore, 1996, pp. 82–87.
16. Kinsey, Wardell, and Martin, 1948, pp. 651.
17. Hooker, 1967, pp. 167–84.
18. Small, 1995, p. 358.

Chapter 17: Same Sex/Hermaphrodite

1. Quoted in Hrdy, 1999, p. 205.
2. Woolf, 1929, p. 98.
3. Hamilton, 1964, pp. 1–52.
4. Trivers, 1971, pp. 35–37.
5. Wilson, 1978.
6. Aiello and Dunbar, 1993, pp. 184–93.
7. Allen and Gorski, 1999, pp. 97–104; Holloway and De Lacoste–Utamsing, 1982, pp. 1431–32.

8. LeVay, 1994, p. 102.
9. Nathans, Thomas, and Hogness, 1986, pp. 193–202.
10. Corballis, 2002, p. 172.
11. Corballis, 1991, p. 94.
12. Lindsay, 1996, p. 24.
13. Falk, 1990, pp. 333–81.
14. Crawford and Marsh, 1989, p. 157.
15. Kimura, 1999, p. 169.
16. Grahn, 1993, p. 202.
17. Potts and Short, 1999, p. 76.
18. Small, 1995, p. 163.

Chapter 18: Mortality/Angst

1. Appel, 1967, pp. 140–41.
2. Becker, 1973, p. 283.
3. Moussaieff Masson and McCarthy, 1995, p. 74.
4. Leakey and Lewin, 1992, p. 303.
5. Moussaieff Masson and McCarthy, 1995, p. 95.
6. Becker, 1973, p. 87.
7. Herodotus, 1954, pp. 460–61.
8. Quoted in Becker, 1973, p. 25.
9. Mithin, 1996, p. 152.
10. Piaget, 1974; Gould, 1977.
11. Bronson, 1982, p. 100.
12. White, 1993a; White, 1993b.
13. Wexler, 1992, pp. 2820–25.

Chapter 19: Superstition/Laughter

1. Barlow, 1958, p. 93.
2. Thompson, 1984, p. 134.
3. Ferenczi, 1955, p. 246.
4. Shaler, 1900, p. 22.
5. Calhoun, 1962, pp. 139–46.
6. Quoted in Brown, 1959, p. 100.
7. Ortega y Gasset, 1957, pp. 156–57.
8. Camus, 1955, p. 3.
9. Quoted in Potts and Short, 1999, p. 189.
10. De Waal, 1996.
11. Rousseau, 1964 (1755), p. 141.

Chapter 20: Father/Mother

1. Quoted in Fisher, 1992, p. 47.
2. Batten, 1992, p. 186.
3. Mead, 1967, p. 192.

4. McBrearty and Brooks, 2000, pp. 453–563.

5. Anthropologist Richard Klein proposes that the Creative Explosion resulted from a brain mutation that rather abruptly changed the way *Homo sapiens* thought. The mutation would have had to have occurred prior to the last significant exodus of *sapients* out of Africa, and then would have had to have lain relatively dormant for approximately 10,000 more years before manifesting in the Creative Explosion. The 10,000-year unexpressed mutation could be the only way Aborigines in Australia could have paralleled the development of the Cro-Magnons in southern France. Recent research has reduced the age of the oldest Australian aboriginal, but it still remains at somewhere around 40,000 years ago. Klein's work draws on the conclusions of Ingman et al., who trace the DNA in everyone living today back to a small group of *Homo sapiens* that left Africa 52,000 years ago.

Klein and others propose that it was this group of very recent immigrants that replaced the older *Homo sapiens,* and everyone alive today derives from this last group. Many experts disagree with Klein's theory, and his proposition of a sudden brain mutation that increased awareness sufficiently to lead to art, music, and burials is as plausible, in my opinion, as my theory of our ancestors' rapid maturation of consciousness due to multiple insights about the human condition. (Ingman, Kaessmann, Pääbo, and Glyllensten, 2000, pp. 708–13; Klein, 2002.)

6. Malinowski, 1929, p. 242.

7. Walker, 1983, p. 680.

8. Holmberg, 1950, p. 73.

9. Brown, 1959, p. 185.

10. Quoted in ibid., p. 107.

11. Ridley, 1993, p. 226.

12. Herodotus, 1954, p. 111.

13. Walker, 1983, p. 709.

Chapter 21: Incest/Dowries

1. Tiger and Fox, 1971, p. 261.

2. Schull and Neel, 1965.

3. Morton and Neel, 1967, pp. 55–62.

4. Lévi-Strauss, 1969, p. 136.

5. Canetti, 1963.

6. Commager, 1977, p. 8.

Chapter 22: Wife/Husband

1. Quoted in Batten, 1992, p. 22.

2. Lewis, Amini, and Lannon, 2000, p. 89.

3. Quoted in ibid., p. 204.

4. Ibid., p. 144.

Chapter 23: Misogyny/Patriarchy

1. Hrdy, 1981, p. 14.

2. Quoted in Coutinho, 1999, p. 164.

3. Quoted in Symons, 1979, p. 253.

4. Innumerable artists believed that capturing the love between a mother and her son and re-creating it on canvas or in marble represented the greatest challenge for their artistic skills. Painters and sculptors of the highest renown have filled museums and private collections with their attempts to distill this essence. The Madonna with her child ranks as the single most familiar theme in Western art. Note that a clear representation of the lad's dad is missing from every single one of these paintings, making Mary history's most famous single mom. Lest someone claim that the second commandment forbids the depiction of the Father, the numerous portrayals of Him engaging in other activities—for example, sparking Adam's finger—make His failure to appear in Christianity's most important family portrait all the more puzzling.

And where is Joseph, Jesus' surrogate father? The gospels tell us that Jesus learned the trade of carpentry from his mortal father. Undoubtedly, the kind and gentle Joseph would have taught his son many other useful things. One would expect that more than a handful of artists would have dared to depict the nuclear family from Bethlehem. Joseph's virtual and Yahweh's complete absence from any artwork portraying mother and son is all the more remarkable in a religion that claims that the Holy Trinity is a father, a son, and . . . a Holy Ghost, or, as in the more modern parlance, the Holy Spirit. This third entity is, most tellingly, not the Holy Mother.

Long before Christianity, the image of a solitary mother holding her son on her lap either just following his birth or immediately following his death had been a central theme in many cultures. A powerful mother embracing her tiny manchild is universal. Tiamat and Kingu, Isis and Horus, Aphrodite and Adonis, Cybele and Attis, and other mother-and-son combinations are a constant thread connecting all the generations back into the miasmic bogs of prehistory.

5. Beauvoir, 1953, p. 239.

6. Tanner and Zihlman, 1976, pp. 585–608.

7. Ortner, 1974, pp. 67–88.

8. Coutinho, 1999, p. 6.

9. Buckley and Gottlieb, 1988, pp. 3–50.

10. Sagan and Druyan, 1992, p. 223.

11. Fisher, 1999, p. 258.

12. Ibid., p. 264.

13. Buss, 1994, pp. 22–26.

14. Ellis, 1992, pp. 241–42.

15. Engels, 1972 (1884), p. 461.

16. Fox, 1972, p. 292.

Chapter 24: Unknown Mother/African Eve/Modern Woman

1. Quoted in Brown, 1959, p. 134.

2. Konner, 2002, pp. xix–xx.

3. Morgan, 1972, p. 155.
4. Daly and Wilson, 1978, p. 303; Ford and Beach, 1951, p. 86.
5. Ridley, 1993, pp. 294.
6. Morgan, 1990, p. 109.
7. Ginsburg, Jurenovski, and Jamieson, 1982, pp. 1079–82.
8. Frisch, 1984, pp. 161–88.
9. Goodall, 1986, p. 484.
10. Smuts et al., 1987, pp. 385–99.
11. Quoted in Mithin, 1996, p. 123.
12. Bickerton, 1995, p. 65.
13. *Washington Post,* 2002, p. AO1.

Bibliography

Ackerman, Diane. 1994. *A Natural History of Love.* New York: Random House.

Adams, Carol J. 1992. *The Sexual Politics of Meat: A Feminist-Vegetarian Critical Theory.* New York: Continuum.

Aiello, L., and R. I. M. Dunbar. 1993. "Neocortex Size, Group Size, and the Evolution of Language," *Current Anthropology* 34: 184–93.

Aiello, L., and P. Wheeler. 1995. "The Expensive Tissue Hypothesis," *Current Anthropology* 36 (no. 2): 199–221.

Alexander, Richard. 1990. *How Did Humans Evolve?* Special Publication no. 1. Ann Arbor: University of Michigan.

Alexander, Richard, and K. M. Noonan. 1979. "Concealment of Ovulation, Parental Care, and the Human Social Evolution," in *Evolutionary Biology and Human Social Behavior,* ed. N. Chagnon and W. Irons (North Scituate, Mass.: Duxbury), pp. 436–53.

Allen, L. S., and R. A. Gorski. 1991. "Sexual Dimorphism and the Anterior Commissure and the Massa Intermedia of the Human Brain," *Journal of Comparative Neurology* 312: 97–104.

Angier, Natalie. 1999. *Woman: An Intimate Geography.* New York: Houghton Mifflin.

Appel, Alfred. 1967. "An Interview with Nabokov," *Wisconsin Studies in Contemporary Literature* 8 (Spring): 140–41.

Ardrey, Robert. 1961. *African Genesis.* New York: Bantam.

Ardrey, Robert. 1976. *The Hunting Hypothesis.* New York: Bantam.

Augustine. 1972. *City of God.* Trans. Henry Betterson. London: Penguin.

Avers, C. J. 1974. *Biology of Sex.* New York: John Wiley and Sons.

Baker, R. R., and M. A. Bellis. 1988. "Kamikaze Sperm in Mammals?" *Animal Behavior* 36: 937–80.

Baker, R. R., and M. A. Bellis. 1993. "Human Sperm Competition: Ejaculate Adjustment by Males and the Function of Masturbation," *Animal Behavior* 46: 861–85.

Barkow, Jerome; Leda Cosmides; and John Tooby. 1992. *The Adapted Mind: Evolutionary Psychology and the Generation of Culture.* New York: Oxford University Press.

Barlow, Nora, ed. 1958. *The Autobiography of Charles Darwin.* New York: Harcourt Brace.

Batten, Mary. 1992. *Sexual Strategies: How Females Choose Their Mates.* New York: G. P. Putnam's Sons.

Beauvoir, Simone de. 1953. *The Second Sex.* London: Penguin.

Becker, Ernest. 1973. *Escape from Evil.* New York: Free Press.

Bellow, Saul. 1973. *Humboldt's Gift.* New York: Viking.

Benedek, T., and B. B. Rubinstein. 1939. "The Correlations Between Ovarian Activity and Psychodynamic Processes: I. The Ovulative Phase," *Psychosomatic Medicine* 1 (no. 2, April): 245–70; "II. The Menstrual Phase," ibid. 1 (no. 2, October): 461–85.

Benshoof, L., and R. Thornhill. 1979. "The Evolution of Monogamy and Concealed Ovulation in Humans," *Journal of Social and Biological Structures* 2: 95–106.

Bermant, G. 1976. "Sexual Behavior: Hard Times with the Coolidge Effect," in M. H. Siegel and H. P. Zeigler, eds., *Psychological Research: The Inside Story*. New York: Harper & Row, pp. 76–103.

Bettelheim, Bruno. 1954. *Symbolic Wounds: Puberty Rites and the Envious Male.* New York: Free Press.

Bickerton, Derek. 1990. *Language and Species.* Chicago: University of Chicago Press.

Bickerton, Derek. 1995. *Language and Human Behavior.* Seattle: University of Washington Press.

Biesele, Megan. 1993. *Women Like Meat: The Folklore and Foraging Ideology of the Kalahari Ju/'hoan.* Bloomington: Indiana University Press.

Blanchard, R. 1997. "Birth Order and Sibling Sex Ratio in Homosexual Versus Heterosexual Males and Females," *Annual Review of Sex Research* 8: 27–67.

Blum, Deborah. 1997. *Sex on the Brain: The Biological Differences Between Men and Women.* New York: Viking.

Bodger, Joan. 2000. *The Crack in the Teacup.* Toronto: McClelland & Stewart.

Bogin, B. 1988. *Patterns of Human Growth.* New York: Cambridge University Press.

Bowlby, John. 1972. *Attachment,* vol. I. Middlesex: Penguin (originally published 1969).

Brabin, B. 1999. "Iron Pots for Cooking: Wishful Thinking or Traditional Common Sense," *Lancet* 353 (Feb. 27): 690–91.

Brain, C. K. 1981. *The Hunters or the Hunted?* Chicago: University of Chicago Press.

Briffault, Robert. 1927. *The Mothers,* 3 vols. London: Allen & Unwin.

Bronson, Gordon. 1982. "Structures States and Characteristics of the Human Nervous System at Birth," in Peter Stratton, ed., *Psychobiology of the Human Newborn.* New York: John Wiley and Sons, p. 100.

Brown, Norman O. 1959. *Life Against Death.* New York: Random House.

Brownmiller, Susan. 1976. *Against Our Will: Men, Women and Rape.* New York: Simon & Schuster.

Bruner, A. B., et al. 1996. "Randomized Study of Cognitive Effects of Iron Supplements in Non-Anemic Iron Deficient Girls," *Lancet* 348: 992–96.

Buckley, Thomas, 1982. "Menstruation and the Power of the Yurok Women: Methods of Cultural Reconstruction," *American Ethnologist* 9: 47–60.

Buckley, Thomas, and Alma Gottlieb, eds. 1988. *Blood Magic: The Anthropology of Menstruation.* Berkeley: University of California Press.

Burley, Nancy. 1977. "Parental Investment, Mate Choice, and Mate Quality," *Proceedings of the National Academy of Sciences* 74: 3476–79.

Buss, David M. 1994. *The Evolution of Desire.* New York: Basic Books.

Byrne, Richard, and Andrew Whiten, eds. 1988. *Machiavellian Intelligence: Social Expertise and the Evolution of Intellect in Monkeys, Apes, and Humans.* Oxford: Clarendon Press.

Calhoun, John B. 1962. "Population Density and Social Pathology," *Scientific American* 206 (no. 2): 139–46.

Calloway, D. H., and M. S. Kunzer. 1982. "Menstrual Cycles' Protein Requirements of Women," *Journal of Nutrition* 112: 356.

Calvin, William. 1983. "A Stone's Throw and Its Launch Window: Timing Precision and Its Implications for Language and Hominid Brains," *Journal of Theoretical Biology* 104: 121–35.

Camus, Albert. 1955. *The Myth of Sisyphus and Other Essays.* Trans. Justin O'Brien. New York: Vintage.

Canetti, Elias. 1963. *Crowds and Power.* New York: Viking.

Cann, Rachel; M. Stoneking; and A. C. Wilson. 1987. "Mitochondrial DNA and Human Evolution," *Nature* 325: 32–36.

Carter, Rita. 1998. *Mapping the Mind.* London: Weidenfeld & Nicolson.

Cavalli-Sforza, Luigi Luca. 2000. *Genes, People and Languages.* Berkeley: University of California Press.

Chagnon, N. A. 1968. *Yanomamo: The Fierce People.* New York: Holt, Rinehart and Winston.

Christiansen, M. H. 1994. "Infinite Languages, Finite Minds: Connectionism, Learning, and the Linguistic Structures" [Ph.D. Thesis, University of Edinburgh], in R. Dunbar, C. Knight, and C. Power, eds., *The Evolution of Culture.* New Brunswick, N.J.: Rutgers University Press, 1999, p. 187.

Christiansen, Morton. Unpublished manuscript. "Language as Organism: Implications for the Evolution and Acquisition of Language."

Comfort, Alex. 1967. *Nature and Human Nature.* New York: Harper and Row.

Comfort, Alex, ed. 1972. *The Joy of Sex.* New York: Simon & Schuster.

Commager, Henry Steele. 1977. *The Empire of Reason: How Europe Imagined and America Realized the Enlightenment.* New York: Doubleday.

Coolidge, Frederick L., and Thomas Wynn. 2001. "Executive Functions and the Frontal Lobes and the Evolutionary Ascendancy of the Homo Sapiens," *Cambridge Archaeological Journal* 11 (no. 2, Oct.): 255–60.

Corballis, Michael. 2002. *From Hand to Mouth: The Origins of Language.* Princeton: Princeton University Press.

Corballis, Michael C. 1991. *The Lopsided Ape.* Oxford: Oxford University Press.

Coren, Stanley. 1992. *The Left-Hander Syndrome: The Causes and Consequences of Left-Handedness.* New York: Vintage.

Coutinho, Elsimar M., with Sheldon J. Segal. 1999. *Is Menstruation Obsolete?* Oxford: Oxford University Press.

Crawford, M., and D. Marsh. 1989. *The Driving Force.* London: Heinemann.

Crawford, Michael, and Sheliagh Crawford. 1972. *What We Eat Today.* London: Spearman.

Cutler, W. B.; C. R. Garcia; and A. M. Krieger. 1980. "Sporadic Sexual Behavior and Menstrual Cycle Length in Women," *Hormones and Behavior* 14: 163–72.

Dalton, Katherina. 1964. *The Premenstrual Syndrome.* London: Heinemann.

Dalton, Katherina. 1969. *The Menstrual Cycle.* Harmondsworth, G.B.: Penguin.

Dalton, Katherina. 1975. "Paramenstrual Baby Battering," *British Medical Journal* 3 (May): 279.

Daly, Martin, and Margot Wilson. 1978. *Sex, Evolution, and Behavior.* New York: Wadsworth.

Daly, Martin, and Margot Wilson. 1988. *Homicide.* New York: Aldine, Hawthorne.

Daly, Martin, and Margot Wilson. 1998. *The Truth About Cinderella: A Darwinian View About Parental Love.* New Haven: Yale University Press.

Damasio, Antonio. 1994. *Descartes' Error: Emotion, Reason, and the Human Brain.* New York: Avon.

Darwin, Charles. 1981 (1871). *The Descent of Man: And Selection in Relation to Sex.* Princeton: Princeton University Press.

Davis, D. E. 1964. "The Physiological Analysis of Aggressive Behavior," in W. Etkin, ed., *Social Behavior and Organization Among Vertebrates.* Chicago: University of Chicago Press.

Dawkins, Richard. 1976. *The Selfish Gene.* Oxford: Oxford University Press.

Dawkins, Richard. 1995. *River out of Eden: A Darwinian View of Life.* New York: Basic Books.

Deacon, Terence. 1997. *The Symbolic Species.* New York: Norton.

Delaney, Janice; Mary Jane Lupton; and Emily Toth. 1979. *The Curse: A Cultural History of Menstruation.* Champaign: University of Illinois Press.

de Waal, Frans. 1996. *Good Natured: The Origins of Right and Wrong in Humans and Other Animals.* Cambridge: Cambridge University Press.

Diamond, Jared M. 1992. *The Third Chimpanzee.* New York: HarperCollins.

Diamond, Jared M. 1997. *Why Is Sex Fun?* New York: Basic Books.

Dillard, Annie. 1974. *Pilgrim at Tinker Creek.* New York: HarperCollins.

Donald, Merlin. 1991. *Origins of the Modern Mind: Three Stages in the Evolution of Culture and Cognition.* Cambridge, Mass.: Harvard University Press.

Drabble, Margaret. 1978. *The Wise Wound: Menstruation and Every Woman.* Bridgend, G.B.: WBC Book Manufacturers.

Dunbar, Robin. 1996. *Grooming, Gossip, and the Evolution of Language.* Cambridge, Mass.: Harvard University Press.

Dunbar, Robin; Chris Knight; and Camilla Power, eds. 1999. *The Evolution of Culture.* New Brunswick, N.J.: Rutgers University Press.

Dunbar, Robin I. M. 1988. *Primate Social Systems.* London and Sydney: Croom Helm.

Durkheim, Emile. 1965 (1912). *The Elementary Forms of Religious Life.* New York: Free Press.

Eliot, George. 1986 (1871). *The Mill on the Floss.* London: Folio Society.

Ellis, B. J. 1992. "The Evolution of Sexual Attraction: Evaluative Mechanisms in Women," in J. H. Barkow, L. Cosmides, and J Tooby, eds., *The Adapted Mind: Evolutionary Psychology and the Generation of Culture.* New York: Oxford University Press.

Ellison, Peter T. 2001. *On Fertile Ground.* Cambridge, Mass.: Harvard University Press.

Engels, Frederick. 1972 (1884). *The Origin of the Family, Private Property, and the State.* New York: Pathfinder Press.

Falk, Dean. 1990. "Brain Evolution in Homo 'The Radiator Theory,'" *Behavioral and Brain Sciences* 13: 333–81.

Ferenczi, S. 1955. *Final Contributions to the Problems and Methods of Psychoanalysis.* Ed. M. Balint, trans. E. Mosbacher and others. London: Hogarth.

Fisher, Helen. 1982. *The Sex Contract.* New York: William Morrow.

Fisher, Helen. 1992. *Anatomy of Love: A Natural History of Mating, Marriage, and Why We Stray.* New York: Ballantine.

Fisher, Helen. 1999. *The First Sex: The Natural Talents of Women and How They Are Changing the World.* New York: Random House.

Fisher, Ronald. 1930. *The Genetical Theory of Natural Selection.* Oxford: Clarendon Press.

Fitch, T. 2002. *Proceedings of the Royal Society* 268 (Mar.): 1669.

Ford, C. S., and F. A. Beach. 1951. *Patterns of Sexual Behavior.* New York: Harper and Row.

Fox, R. 1972. "Alliance and Constraint: Sexual Selection in the Evolution of Human Kinship Systems," in B. Campbell, ed., *Sexual Selection and the Descent of Man.* Chicago: Aldine.

Frankl, Victor E. 1959. *Man's Search for Meaning.* New York: Pocket Books.

Frazer, James G. 1960. *The Golden Bough: A Study in Magic and Religion.* London: Macmillian.

Frisch, Karl von. 1967. *The Dance: Language and Orientation of Bees.* Berlin: Dadant.

Frisch, Rose. 1984. "Body Fat, Puberty, and Fertility," *Biology Review* 59: 161–88.

Freud, Sigmund. 1930. *Civilization and Its Discontents.* Trans. Joan Riviere. New York: Hogarth Press and Institute of Psycho-Analysis.

Furuichi, Takeshi. 1987. "Sexual Swelling, Receptivity, and Grouping of Wild Pygmy Chimpanzee Females at Wamba, Zaire," *Primates* 28: 309–18.

Gardiner, D. 1949. "The Fate of the Foreskin: A Study of Circumcision," *British Medical Journal* 2: 1433–37.

Garner, Martin. 1964. *The New Ambidextrous Universe.* New York: W. H. Freeman.

Geschwind, N., and A. M. Galabruda. 1985. "Cerebral Lateralization: Biological Mechanisms, Associations, and Pathology, I. A Hypothesis and a Program for Research," *Archives of Neurology* 42: 428–59.

Ghiselin, M. T. 1969. "The Evolution of Hermaphrodism Among Animals," *Quarterly Review of Biology* 44: 189–208.

Giedd, Jay, et al. 1999. "Brain Development During Childhood and Adolescence: A Longitudinal MRI Study," *Nature Neuroscience* 2: 861.

Ginsburg, N.; M. Jurenovski; and J. Jamieson. 1982. "Sex Differences in Critical Flicker Frequency," *Perceptual & Motor Skills* 54: 1079–82.

Gladwell, Malcolm. 2000. "John Rock's Error," *New Yorker,* March 13, pp. 52–56.

Goldsmith, Judith. 1990. *Childbirth Wisdom*. Brookline, Mass.: East West
    Health Books.
Goodall, Jane. 1986. *The Chimpanzees of Gombe: Patterns of Behavior*. Cam-
    bridge, Mass.: Belknap Press of Harvard University Press.
Gould, S. 1977. *Ontogeny & Phylogeny*. Cambridge, Mass.: Harvard University
    Press.
Gould, S. J. 1983. *The Panda's Thumb: More Reflections on Natural History*. New
    York: Norton.
Gould, S. J., 1993. *Eight Little Piggies*. London: Jonathan Cape.
Gould, S. J., and N. Eldrege. 1993. "Punctuated Equilibrium Comes of Age,"
    Nature 366 (Nov. 18): 223–27.
Gould, S. J., and R. C. Lewontin. 1979. "The Spandrels of San Marcos and
    Panglossian Paradigm," *Proceedings of the Royal Society of London*, ser. B,
    205: 281–88.
Grahn, Judy. 1993. *Blood, Bread, and Roses*. Boston: Beacon Press.
Gross, D. R. 1975. "Protein Capture and Cultural Development in the Amazon
    Basin," *American Anthropologist* 77: 526–49.
Gunn, D. L.; P. M. Jenkin; and A. L. Gunn. 1937. "Menstrual Periodicity: Sta-
    tistical Observations on a Large Sample of Normal Cases," *Journal of
    Obstetrics and Gynecology of the British Empire* 44: 839.
Hall, K. R. L., and I. Devore. 1965. "Baboon Social Behavior," in I. Devore,
    ed., *Primate Behavior*. New York: Holt, Rinehart and Winston, pp.
    53–110.
Halterman, Jill. 2001. "Iron Deficiency and Cognitive Achievement Among
    High School Aged Children and Adolescents in the United States,"
    Pediatrics 107(6): 1381–86.
Hamer, D. H. S.; V. L. Hu; N. Magnuson; and M. L. Pattatucci. 1993. "A Link-
    age Between DNA Markers and the X Chromosomes and Male Sexual
    Orientation," *Science* 200: 1405–9.
Hamilton, W. D. 1964. "The Genetical Evolution of Social Behavior," *Journal of
    Theoretical Biology* 7: 1–52.
Hanbury-Tenison, R. 1982. *Aborigines of the Amazon Rain Forest: The Yanomami*.
    Amsterdam: Time-Life Books.
Haraway, Donna. 1989. *Primate Visions: Gender, Race and Nature in the World of
    Modern Science*. New York: Routledge.
Hardy, Alister. 1960. "Was Man More Aquatic in the Past?," *New Scientist* 17
    (March): 642–45.
Harrington, J. P. 1933. *Annotations in G. Boscana Chingchinich*. Ed. P. Townsend
    Hanna. Glendale, Calif.: Clark.
Hawkes, Kristen. 1991. "Showing Off: Tests of an Hypothesis about Men's For-
    aging Goals," *Ethology and Sociobiology* 12: 29–54.
Hawkes, Kristin. 1997. "Hazda Women's Time Allocation, Offspring Provision-
    ing, and the Evolution of Long Postmenopausal Life Spans," *Current
    Anthropology* 38: 551–77.

Herodotus. 1954. *The Histories.* Trans. Aubrey de Selincourt. New York: Penguin.

Hill, E., and P. Wenzel. 1981. "Variation in Ornamentation and Behavior for Females at Differing Menstrual Phases," paper presented to the Animal Behavior Society, Knoxville, Tenn.

Hines, T. 2001. "The G-Spot: A Modern Gynecologic Myth," *American Journal of Obstetrics and Gynecology* 185 (no. 2): 359–62.

Hoffman, M. 1968. *The Gay World.* New York: Basic Books.

Holloway, R. L. 1983. "Human Paleontological Evidence Relevant to Language Behavior," *Human Neurobiology* 2: 105–14.

Holloway, Ralph L., and Christine De Lacoste–Utamsing. 1982. "Sexual Dimorphism in the Human Corpus Callosum," *Science* 216: 1431–32.

Holmberg, A. R. 1950. *Nomads of the Long Bow: The Siriono of Eastern Bolivia.* Washington, D.C.: United States Government Printing Office.

Hooker, Evelyn. 1967. "The Homosexual Community," in *Sexual Deviance,* ed. J. H. Gagnon and W. Simon (New York: Harper and Row), pp. 167–84.

Hoppe, Klaus D., ed. 1988. *The Psychiatric Clinics of North America.* Philadelphia: W. B. Saunders.

Houppert, Karen. 2000. *The Curse: Confronting the Last Unmentionable Taboo, Menstruation.* New York: Farrar, Straus & Giroux.

Hrdy, Sarah Blaffer. 1981. *The Woman That Never Evolved.* Cambridge, Mass.: President and Fellows of Harvard College.

Hrdy, Sarah Blaffer. 1999. *Mother Nature: A History of Mothers, Infants, and Natural Selection.* New York: Pantheon.

Humphrey, Nicholas. 1992. *A History of the Mind: Evolution and the Birth of Consciousness.* New York: Simon & Schuster.

Huxley, Aldous. 1990. *The Doors of Perception and Heaven and Hell.* New York: HarperCollins.

Huxley, Julian. 1943. *The Uniqueness of Man.* London: Chatto and Windus.

Ingman, M.; H. Kaessmann; S. Pääbo; and U. Glyllensten. 2000. "Mitochondrial Genome Variation and the Origin of Modern Humans," *Nature* 408: 708–13.

Ivry, Richard B., and Lynn C. Robertson. 1999. *The Two Sides of Perception.* Cambridge, Mass.: MIT Press.

Jacobs, A.; E. Butler; and M. Blanche. 1965. "Menstrual Blood Loss in Iron Deficiency Anemia," *Lancet,* Aug. 28, pp. 1102–1107.

Janowsky, David S.; Roderick Gorney; and Bret Kelley. 1966a. "The Curse—Vicissitudes and Variations of the Female Fertility Cycle, Part I, Psychiatric Aspects," *Psychosomatics* 7 (July–Aug.): 242–47.

Janowsky, David S.; Roderick Gorney; and Bret Kelley. 1966b. "The Curse—Vicissitudes and Variations of the Female Fertility Cycle, Part II, Psychiatric Aspects," *Psychosomatics* 7 (Sept.–Oct.): 283–87.

Jesperson, O. 1922. *Language: Its Nature, Development, and Origin.* London: G. Allen & Unwin.

Johanson, D. C., and M. A. Edey. 1981. *Lucy: The Beginnings of Humankind.* London: Granada.

Jolly, Allison. 1999. *Lucy's Legacy.* Cambridge, Mass.: Harvard University Press.

Jong, Erica. 1973. *Fear of Flying.* New York: Holt, Reinhart and Winston.

Jung, Carl. 1964. *Man and His Symbols.* New York: Dell.

Kaplan, Hilliard. 1994. "Evolutionary Wealth Flows Theories of Fertility: Empirical Tests and New Models," *Population and Development Review* 20 (no. 4): 753–91.

Kimura, Doreen. 1979. "Neuromotor Mechanisms in the Evolution of Human Communication," in H. A. Whitaker and H. Whitaker, eds., *Current Trends in Neurolinguistics.* New York: Academic Press.

Kimura, Doreen. 1999. *Sex and Cognition.* Cambridge, Mass.: MIT Press.

Kingdon, Jonathan. 1993. *Self-Made Man: Human Evolution from Eden to Extinction?* New York: John Wiley and Sons.

Kinsey, A. C.; W. B. Pomeroy; C. E. Martin; and P. H. Gebhard. 1953. *Sexual Behavior in the Human Female.* Philadelphia: Saunders.

Kinsey, Alfred; Wardell Pomeroy; and Clyde Martin. 1948. *Sexual Behavior in the Human Male.* Philadelphia: Saunders.

Klein, Richard, with Blake Edgar. 2002. *The Dawn of Human Culture.* New York: John Wiley and Sons.

Knight, Chris. 1991. *Blood Relations: Menstruation and the Origin of Culture.* New Haven: Yale University Press.

Konner, Melvin. 2002. *The Tangled Wing: Biological Constraints on the Human Spirit.* New York: Times Books/Henry Holt.

Kruijver, Frank; Jiang-Ning Zhou; Chris Pool; Michael Hofman; Louis Gooren; and Dick Swaab. 2000. "Male to Female Transsexual Individuals Have Female Neuron Numbers in the Central Subdivision of the Bed Nucleus of the Stria Terminalis," *Journal of Clinical Endocrinology and Metabolism* 85 (5): 2034–41.

Kruuk, Hans. 1972. *The Spotted Hyena.* Chicago: University of Chicago Press.

Labrecque, M.; E. Eason; and S. Marcoux. 2001. "Perineal Massage in Pregnancy," *British Medical Journal* 29 (Sept.): 323.

Ladas, Alice Kahn; Beverly Whipple; and John D. Perry. 1983. *The G Spot.* New York: Dell.

Landes, R. 1938. *The Ojibwa Woman.* New York: Columbia University Press.

Laporte, L. F., and A. Zihlman. 1983. "Plates, Climates, and Hominid Evolution," *South African Journal of Science* 79: 96–110.

Larrington, Carolyn, ed. 1992. *The Feminine Companion to Mythology.* New York: HarperCollins.

Laumann, E. O.; J. H. Gagnon; R. T. Michael; and S. Michaels. 1994. *The Social Organization of Sexuality: Sexual Practices in the United States.* Chicago: University of Chicago Press.

Law, Sung Ping. 1986. "The Regulation of the Menstrual Cycle and Its Relationship to the Moon," *Acta Obstetricia et Gynecologica Scandinavica* 65: 45–48.

Leakey, Richard E., and Roger Lewin. 1978. *People of the Lake: Mankind and Its Beginnings.* New York: Anchor/Doubleday.

Leakey, Richard, and Roger Lewin. 1992. *Origins Reconsidered: In Search of What Makes Us Human.* New York: Dell.

LeDoux, Joseph. 1998. *The Emotional Brain.* New York: Touchstone.

Lee, Richard B. 1988. "Reflections on Primitive Communism," in T. Ingold, D. Riches, and J. Woodburn, eds., *Hunters and Gatherers. 1: History, Evolution, and Social Changes.* Chicago: Aldine, pp. 252–68.

Lee, Richard B., and Irven deVore, eds. 1968. *Man the Hunter.* New York: Aldine de Gruyter.

Leonard, W. R., and M. L. Robertson. 1994. "Evolutionary Perspectives on Human Nutrition," *American Journal of Human Biology* 6 (no. 1): 77–88.

LeVay, Simon. 1991. "A Difference in Hypothalamic Structure Between Heterosexual and Homosexual Men," *Science* 253: 1034–37.

LeVay, Simon. 1994. *The Sexual Brain.* Cambridge, Mass.: MIT Press.

Lévi-Strauss, Claude. 1969. *The Elementary Structures of Kinship.* London: Eyre & Spottiswoode.

Lewis, Thomas; Fari Amini; and Richard Lannon. 2000. *A General Theory of Love.* New York: Random House.

Lieberman, Philip. 1984. *The Biology and Evolution of Language.* Cambridge, Mass.: Harvard University Press.

Lindsay, Rae. 1996. *Left Is Right.* Englewood Cliffs, N.J.: Gilmour House.

Loebenstein, Rabbi Yosef, trans. 1983. *Kitzur Dinei Taharah: A Summary of the Niddah Laws Following the Rulings of the Rebbes of Chabad.* Brooklyn, N.Y.: Kehot Publication Society.

Lorenz, Konrad. 1966. *On Aggression.* Trans. Marjorie Kerr Wilson. New York: Bantam.

Lovejoy, C. O. 1981. "The Origin of Man," *Science* 211: 341–50.

Low, Bobbi S. 2000. *Why Sex Matters: A Darwinian Look at Human Behavior.* Princeton: Princeton University Press.

Lucretius. 1950. *On the Nature of Things.* Trans. Willa Ellery Leonard. New York: Dutton.

MacCormack, Carol P., and Marilyn Strathern. 1980. *Nature, Culture, and Gender.* Cambridge, Mass.: Cambridge University Press.

MacLean, Paul D. 1990. *The Triune Brain in Evolution: Role in Paleocerebral Functions.* New York: Plenum Press.

Maimonides. 1956. *A Guide for the Perplexed.* Trans. M. Friedlander. New York: Dover.

Malinowski, Bronislaw. 1929. *The Sexual Life of Savages in North-Western Melanesia.* New York: Halcyon House.

Manniche, Lisa. 1987. *Sexual Life in Ancient Egypt.* London: KPI.

March, K. S. 1980. "Deer, Bears, and Blood: A Note on Non Human Animal Response to Menstrual Odor," *American Anthropology* 82: 125–27.

Markee, J. E. 1940. "Menstruation in the Introcular Endometrial Transplants in the Rhesus Monkey," *American Journal of Obstetrics and Gynecology* 131 (no. 5): 558–59.

Marler, Peter, and J. William Hamilton III. 1966. *Mechanisms of Animal Behavior.* New York: John Wiley and Sons.

Marshack, Alexander. 1964. "Lunar Notation on Upper Paleolithic Remains," *Science* 146: 743–45.

Marshack, Alexander. 1972. "Cognitive Aspects of Upper Paleolithic Engraving," *Current Anthropology* 13 (nos. 3–4): 445–77.

Marshall, D. S. 1971. "Sexual Behavior on Mangaia," in D. S. Marshall and R. C. Suggs, eds., *Human Sexual Behavior.* New York: Basic Books, pp. 103–62.

Martin, T. C., and L. L. Bumpass. 1989. "Recent Trends in Marital Disruption," *Demography* 26: 37–51.

Masters, W. H., and V. E. Johnson. 1966. *Human Sexual Response.* Boston: Little, Brown.

Mayer, P. J. "Evolutionary Advantage of the Menopause," *Human Ecology* 10: 477–93.

McBrearty, S., and A. S. Brooks. 2000. "The Revolution That Wasn't: A New Interpretation of the Origin of Modern Human Behavior," *Journal of Human Evolution* 39: 453–563.

McClintock, Barbara K. 1971. "Menstrual Synchrony and Suppression," *Nature* 229: 244–45.

McCrone, John. 1991. *The Ape That Spoke: Language and the Evolution of the Human Mind.* New York: William Morrow.

Mead, Margaret. 1967. *Male and Female: A Study of the Sexes in a Changing World.* New York: William Morrow.

Menaker, W., and A. Menaker. 1959. "Lunar Periodicity in Human Reproduction: A Likely Unit of Biological Time," *American Journal of Obstetrics and Gynecology* 77: 905–14.

Miller, G. F. 1992. "Sexual Selection for Protean Expressiveness: A New Model of Hominid Encephalization," paper delivered to the Fourth Annual meeting of the Human Behavior and Evolution Society, Albuquerque, N.M., July 22–26.

Miller, Geoffrey. 2001. *The Mating Mind: How Sexual Choice Shaped the Evolution of Human Nature.* New York: Anchor.

Milton, K. 1984. "The Role of Food-Processing Factors in Primate Food Choice," in P. S. Rodman and J. G. H. Cant, eds., *Adaptations for Foraging in Nonhuman Primates: Contributions to an Organismal Biology of Prosimians, Monkeys, and Apes.* New York: Columbia University Press, pp. 249–79.

Mithin, Stephen. 1996. *The Prehistory of the Mind.* London: Thames and Hudson.

Moir, Anne, and David Jessel. 1989. *Brain Sex: The Real Difference Between Men and Women.* New York: Dell.

Mondimore, Francis Mark. 1996. *A Natural History of Homosexuality.* Baltimore: Johns Hopkins University Press.

Money, J., and A. A. Eberhardt. 1972. *Man & Woman: Boy & Girl.* Baltimore: Johns Hopkins University Press.

Montaigne, Michel de. 1958. *The Complete Essays of Montaigne.* Trans. D. M. Frame. Stanford: Stanford University Press.

Morbeck, Mary Ellen; Alison Galloway; and Adrienne Zihlman. 1997. *The Evolving Female: A Life History Perspective.* Princeton: Princeton University Press.

Morgan, Elaine. 1972. *The Descent of Woman.* New York: Bantam.

Morgan, Elaine. 1990. *The Scars of Evolution.* Oxford: Oxford University Press.

Morgan, Elaine. 1997. *The Aquatic Ape Hypothesis.* London: Souvenir Press.

Morris, Desmond. 1967. *The Naked Ape.* New York: Dell.

Morton, S. Adams, and James V. Neel. 1967. "Children of Incest," *Pediatrics* 40: 55–62.

Moussaieff Masson, Jeffrey, and Susan McCarthy. 1995. *When Elephants Weep: The Emotional Lives of Animals.* New York: Dell.

Myers-Helfgott, M. G., and A. W. Helfgott. 1999. "Routine Use of Episiotomy in Modern Obstetrics: Should It Be Performed?" *Obstetrics and Gynecology Clinics North America* 26: 305–25.

Nathans, J.; D. Thomas; and D. S. Hogness. 1986. "Molecular Genetics of Human Color Vision: The Genes Encoding Blue, Green, and Red Pigments," *Science* 232: 193–202.

Nicholson, Paula. 1995. "The Menstrual Cycle, Science and Femininity: Assumptions Underlying Menstrual Cycle Research," *Social Science and Medicine* 41: 779–84.

Norwitz, E. R.; D. Schust; and S. J. Fisher. 2001. "Implantation and the Survival of Early Pregnancy," *New England Journal of Medicine* 19: 1400–1408.

Ornstein, Robert E. 1972. *The Psychology of Consciousness.* New York: W. H. Freeman.

Ortega y Gasset, J. 1957. *The Revolt of the Masses.* New York: Norton.

Ortner, Sherry B. 1974. "Is the Female to Male as Nature Is to Culture?" in *Women, Culture, and Society,* ed. M. Z. Rosaldo and L. Lamphere (Stanford: Stanford University Press), pp. 67–88.

Owen, Lara. 1998. *Honoring Menstruation.* Freedom, Calif.: Crossing Press.

Page, E. W. 1939. "The Relation Between Hydatid Moles, Relative Ischemia of the Gravid Uterus and the Placental Origin of Elampsia," *American Journal of Obstetrics and Gynecology* 37: 291–93.

Pert, Candace B. 1997. *Molecules of Emotion: Why You Feel the Way You Feel.* New York: Simon & Schuster.

Pfeiffer, John E. 1982. *The Creative Explosion: An Inquiry into the Origins of Art and Religion.* New York: Harper and Row.

Piaget, J. 1974. *Biology and Knowledge.* Trans. B. Walsh. Chicago: University of Chicago Press.

Pinker, Steven. 1994. *The Language Instinct.* New York: William Morrow.

Pinker, Steven. 1997. *How the Mind Works.* London: Penguin.

Pinker, Steven, and Paul Bloom. 1990. "Natural Language and Natural Selection," *Behavioral and Brain Sciences* 13: 707–84.

Pliny, The Elder. 1963. *The Natural History.* Book 28, chapter 23. Trans. W. H. S. Jones. Cambridge, Mass.: Harvard University Press.

Potts, Malcolm, and Roger Short. 1999. *Ever Since Adam and Eve: The Evolution of Human Sexuality.* Cambridge: Cambridge University Press.

Power, Camilla. 1999. "Beauty Magic," in Robin Dunbar, Chris Knight, and Camilla Power, eds., *The Evolution of Culture.* New Brunswick, N.J.: Rutgers University Press, pp. 92–112.

Profet, Margie. 1993. "Menstruation as a Defense Against Pathogens Transported by Sperm," *Quarterly Review of Biology* 68: 335–85.

Ramachandran, V. S., and Sandra Blakeslee. 1998. *Phantoms in the Brain.* New York: William Morrow.

Reyes, F. I.; J. S. D. Winter; and C. Faiman. 1978. "Studies on Human Sexual Development, I. Fetal Gonadal and Adrenal Sex Steroids," *Journal of Clinical Endocrinology and Metabolism* 74: 74–78.

Rich, Adrienne. 1978. *Twenty-One Love Poems: The Dream of a Common Language.* New York: Norton.

Richards, G. 1987. *Human Evolution.* London and New York: Routledge & Kegan Paul.

Richards, Marcus. 1999. "Lifetime Cognitive Function and the Timing of Natural Menopause," *Neurology* 53: 508.

Ridley, Matt. 1993. *The Red Queen: Sex and the Evolution of Human Nature.* New York: Penguin.

Ridley, Matt. 1996. *The Origins of Virtue: Human Instincts and the Evolution of Cooperation.* New York: Penguin.

Ridley, Matt. 1999. *Genome: The Autobiography of a Species in 23 Chapters.* New York: HarperCollins.

Rodwell-Wilton, Sue. 1985. *Nutrition and Dietary Therapy.* St. Louis: Times Mirror.

Rosenberg, Karen. 1988. "The Functional Significance of Neanderthal Pubic Length," *Current Anthropology* 29: 611.

Rosenberg, Karen R., and Wenda R. Trevathan. 1996. "Bipedalism and Human Birth: The Obstetrical Dilemma Revisited," *Evolutionary Anthropology* 4 (no. 5): pp. 161–68

Rosenblum, A. 1976. *The Natural Birth Control Book.* Philadelphia: Aquarian Research Foundation.

Rousseau, Jean-Jacques. 1964 (1755). *The First and Second Discourses.* Ed. R. D. Masters. New York: St. Martin's Press.

Sacks, Oliver. 1970. *The Man Who Mistook His Wife for a Hat.* New York: Touchstone.

Sagan, Carl, and Ann Druyan. 1992. *Shadows of Forgotten Ancestors: A Search for Who We Are.* New York: Random House.

Sahlins, Marshall D. 1960. "The Origin of Society," *Scientific American* 203 (no. 3): 76–87.

Santayana, George. 1988. *The Works of George Santayana,* Vol. II, *The Sense of Beauty: Being the Outlines of Aesthetic Theory.* Ed. William Holzberger and Herman J. Saatkamp, Jr. Cambridge, Mass.: MIT Press.

Sapolsky, Robert M. 1997. *The Trouble with Testosterone, and Other Essays on the Biology of the Human Predicament.* New York: Simon & Schuster.

Schiffer, Fredric. 1998. *Of Two Minds: The Revolutionary Science of Dual-Brain Psychology.* New York: Simon & Schuster.

Schreiner, Olive. 1978. *Women and Labour.* London: Virago.

Schull, William J., and James V. Neel. 1965. *The Effects of Inbreeding on Japanese Children.* New York: Harper and Row.

Shaler, N. S. 1900. *The Individual: A Study of Life and Death.* New York: Appleton.

Sherfey, Mary Jane. 1966. *The Nature and Evolution of Female Sexuality.* New York: Random House.

Sherman, Paul. 1998. "The Evolution of Menopause," *Nature* 392: 759–61.

Shuttle, Penelope, and Peter Redgrove. 1999. *The Wise Wound: Menstruation and Everywoman.* London and New York: Marion Boyars.

Sibley, C. G., and J. E. Ahlquist. 1984. "The Phylogeny of the Hominoid Primates, as Indicated by DNA-DNA Hybridization," *Journal of Molecular Evolution* 26: 99–121.

Siskind, Janet. 1973. *To Hunt in the Morning.* New York: Oxford University Press.

Small, Meredith F. 1995. *What's Love Got to Do With It?* New York: Anchor.

Smuts, Barbara B., and J. M. Wantanabe. 1990. "Social Relationships and Ritualized Greetings in Male Baboons," *International Journal of Primatology* 11: 147–72.

Smuts, Barbara B., et al., eds. 1987. *Primate Societies.* Chicago: University of Chicago Press.

Sowell, E. R. 1999. "In Vivo Evidence for Post-Adolescent Brain Maturation," *Nature Neuroscience* 2: 859.

Speth, J. D. 1987. "Early Hominid Subsistence Strategies in Seasonal Habitats," *Journal of Archeological Science* 14: 13–29.

Spock, Benjamin. 1989. "Circumcision: It's Not Necessary," *Redbook* (April).

Springer, Sally P., and George Deutsch. 1981. *Left Brain, Right Brain.* New York: W. H. Freeman.

Stahl, A. B. 1984. "Hominid Dietary Selection Before Fire," *Current Anthropology* 25: 151–68.

Stanford, Craig B. 1999. *The Hunting Apes: Meat Eating and the Origins of Human Behavior.* Princeton: Princeton University Press.

Strassman, B. I. 1996. "The Evolution of Endometrial Cycles and Menstruation," *Quarterly Review of Biology* 7 (no. 2): 181–220.

Strassman, B. I. 1998. "Predictors of Fecundability and Conception Waits Among the Dogan of Mali," *American Journal of Physical Anthropology* 105 (no. 22): 167–84.

Stringer, C. B., and P. Andrews. 1988. "Genetic and Fossil Evidence for the Origins of Modern Humans," *Science* 239: 1263–68.

Swindler, Daris R. 1998. *Introduction to the Primates.* Seattle: University of Washington Press.

Symons, Donald. 1979. *The Evolution of Human Sexuality.* New York: Oxford University Press.

Tanner, N. M., and A. L. Zihlman. 1976. "Women in Evolution, Part 1: Innovation and Selection in Human Origins," *Signs: Journal of Women, Culture, and Society* 1: 585–608.

Tattersall, Ian. 1998. *Becoming Human: Evolution and Human Uniqueness.* Orlando: Harcourt Brace.

Thatcher, R. W.; R. A. Walker; and S. Guidice. 1987. "Human Cerebral Hemispheres Develop at Different Rates and Ages," *Science* 236: 1110–13.

Thompson, William Irwin. 1984. *The Time Falling Bodies Take to Light.* New York: St. Martin's Press.

Thornhill, N. W., and R. Thornhill. 1990a. "An Evolutionary Analysis of Psychological Pain Following Rape, 1. The Effects of the Victim's Age and Marital Status," *Ethology and Sociobiology* 11: 155–76.

Thornhill, N. W., and R. Thornhill. 1990b. "An Evolutionary Analysis of Psychological Pain Following Rape, 2. The Effects of Stranger, Friend, and Family Member Offenders," *Ethology and Sociobiology* 11: 177–93.

Thornhill, Randy, and Craig T. Palmer. 2000. *A Natural History of Rape.* Cambridge, Mass.: MIT Press.

Tiger, Lionel, and Robin Fox. 1971. *The Imperial Animal.* New York: Dell.

Trevathan, Wenda. 1987. *Human Birth: An Evolutionary Perspective.* New York: Aldine de Gruyter.

Trinkhaus, E. 1984. "The Neanderthal Pubic Morphology and Gestation Length," *Current Anthropology* 25: 509–14.

Trivers, R. L. 1971. "The Evolution of Reciprocal Altruism," *Quarterly Review of Biology* 46: 35–57.

Turke, Paul W. 1984. "Effects of Ovulatory Concealment and Synchrony on Protohominid Mating Systems and Parental Roles," *Ethology and Sociobiology* 5: 33–44.

Tutin, C. E. G. 1979. "Mating Patterns and Reproductive Strategies in a Community of Wild Chimpanzees," *Behavioral Ecology and Sociobiology* 6: 29–38.

Twain, Mark. 1938. *Letters from the Earth.* Ed. Bernard De Voto. New York: Crest.

Underhill, P. A.; P. J. Oefner; et al. 2000. "Y Chromosome Sequence Variation and the History of Human Populations," *Nature Genetics* 26: 358–61.

Van Valen, L. 1973. "A New Evolutionary Law," *Evolutionary Theory* 1: 1–30.

Vollman, R. F. 1968. "The Length of the Premenstrual Phase by Age of Women," *Proceedings of the Fifth World Congress on Fertility and Sterility,* Stockholm, *Excerpta Medica International Congress Series* 133: 1171–75.

Walker, A. C., and R. E. Leakey. 1978. "The Hominids of East Turkana," *Scientific American* 239 (no. 8): 54–66.

Walker, Barbara G. 1983. *The Woman's Encyclopedia of Myths and Secrets.* San Francisco: Harper.

Wallis, J. 1997. "A Survey of Reproductive Parameters in the Free Ranging Chimpanzees of Gombe National Park," *Journal of Reproduction and Fertility* 109: 297–307.

Washburn, S. L. 1960. "Tools and Human Evolution," *Scientific American* 203 (no. 3): 60–75.

Washburn, S. L. 1973. "Human Evolution: Science or Game," *Yearbook of Physical Anthropology* 17: 67–70.

Weinstock, Eugene. 1947. *Beyond the Last Path.* New York: Boni and Gaer.

Wexler, N. S. 1992. "The Tiresias Complex: Huntington's Disease as a Paradigm for Testing Late Onset Disorders," *FASEB Journal* 6: 2820–25.

Wheeler, P. 1994. "The Thermoregulatory Advantages of Heat Storage and Shade Seeking Behavior to Hominids Foraging in Equatorial Savannah Environments," *Journal of Human Evolution* 24 (no. 4): 339–50.

White, R. 1989. "Toward a Contextual Understanding of the Earliest Body Ornaments," in *The Emergence of Modern Humans: Biocultural Adaptations in the Later Pleistocene,* ed. E. Trinkhaus (Cambridge: Cambridge University Press), pp. 211–31.

White, R. 1993a. "The Dawn of Adornment," *Natural History* 5: 62–66.

White, R. 1993b. "Technological and Social Dimensions of 'Aurignacian-Age' Body Ornaments Across Europe," in H. Knecht, A. Pike-Tay, and R. White, eds., *Before Lascaux: The Complex Record of the Upper Paleolithic.* Boca Raton: CRC, pp. 277–99.

Williams, G. C. 1975. *Sex and Evolution.* Princeton: Princeton University Press.

Wilson A. N. 1997. *Paul: The Mind of the Apostle.* New York: Fawcett Columbine.

Wilson, E. O. 1978. *On Human Nature.* Cambridge, Mass.: Harvard University Press.

Wilson, Frank R. 1998. *The Hand: How Its Use Shapes the Brain, Language, and Human Culture.* New York: Pantheon.

Winter, Ernst F., trans. and ed. 1961. *Erasmus, D., and Luther, M.: Discourses on Free Will.* New York: Frederick Ungar.

Witelson, Sandra. 1982. "Sexual Dimorphism in the Human Corpus Callosum," *Science* 216: 1431–32.

Woolf, Virginia. 1929. *A Room of One's Own.* New York: Harcourt Brace.

Wrangham, Richard. 2001. *Tree of Origin: What Primate Behavior Can Tell Us About Human Social Evolution.* Ed. Frans B. M. de Waal. Cambridge, Mass.: Harvard University Press.

Wrangham, Richard, and Dale Peterson. 1996. *Demonic Males: Apes and the Origins of Human Violence.* New York: Houghton Mifflin.

Zahavi, A., and A. Zahavi. 1997. *The Handicap Principle: A Missing Piece of Darwin's Puzzle.* Oxford: Oxford University Press.

Zerries, O. 1968. "Primitive Religions of South America and the West Indies," in *Pre-Columbian American Religions,* ed. W. Krickberg, H. Thimborn, W. Muller, and O. Zerries (London: Weidenfeld & Nicolson), pp. 230–310.

Zhilman, Adrienne. 1981. "Women as Shapers of Human Adaptation," in *Woman: The Gatherer,* ed. F. Dahlberg (New Haven: Yale University Press), pp. 75–120.

Zuckerman, Z. 1932. *The Social Life of Monkeys and Apes.* London: Butler and Tanner.

Illustration Credits

Page 2 Henri Rousseau, *The Dream,* 1910. Oil on canvas. Museum of Modern Art, New York, New York. Digital image © The Museum of Modern Art/Licensed by Scala/Art Resource, New York.

8 René Magritte, *Woman-Bottle,* 1945. © 2003 c. Herscovici, Brussels/Artists Rights Society (ARS), New York. © Christie's Images/Corbis.

10 *The Goddess Tlazolteotl Giving Birth to the Sun God.* Aztec Dumbarton Oaks Pre-Columbian Collection, Washington, D.C.

21 Nude Study # 264 from *Nude Body Nude* by Howard Schatz (HarperCollins Publishers). © Schatz/Ornstein 2000.

22 Georges Lacombe, *Isis,* 1895. Polychrome wood. Photo: C. Jean. Musée d'Orsay, Paris, France. © Réunion des Musées Nationaux/Art Resource, New York.

38 Nat Farbman, *Bushman Throwing His Spear at a Winded Gemsbok,* Botswana, 1947. Nat Farbman/TimePix.

44 Sandro Botticelli, *The Birth of Venus,* 1486. Uffizi, Florence, Italy. © Erich Lessing /Art Resource, New York.

56 Leon-Maxime Faivre, *Two Mothers* (detail), 1888. Louvre, Paris, France. Photo: Hervé Lewandowski. © Réunion des Musées Nationaux/Art Resource, New York.

61 Artist: Mark Reynolds.

68 Knot # 108 from *Body Knots* by Howard Schatz (Rizzoli Publishers International). © Schatz/Ornstein 2000.

84 Leonardo da Vinci, *The Virgin and Child with Saint Anne,* 1508. Oil on wood. Louvre, Paris, France. © Erich Lessing/Art Resource, New York.

100 Edward Landseer, *The Hunting of Chevy Chase,* 1825–26. Birmingham Museums and Art Gallery.

116 Pieter Bruegel, *Return of the Hunters,* 1565. Oil on panel. Kunsthistorisches Museum, Vienna.

136 Edvard Munch, *Puberty,* 1894. © 2003 The Munch Museum/The Munch-Ellingsen Group/Artists Rights Society (ARS), New York. © National Gallery, Norway.

148 René Magritte, *Le Viol,* 1934. The Menil Collection, Houston. 2003 c. Herscovici, Brussels/Artists Rights Society (ARS), New York.

164 An illustration by Florence Harrison of a poem by Tennyson from *Guinevere and other poems, by Alfred Lord Tennyson; illustrated by Florence Harrison* (London: Blackie and Son Ltd., 1912, EB Children's Lit. PR5551 1912). Special Collections Research Center, University of Chicago Library.

183 Drawing by Mathieu Hunter.

184 The Venus of Laussel. Paleolithic sculpture. Musée d'Aquitaine, Bordeaux, France. © Scala/Art Resource, New York.

186 Nat Farbman, *Bushman Children Sitting Around Their Chief as He Acts Out a Story,* Botswana, 1947. Nat Farbman/TimePix.

193 Artist: Mark Reynolds.

Index

Page numbers in *italics* refer to illustrations.